Testing matters! It can determine kids' and schools' futures. In a conference at the Mathematical Sciences Research Institute, mathematicians, math education researchers, teachers, test developers, and policymakers gathered to work through critical issues related to mathematics assessment. They examined:

- the challenges of assessing student learning in ways that support instructional improvement;
- ethical issues related to assessment, including the impact of testing on urban and high-poverty schools;
- the different (and sometimes conflicting) needs of the different groups; and
- different frameworks, tools, and methods for assessment, comparing the kinds of information they offer about students' mathematical proficiency.

This volume presents the results of the discussions. It highlights the kinds of information that different assessments can offer, including many examples of some of the best mathematics assessments worldwide. A special feature is an interview with a student about his knowledge of fractions, demonstrating what interviews (versus standardized tests) can reveal.

Mathematical Sciences Research Institute
Publications

53

Assessing Mathematical Proficiency

Mathematical Sciences Research Institute Publications

Volumes 1–4 and 6–27 are published by Springer-Verlag

Assessing
Mathematical Proficiency

Edited by

Alan H. Schoenfeld
University of California, Berkeley

Alan H. Schoenfeld, Elizabeth and Edward Conner Professor of Education
School of Education, EMST, Tolman Hall #1670
University of California, Berkeley, CA 94720-1670
alans@berkeley.edu

Silvio Levy (*Series Editor*)
Mathematical Sciences Research Institute
17 Gauss Way, Berkeley, CA 94720
levy@msri.org

The Mathematical Sciences Research Institute wishes to acknowledge support by
the National Science Foundation and the Pacific Journal of Mathematics for the
publication of this series.

CAMBRIDGE UNIVERSITY PRESS
Cambridge, New York, Melbourne, Madrid, Cape Town, Singapore, São Paulo

Cambridge University Press

32 Avenue of the Americas, New York, NY 10013-2473, USA

www.cambridge.org

Information on this title: www.cambridge.org/9780521874922

© Mathematical Sciences Research Institute 2007

First published 2007

Printed in the United States of America

A catalogue record for this book is available from the British Library.

Library of Congress Cataloging in Publication data

Assessing mathematical proficiency / edited by Alan H. Schoenfeld.
    p. cm. – (Mathematical Sciences Research Institute publications ; 53)
    Includes bibliographical references and index.
    ISBN 978-0-521-87492-2 (hardback) – ISBN 978-0-521-69766-8 (pbk)
    1. Mathematics – Study and teaching – United States – Evaluation. 2. Mathe-
matical ability – Testing. I. Schoenfeld, Alan H. II. Title. III. Series.

QA13.A8773 2007
510.71–dc22

                                                                   2007060895

            ISBN   978-0-521-87492-2 hardcover
            ISBN   978-0-521-69766-8 paperback

Assessing Mathematical Proficiency
MSRI Publications
Volume **53**, 2007

Contents

Preface

Mathematics tests — and more broadly, assessments of students' mathematical proficiency — play an extremely powerful role in the United States and other nations. California, for example, now has a High School Exit Examination, known as CAHSEE. If a student does not pass CAHSEE, he or she will not be awarded a diploma. Instead, the four years that the student has invested in high school will be recognized with a certificate of attendance. In many states, annual examinations in mathematics and English Language Arts are used to determine whether students at any grade level will advance to the next grade.

Tests with major consequences like those just described are called high-stakes tests. Such tests raise myriad questions. What kinds of understandings do they test? Do they capture the kinds of mathematical thinking that is important? Are they equitable? Is it fair to let a child's career hinge on his or her performance on one particular kind of assessment? Do such tests reinforce and perhaps exacerbate patterns of discrimination, further penalizing students who already suffer from attending "low-performing" schools? Will they increase dropout rates, because students who see themselves as having no chance of passing a high-stakes exam decide to leave school early? Or, can such tests be levers for positive change, compelling attention to mathematics instruction and helping to raise standards?

These are complex issues, all the more so because of the multiple roles that the assessment of students' mathematical proficiency plays in the U.S. educational system, and the multiple groups that have an interest in it. As the chapters in this volume reveal, many different groups — mathematicians, researchers in mathematical thinking and learning, students and teachers, administrators at various levels (mathematics specialists, principals, school-district and state-level superintendents of education), policy-makers (some of the above; state and federal officials and legislators), test-makers and test-consumers (parents, college admissions officers, and more) all have strong interests in the information that mathematics assessments can provide. At the same time, many of these groups have different needs. Is the result of interest a number (How is this school or this district doing? How does this student rank against others?), or is it a profile (This is what the student knows and can do; these are things he or she needs to work on)? Does it matter whether the score assigned to a student or school is legally defensible? (In some contexts it does, and that imposes serious constraints on the kind of reporting that can be done.)

Beyond these complexities is the fact that many of the groups mentioned in the previous paragraph have little knowledge of the needs of the other groups. For example, what relevance do "reliability" or "validity" have for teachers and mathematicians? There is a good chance that they would not know the technical meanings of the terms — if they have heard them at all. Yet, to the test manufacturer, an assessment without good reliability and validity data is worthless — the makers of the SAT or GRE would no sooner release such a test than a pharmaceutical company would release a drug that had not undergone clinical trials. Similarly, the policy-maker who demands that "high standards" be reflected by an increase in average scores or the percentage of students attaining a one-dimensional passing score may not understand teachers' needs for "formative diagnostic assessments," or how disruptive "teaching to the test" can be if the test is not aligned with an intellectually robust curriculum.

What to assess in mathematics learning is not as simple as it might seem. In the 1970s and 1980s, research on mathematical thinking and learning resulted in a redefinition of thinking and learning in general [Gardner 1985] and of mathematical understanding in particular [DeCorte et al. 1996; Grouws 1992; Schoenfeld 1985]. This refined understanding of mathematical competency resulted in a new set of goals for mathematics instruction, for example those stimulated by the National Council of Teachers of Mathematics' *Curriculum and Evaluation Standards for School Mathematics*, commonly known as the *Standards* [NCTM 1989]. This document, which delineated content and process desiderata for K–12 mathematics curricula, focused on four "process standards" at every grade level: problem solving, reasoning, making mathematical connections, and communicating mathematics orally and in writing. Such learning goals, which continue to play a central role in NCTM's refinement of the *Standards* [NCTM 2000], pose a significant challenge for assessment.

Although it may seem straightforward to measure how much algebra, or geometry, or probability a student understands, the issues involved in obtaining accurate measurements of students' content understandings are actually complex. Measuring students' abilities to solve problems, reason, and make mathematical connections is much more difficult. There are myriad technical and mathematical issues involved. Which mathematics will be assessed: Technical skills? Conceptual understanding? For purposes of reliable scoring, when are two problems equivalent? How can one compare test scores from year to year, when very different problems are used? (If similar problems are used year after year, teachers and students learn what they are, and students practice them: problems become exercises, and the test no longer assesses problem solving.)

One measure of the complexities of the issue of assessment in general, and mathematics assessment in particular, is the degree of attention given to the topic by the National Research Council (NRC). The NRC's Board on Testing and Assessment has published a series of general reports on the issue; see, for example, *High Stakes: Testing for Tracking, Promotion, and Graduation* [NRC 1999]. Among the NRC publications focusing specifically on assessments that capture students' mathematical understandings in accurate ways are *Keeping Score* [Shannon 1999], *Measuring Up* [NRC 1993a], *Measuring What Counts* [NRC 1993b]. A more recent publication, *Adding It Up* [NRC 2001] provides a fine-grained portrait of what we might mean by mathematical proficiency. All of these volumes, alongside the large literature on mathematical thinking and problem solving, and volumes such as NCTM's *Principles and Standards* [NCTM 2000], point to the complexity of teaching for, and assessing, mathematical proficiency.

Added to the intellectual challenges are a series of social, political, and ethical challenges. A consequence of the "standards movement" has been the establishment of high-stakes accountability measures — tests designed to see whether students, schools, districts, and states are meeting the standards that have been defined. Under the impetus of the No Child Left Behind legislation [U.S. Congress 2001], all fifty states have had to define standards and construct assessments to measure progress toward them. The consequences of meeting or not meeting those standards have been enormous. For students, failing a test may mean being held back in grade or being denied a diploma. Consequences for schools are complex. As has been well documented, African Americans, Latinos, Native Americans, and students in poverty have tended to score much worse on mathematical assessments, and to have higher dropout rates, than Whites and some Asian groups. One mechanism for compelling schools to focus their attention on traditionally lower-performing groups has been to disaggregate scores by groups. No longer can a school declare that it is performing well because its average score is good; all subgroups must score well. This policy raises new equity issues: schools with diverse populations now face more stringent requirements than those with homogeneous populations, and dire consequences if they fail to meet the requirements. Even before the advent of the No Child Left Behind legislation, various professional societies cautioned about the use of a single test score as the sole determinant of student success or failure: see, for example, the position paper by the American Educational Research Association [AERA 2000] and the updated paper by the National Council of Teachers of Mathematics [NCTM 2006].

There are also issues of how policy drives curricula. With high-stakes testing in place, many schools and districts take the conservative route, and teach to the test. If the assessments are robust and represent high standards, this can be a good thing. However, if the standards are high, schools run the risk of not making "adequate yearly progress" toward proficiency for all students, and being penalized severely. If standards are set lower so that they are more easily attainable, then education is weakened — and "teaching to the (narrow) test" may effectively lower standards, and limit what students learn [Shepard 2001].

In short, the issues surrounding mathematics (and other) assessments are complex. It may be that tests are simply asked to do too much. As noted above, many of the relevant stakeholders tend to have particular needs or interests in the assessments:

- Mathematicians want students to experience the power, beauty, and utility of mathematics. Accordingly, the mathematics represented in the tests should be important and meaningful, and not be weakened by technical or legal concerns (e.g., psychometric issues such as reliability, validity, and legal defensibility).

- Researchers, teachers, principals, and superintendents want tests that provide meaningful information about the broad set of mathematical skills and processes students are supposed to learn, and that provide diagnostic information that helps to improve the system at various levels.

- Psychometricians (and more broadly, test developers) want tests to have the measurement properties they are supposed to have — that the problems and their scoring capture what is important, that different versions of a test measure the same underlying skills, that they are fair in a substantive and legal sense.

- Parents and teachers want information that can help them work with individual students. Teachers also want information that can help them shape their instruction in general.

- Policy-makers want data that says how well their constituents are moving toward important social and intellectual goals.

- Cost, in terms of dollars and time, is a factor for all stakeholders.

Some of the goals and needs of some constituencies tend to be in conflict with each other. For example, multiple-choice tests that focus largely on skills can be easily graded, tend to be easy to construct in ways that meet psychometric criteria, and to provide aggregate statistics for purposes of policy — but, they are unlikely to capture the desired spectrum of mathematics, and provide little or no useful diagnostic information. Moreover, if there is a large amount of "teaching to the test," such tests may distort the curriculum.

Assessments that focus on a broad range of skills turn out to be expensive to grade, and are much more difficult to construct in ways that meet psychometric criteria — but, they can provide diagnostic information that is much more useful for teachers, and may be used for purposes of professional development as well. In addition, the vast majority of stakeholders in assessment are unaware of the complexities involved or of the needs of the other constituencies. That can lead to difficulties and miscommunication.

In short, different groups with a stake in the assessment of students' mathematical thinking and performance need to understand the perspectives and imperatives of the others. They need to seek common ground where it exists, and to recognize irreconcilable differences where they exist as well. To this end, in March 2004 the Mathematical Sciences Research Institute brought together a diverse collection of stakeholders in mathematics assessment to examine the goals of assessment and the varied roles that it plays. Major aims of the conference were to:

- Articulate the different purposes of assessment of student performance in mathematics, and the sorts of information required for those purposes.
- Clarify the challenges of assessing student learning in ways that support instructional improvement.
- Examine ethical issues related to assessment, including how assessment interacts with concerns for equity, sensitivity to culture, and the severe pressures on urban and high-poverty schools.
- Investigate different frameworks, tools, and methods for assessment, comparing the kinds of information they offer about students' mathematical proficiency.
- Compile and distribute a list of useful and informative resources about assessment: references, position papers, and sources of assessments, etc.
- Enlarge the community of mathematicians who are well informed about assessment issues and interested in contributing to high-quality assessments in mathematics.
- Articulate a research and development agenda on mathematics assessment.

This book is the product of that conference. Here is what you will find within its pages.

Section 1 of the book provides an orientation to issues of assessment from varied perspectives. In Chapter 1, Alan H. Schoenfeld addresses three sets of issues: Who wants what from assessments? What are the tensions involved? What other issues do we confront? This sets the stage for the contributions that follow. In Chapter 2, Judith A. Ramaley frames the issues of assessment in a broad way, raising questions about the very purposes of education. The underlying philosophical issue is, "Just what are our goals for students?" Then, as a corollary activity, how do we know if we are meeting them? In Chapter 3, Susan Sclafani describes the current political context and national goals following from the No Child Left Behind Act (NCLB). The goals of NCLB are that all students will develop "fundamental knowledge and skills in mathematics, as well as in all core subjects." The mechanism for achieving these goals is premised on four basic principles: accountability for results, local control and flexibility, choice, and research-based practice. By virtue of these three introductory chapters, Section 1 identifies some of the major constituencies and perspectives shaping the varied and sometimes contradictory approaches to mathematics assessment found in the United States today.

Different notions of mathematical proficiency may underlie different approaches to assessment. Section 2 offers two perspectives regarding the nature of mathematical proficiency. Chapter 4, by R. James Milgram, describes one mathematician's perspective on what it means to do mathematics, and the implications of that perspective for teaching and assessing mathematics and understanding mathematics learning. Chapter 5, by Alan H. Schoenfeld, provides a top-level view of the past quarter-century's findings of research in mathematics education on mathematical thinking and learning. This too has implications for what should be taught and assessed. A comparison of the two chapters reveals that even the "basics" — views of the nature of mathematics, and thus what should be taught and assessed — are hardly settled.

Section 3 gets down to the business of assessment itself. At the conference organizers' request, the chapters in this and the following section contain numerous illustrative examples of assessment tasks. The organizers believe that abstract generalizations about mathematical competencies can often be misunderstood, while specific examples can serve to demonstrate what one cares about and is trying to do. Chapter 6, by Hugh Burkhardt, gives the lay of the land from the point of view of an assessment developer. It offers some design principles for assessment, discusses values (What should we care about?), the dimensions of performance in mathematics, and issues of the quality and cost of assessments. As in the chapters that follow, the examples of assessment items contained in the chapter show what the author values. There are modeling problems, and tasks that take some time to think through. Burkhardt has argued that "What You Test Is What You Get," and that in consequence assessment should not only capture what is valued but should help focus instruction on what is valued. Chapter 7, by Jan de Lange, lays out some of the foundations of the International Organisation for Economic Co-operation and Development Program for International Student Assessment (PISA) mathematics assessments. There is an interesting non-correlation between national scores on the PISA exams and scores on exams given in the Trends in International Mathematics and Science

Study, precisely because of their different curricular and content foci. Thus, de Lange's chapter, like the others, underscores the point that curriculum and assessment choices are matters of values. In Chapter 8, Bernard Madison expands the scope of assessment yet further. Madison reflects on the kinds of reasoning that quantitatively literate citizens need for meaningful participation in our democratic society. His examples, like others in this section, will certainly challenge those who think of mathematics assessments as straightforward skills-oriented (often multiple-choice) tests. In Chapter 9, Richard Askey provides a variety of assessment tasks, dating all the way back to an 1875 examination for teachers of mathematics in California. Some of Askey's historical examples are remarkably current — and, Askey would argue, more demanding than what is required of teachers today. In Chapter 10, David Foster, Pendred Noyce, and Sara Spiegel address the issue of teacher professional development through the use of student assessment. They describe the work of the Silicon Valley Mathematics Initiative, which uses well-designed assessments as a mechanism for focusing teachers' attention on the mathematics that students are to learn, and the kinds of understandings (and misunderstandings) that students develop.

Section 4 focuses on algebra. In Chapter 11, William McCallum takes a broad view of the "assessment space." McCallum gives examples of more advanced tasks that can be used to assess different strands of proficiency such as conceptual understanding and strategic competence. In Chapter 12, David Foster identifies some of the core skills underlying algebraic understandings — examples of algebraic "habits of mind" such as abstraction from computation, rule-building, and constructing and inverting processes ("doing-undoing"). Foster also discusses foundational concepts for young students transitioning to algebra, such as understanding and representing the concepts of equality. And he provides examples of assessment items focusing on these critically important topics. Chapter 13, by Ann Shannon, looks at issues of problem context in ways consistent with the approaches of Burkhardt and de Lange. Shannon discusses tasks that focus on understanding linear functions. Her chapter shows how very different aspects of reasoning can be required by tasks that all ostensibly deal with the same concept. Taken together, these chapters (and some examples from Section 3 as well) provide rich illustrations of the broad range of algebraic competency — and thus of things to look for in assessing it.

Section 5 focuses on fractions, with a slightly different purpose. The idea here is to see what kinds of information different kinds of assessments can provide. In Chapter 14, Linda Fisher shows how a close look at student responses to somewhat complex tasks provides a structured way to find out what one's students are understanding (and not). This kind of approach provides substantially more information than simple test results. Yet, as we see in Chapter 15, there is no substitute for a thoughtful conversation with a student about what he or she knows. Chapter 15 provides the transcript of an interview with a student conducted by Deborah Ball at the conference (a video of the interview is available online). In Chapter 16, Alan H. Schoenfeld reflects on the nature and content of the interview in Chapter 15. He considers the detailed information about student understanding that this kind of interview can reveal, and what this kind of information implies — about the nature of learning, and about how different assessments can reveal very different things about what students understand.

Section 6 takes a more distanced view of the assessment process. To this point (with the exception of Section 1) the volume has focused on issues of which mathematics is important to assess, and what kinds of tasks one might give students to assess their mathematical knowledge. Here we open up to the issue of societal context — the fact that all assessment takes place within a social, political, and institutional set of contexts and constraints, and that those constraints and contexts shape what one can look for and what one can see as a result. In Chapter 17, Michèle Artigue presents a picture of the assessment system in France — a system very different from that in the U.S. (the educational system is much more centralized) and in which attempts at providing information about student understanding seem much more systematic than in the U.S. As always, a look outside one's own borders helps to calibrate what happens within them. In Chapter 18, Mark Wilson and Claus Carstensen take us into the realm of the psychometric, demonstrating the issues that one confronts when trying to build assessment systems that are capable of drawing inferences about student competence in particular mathematical domains. In Chapter 19, Lily Wong Fillmore describes the complexities of assessment when English is not a student's native language. If a student is not fluent, is his or her failure to solve a problem a result of not understanding the problem (a linguistic issue) or of not understanding the mathematics? Judit Moschkovich addresses a similar issue in Chapter 20, with a close examination of bilingual students' classroom work. Here she shows that the students have a fair amount of conceptual competency, while not possessing the English vocabulary to appear (to an English speaker) that they are very competent. This too raises the question of how one can know what a student knows — the real purpose of assessment. In Chapter 21, Elizabeth Taleporos takes us outside the classroom and into the realm of politics. The issue: How does one move a system toward looking at the right things? Taleporos describes her experiences in New York City. In Chapter 22, Elizabeth Stage discusses the systemic impact of assessments in California. Stage examines the ways in which the publicly released items on the California assessments shaped what teachers taught — sometimes for good, when testing identified consistent and important weaknesses in students' knowledge; sometimes for ill, when teaching to the test resulted in a focus on less than essential mathematics. This brings us full circle, in that these issues address some of the critical concerns in Section 1, such as concerns about assessments mandated by the No Child Left Behind legislation.

At the end of the conference, participants reflected on what they had experienced and what they (and the field) needed to know. Eight working groups at the conference were charged with formulating items for a research agenda on the topic of the conference. The product of their work, a series of issues the field needs to address, is presented as an Epilogue. There is great variety in the chapters of this book, reflecting the diverse perspectives and backgrounds of the conference participants. As mentioned before, that mixing was intentional. The conference participants learned a great deal from each other. The organizers hope that some of the "lessons learned" have made their way into this volume.

References

[AERA 2000] American Educational Research Association, Position statement concerning high-stakes testing in PreK-12 education, July 2000. Available at http://edtech.connect.msu.edu/aera/about/policy/stakes.htm. Retrieved 7 Feb 2006.

[DeCorte et al. 1996] E. DeCorte, B. Greer, and L. Verschaffel, "Mathematics teaching and learning", pp. 491–549 in *Handbook of educational psychology*, New York: Macmillan, 1996.

[Gardner 1985] H. Gardner, *The mind's new science: A history of the cognitive revolution*, New York: Basic Books, 1985.

[Grouws 1992] Grouws, D. (editor), *Handbook for research on mathematics teaching and learning*, New York: Macmillan, 1992.

[NCTM 1989] National Council of Teachers of Mathematics, *Curriculum and evaluation standards for school mathematics*, Reston, VA: Author, 1989.

[NCTM 2000] National Council of Teachers of Mathematics, *Principles and standards for school mathematics*, Reston, VA: Author, 2000.

[NCTM 2006] National Council of Teachers of Mathematics, Position statement on high stakes testing, January 2006. Available at http://www.nctm.org/about/position_statements/highstakes.htm. Retrieved 24 Jun 2006.

[NRC 1993a] National Research Council (Mathematical Sciences Education Board), *Measuring up: Prototypes for mathematics assessment*, Washington, DC: National Academy Press, 1993.

[NRC 1993b] National Research Council (Mathematical Sciences Education Board), *Measuring what counts: A conceptual guide for mathematics assessment*, Washington, DC: National Academy Press, 1993.

[NRC 1999] National Research Council (Committee on Appropriate Test Use: Board on Testing and Assessment — Commission on Behavioral and Social Sciences and

Education), *High stakes: Testing for tracking, promotion, and graduation*, edited by J. P. Heubert and R. M. Hauser, Washington, DC: National Academy Press, 1999.

[NRC 2001] National Research Council (Mathematics Learning Study: Center for Education, Division of Behavioral and Social Sciences and Education), *Adding it up: Helping children learn mathematics*, edited by J. Kilpatrick et al., Washington, DC: National Academy Press, 2001.

[Schoenfeld 1985] A. H. Schoenfeld, *Mathematical problem solving*, Orlando, FL: Academic Press, 1985.

[Shannon 1999] A. Shannon, *Keeping score*, National Research Council: Mathematical Sciences Education Board, Washington, DC: National Academy Press, 1999.

[Shepard 2001] L. Shepard, Protecting learning from the harmful effects of high-stakes testing, paper presented at the annual meeting of the American Educational Research Association, Seattle, April 2001.

[U.S. Congress 2001] U.S. Congress, H. Res. 1, 107th Congress, 334 Cong. Rec. 9773, 2001. Available at http://frwebgate.access.gpo.gov.

Assessing Mathematical Proficiency
MSRI Publications
Volume 53, 2007

Acknowledgments

A conference and the book it produces are made possible by many contributions of different kinds. To begin, there is the team that conceptualized and organized the whole enterprise. I owe a debt of gratitude to MSRI's Educational Advisory Committee: Bruce Alberts, Michèle Artigue, Deborah Ball, Hyman Bass, David Eisenbud, Roger Howe, Jim Lewis, Robert Moses, Judith Ramaley, Hugo Rossi, and Lee Shulman. A planning team consisting of Deborah Ball, Hyman Bass, David Eisenbud, Jim Lewis, and Hugo Rossi put in many hours organizing and running the conference. David and Hugo were responsible for much of the conference infrastructure, which is typically invisible (when things go well, which they did) but absolutely essential. An equally essential part of the infrastructure is fiscal: I am grateful to the National Science Foundation, the Noyce Foundation, and the Spencer Foundation for their collective underwriting of the conference. Of course, the participants and the authors are who make the conference and the volume what they are. It has been a pleasure to work with them, individually and collectively. Most special thanks are due to Cathy Kessel, who worked with me in editing this volume. Every chapter — every word, every figure — has been reviewed by Cathy, and the manuscript is much improved for her efforts. Thanks as well to Silvio Levy for turning our drafts into technically viable LaTeX, and to Cambridge University Press for converting Silvio's labors into the volume you now hold in your hand.

Alan H. Schoenfeld
Berkeley, CA
June 2006

Section 1
The Big Picture

This introductory section establishes the context for what follows in this volume. In broad-brush terms, it does so by addressing three P's: pragmatics, philosophy, and policy.

Alan H. Schoenfeld opens with a discussion of pragmatics in Chapter 1. A simple question frames his contribution: Who wants what from mathematics assessments? As he shows, the issue is far from simple. It is true that at some level everyone involved in mathematics testing is interested in the same basic issue: What do students know? However, various groups have very different specific interests. For example, a teacher may want to know what his or her particular students need to know in order to improve, while a state superintendent may be most interested in trends and indications of whether various performance gaps are being closed. Other groups have other interests. Those interests are not always consistent or even compatible. Understanding who has a stake in assessment, and what different groups want to learn from it, is part of the big picture — the picture that is elaborated throughout the volume, as representatives of different stakeholder groups describe the kinds of information that they need and that carefully constructed assessments can provide. Schoenfeld also looks at the impact of assessment. Tests reflect the mathematical values of their makers and users. In the United States, tests are increasingly being used to drive educational systems — to measure performance aimed at particular educational goals. This is the case at the national level, where various international assessments show how one nation stacks up against another; at the state level, where individual states define their intended mathematical outcomes; and at the individual student level, where students who do not pass state-mandated assessments may be retained in grade or denied a diploma. Schoenfeld addresses both intended and unintended consequences of assessments.

Judith A. Ramaley's discussion in Chapter 2 addresses the second P, philosophy. As indicated in the previous paragraph, testing reflects one's values and goals. The issue at hand is not only "What is mathematics," but "Which aspects of mathematics do we want students to learn in school?" Is the purpose of schooling (and thus of mathematics instruction in school) to provide the skills needed for successful participation in the marketplace and in public affairs? Is it to come to grips with fundamental issues of truth, beauty, and intellectual coherence? To use some common jargon, these are consequential decisions: the answers to questions of values shape what is taught, and how it is taught. As Ramaley notes, the Greeks' two-thousand-year-old philosophical debates lie at the heart of today's "math wars." But, as she also observes, science brings philosophy into the present: questions of "what counts" depend on one's understanding of thinking and learning, and of the technologies available for assessing it. Discussions of what can be examined, in what ways, bring us firmly into the twenty-first century.

Susan Sclafani's contribution in Chapter 3 brings us into the policy arena. Having certain goals is one thing; working to have a complex system move toward achieving those goals is something else. At the time of the MSRI conference Sclafani served as Assistant Secretary in the Office of Vocational and Adult Education of the U.S. Department of Education. One of her major concerns was the implementation of the No Child Left Behind Act (NCLB), federal legislation that mandates the development and implementation of mathematics assessments in each of the nation's fifty states. The creation of NCLB was a political balancing act, in which the traditional autonomy granted to states on a wide range of issues was weighed against a federal interest in policies intended to have a beneficial impact on students nationwide. How such issues are resolved is of great interest. In NCLB states are given particular kinds of autonomy (e.g., what is tested is, in large measure, up to the states) but they are subject to national norms in terms of implementation. Broadly speaking, NCLB mandates that scores be reported for all demographic groups, including poor students, English as a Second Language students, various ethnic groups, and more. In order for a school to meet its state standard, every demographic group with 30 or more students in the school must meet the standard. In this way, NCLB serves as a policy lever for making sure that under-represented minority groups cannot slip through the cracks.

In sum, then, the philosophical, pragmatic, and policy discussions in the three chapters that follow establish the overarching context for the more detailed discussions in the core of the volume.

Chapter 1
Issues and Tensions in the Assessment of Mathematical Proficiency

ALAN H. SCHOENFELD

Introduction

You'd think mathematics assessment — thought of as "testing" by most people — would be simple. If you want to know what a student knows, why not write (or get an expert to write) some questions about the content you want examined, give those questions to students, see if the answers are right or wrong, and add up a total score? Depending on your predilections (and how much time you have available) you might give a multiple-choice test. You might give an "open answer" or "constructed response" test in which students show their work. You could give partial credit if you wish.

This version of assessment fits with most people's experiences in school, and fits with descriptions of the National Assessment of Educational Progress (NAEP) as "the nation's report card." From this perspective, mathematics assessment — discovering what mathematics a person (typically, a student) knows — seems straightforward.

Would that things were so simple. As this essay and later contributions to this volume will indicate, different groups can have very different views of what "counts," or should count, in mathematics. Assessing some aspects of mathematical thinking can be very difficult — especially if there are constraints of time or money involved, or if the tests have to have certain "psychometric" properties (discussed further in this essay) in order to make sure that the test-makers stand on legally safe ground. Different groups may want different information from tests. And, the tests themselves are not neutral instruments, in the ways that people think of thermometers as being neutral measures of the temperature of a system. In many ways, tests can have a strong impact on the very system they measure.

3

This essay introduces and illustrates such issues. I begin by identifying a range of "stakeholder" audiences (groups who have an interest in the quality or outcomes) for mathematics assessments, and identifying some of the conflicts among their interests. I proceed with a discussion of some of the side effects of certain kinds of large-scale testing. These include: test score inflation and the illusion of competence; curriculum deformation; the stifling of innovation; the disenfranchising of students due to linguistic or other issues; and a possible impact on drop-out rates.

My purpose is to lay out some of the landscape, so that the varied groups with a stake in assessing mathematical proficiency (nearly all of us!) can begin to understand the perspectives held by others, and the kinds of issues we need to confront in order to make mathematics assessments serve the many purposes they need to serve.

Who Wants What from Mathematics Assessments?

Here, in brief, are some assertions about what various communities want from mathematics assessments. It goes without saying that these communities are not monolithic, and that my assertions are simple approximations to complex realities.

Mathematicians. Generically speaking: Mathematicians want assessments to be true to mathematics. From the mathematician's perspective, mathematics assessments should focus on revealing whether and what students understand about mathematically central ideas.

This does not happen automatically. For example, I served for many years on the committee that produced the Graduate Record Examination (GRE) advanced mathematics exam. In a sense, our task was simple: the audience (mathematics professors assessing the potential of applicants to their graduate programs) wants to judge applicants' promise for success in graduate school. This is usually understood as "How much do they know?" or "What problems can they solve?" The paper-and-pencil version of the GRE advanced mathematics exam had 65 multiple-choice questions, which students worked over a three-hour period. Student scores correlated reasonably well with the grades that those students earned during their first year of graduate school — but those grades themselves are not great predictors of future performance in graduate school or beyond, and there has been perennial dissatisfaction with the test because it reveals so little about actual student thinking. Indeed, the Educational Testing Service (ETS) spent some time trying to alter the exam, replacing it with an exam that focused on a deeper examination of what students know. ETS looked at the possibility of putting essay questions on the test to see if students could produce proofs, explain concepts, etc. For a variety of reasons, including cost and the difficulty of creating essay tests with the right psychometric properties, some of which are examined below, ETS abandoned this approach.

Mathematics education researchers. Again, generically speaking: From the perspective of mathematics educators, mathematics assessments should reflect the broad spectrum of mathematical content and processes that comprise what might be called "thinking mathematically." (See my chapter "What is Mathematical Proficiency and How Can It Be Assessed?" later in this volume.) Thus, for example, the National Research Council document *Adding It Up* [NRC 2001, p. 5] describes five interwoven strands of mathematical proficiency:

- *conceptual understanding*: comprehension of mathematical concepts, operations, and relations
- *procedural fluency*: skill in carrying out procedures flexibly, accurately, efficiently, and appropriately
- *strategic competence*: ability to formulate, represent, and solve mathematical problems
- *adaptive reasoning*: capacity for logical thought, reflection, explanation, and justification
- *productive disposition*: habitual inclination to see mathematics as sensible, useful and worthwhile, coupled with a belief in diligence and one's own efficacy.

The National Council of Teachers of Mathematics offers a more fine-grained characterization of desired proficiency in its *Principles and Standards for School Mathematics* [NCTM 2000]. This document argues for competency along these ten dimensions, clustered by content and process:

Content:

    Number and operations
    Algebra
    Geometry
    Measurement
    Data analysis and probability

Process:

    Problem solving
    Reasoning and proof
    Making connections
    Oral and written communication
    Uses of mathematical representation

To give a far too simple example, someone interested in students' ability to operate comfortably with fractions might ask a student to "reduce the fraction 2/8 to lowest terms." Someone interested in whether students understand different representations of fractions, and operate on them, might ask a question such as this:

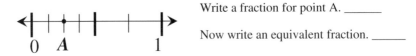

Write a fraction for point A. _____

Now write an equivalent fraction. _____

Part of the interest in this question is whether the student understands that the interval from 0 to 1 must be partitioned into n parts of equal length in order for each sub-interval to have length $1/n$. A student who merely counts parts is likely to say that point A is 2/6 of the way from 0 to 1; that student may well reduce the fraction 2/6 to 1/3. To the mathematics educator, this is evidence that mastery of procedures in no way guarantees conceptual understanding. And, the problem above only examines a small part of conceptual understanding. With their broad interpretation of what it means to understand mathematics and to think mathematically, mathematics education researchers tend to want assessments to cover a very broad range of mathematical content and processes. Moreover, they would want to see students answer analogous questions with different representations — e.g., with a circle partitioned into parts that are not all congruent.

Researchers employ a wide range of assessment techniques. These range from extended interviews with and observations of individual children to the large-scale analysis of standardized test data.

Parents. From parents' point of view, a mathematics assessment should enable parents to understand their children's knowledge and progress (and to help their kids!). Thus, parents tend to want: (a) simple performance statistics that place the student on a continuum (e.g., a grade or a percentile score); and (b) perhaps enough information about what their child is doing well and doing poorly so that they can provide or arrange for help in the areas where it is needed. Homework plays this role to some degree, but assessments can as well.

Policy-makers. By policy-makers I mean those who have a direct say in the ways that schools are run. This includes principals, superintendents, and school boards at the local level; it includes state departments of education and their policy leaders (often elected or appointed state superintendents of education); it includes legislatures at the state and federal level; it includes governors and the president. For example, the No Child Left Behind (NCLB) Act [U.S. Congress 2001], passed by the Congress and signed by the president in 2002, mandated mathematics testing at almost all grades K–12. State bureaucracies, under direction from their state legislatures, established curriculum frameworks, standards, and assessments: Local administrators are responsible for making sure that students meet the standards, as defined by performance on the assessments.

From the policy-maker's perspective, the primary use of assessments is to provide indicators of how well the system is going. The further away instruction the policy-maker is, the less details may matter. That is, a teacher is interested in detailed information about all of his or her students (and in some detail — see below). A principal may want to know about how the school as a whole is doing, how subgroups of students are doing, and perhaps how particular teachers are doing. By this level of the system, then, one number per student (the test score) is as much as can be handled, if not too much; one number per subgroup is more typically used. Statistical trends are more important than what the scores reveal about individuals. As one travels up the political food chain, the unit of analysis gets larger: what counts is how well a school or a district did, the typical question being "Did scores go up?" Test scores may be used to "drive" the system, as in the case of NCLB: each year scores must go up, or there will be serious consequences. Those making the policies may or may or may not know anything about the content of the assessments or what scores mean in terms of what students actually know. Such details can be (and usually are) left to experts.

Publishers and test developers. Commercially developed assessments play a significant and increasing role in schooling at all levels. In the U.S. students have faced the SAT and ACT (for college admission) and the GRE (for admission to graduate school) for decades, but with NCLB, students are assessed annually from grade 3 through 8 and often beyond. These tests are typically developed and marketed by major corporations — Harcourt Brace, CTB McGraw-Hill (CTB used to stand for "Comprehensive Test Bureau"), ETS, the College Board, etc.

What must be understood is that these corporations — indeed, every developer of "high-stakes" assessments for wide distribution and use — must design their tests subject to some very strong constraints. An easily understandable constraint is cost. School districts or other test consumers typically want tests that can be administered and graded at low cost. This is one reason multiple-choice questions are so common. Another constraint is security — the need to administer tests in ways that the potential for cheating is minimized. Thus tests are given under high security conditions. "Objective" machine grading is used not only to lower costs, but to lower the possibility of teachers acting, individually or collectively, to modify papers or scores to their students' advantage. (There is some individual scoring of tests such as the Advanced Placement exams that students take to get college credit for high school courses. However, the scoring is done at centralized locations where teachers do not have a stake in the scores, and the high costs of these exams are borne by the students who pay to take them.)

More important and more constraining, however, are the constraints imposed by the test design system itself. No major commercial test producer in the United States will market a test that fails to meet a series of technical criteria. These are called psychometric criteria. (Psychometrics is the field engaged in the quantification of the measurement of mental phenomena.) There are technical terms called "reliability," "construct validity," "predictive validity," and "test comparability" that play little or no formal role at the classroom level, but that are essential considerations for test designers. Indeed, the American Educational Research Association, American Psychological Association, and National Council on Measurement in Education have jointly issued sets of criteria that standardized tests should meet, in their (1999) volume *Standards for Educational and Psychological Testing*. Relevant issues include: Will a student get the same score if he or she takes the test this month or next? Is the same content covered on two versions of a test, and does it reflect the content that the test says it covers? Will a score of 840 on this year's GRE exam mean the same thing as a score of 840 on next year's GRE? Such considerations are essential, and if a test fails to meet any of these or related criteria it will not be used. Among other things, test producers must produce tests that are legally defensible. If a test-taker can demonstrate that he or she suffered lost opportunities (e.g., failed to graduate, or was denied admission to college or graduate school) because of a flaw in a test or inconsistencies in grading, the test-maker can be sued, given that the test was used for purposes intended by the test-maker. Thus from the perspective of test-makers and publishers, the psychometric properties of a mathematics test are far more important than the actual mathematical content of the test.

Here is one way the tension between "testing that is affordable and meets the relevant psychometric criteria" and "testing that informs teaching and careful decision-making" plays out. Almost all teachers (from elementary school through graduate school) will say that "performance tasks" (asking the student to do something, which might be to build a mathematical model or write a proof, and then evaluating that work) provide one of the best contexts for understanding what students understand. Such tasks are found very rarely on high-stakes exams. This is partly because of the cost of grading the tasks, but also because it is very difficult to make the grading of student work on open-ended tasks consistent enough to be legally bullet-proof. The kinds of tasks that K–12 teachers and college and university faculty put on their in-class exams rarely meet the psychometric criteria that are necessary to satisfy the technical departments of the publishers. Thus high-stakes exams tend to be machine graded. This has consequences. Machine-graded exams tend to focus largely on skills, and do not emphasize conceptual understanding and problem solving to the degree that many would like.

Teachers. From the teacher's perspective, assessment should help both student and teacher to understand what the student knows, and to identify areas in which the student needs improvement. In addition, assessment tasks should have curricular value. Otherwise they steal time away from the job of teaching.

Teachers are assessing students' proficiency all the time. They have access to and use multiple assessments. Some assessment is done informally, by observation in the classroom and by grading homework; some is done one-on-one, in conversations; some is done via quizzes and in-class tests; and some is done by formal assessments. But, if the formal assessments deliver just scores or percentile ratings, that information is of negligible use to teachers. If the results are returned weeks or months after the test is taken, as is often the case with high-stakes assessments, the results are of even less value. Moreover, if a teacher spends a significant amount of time on "test prep" to prepare students for a high-stakes test, then (depending on the nature of the mathematics that appears on the test) the exam — though perhaps low-cost in terms of dollars — may actually come at a significant cost in terms of classroom opportunities to learn. (See the issue of curriculum deformation, below.)

Professional development personnel. For the most part, professional development (PD) personnel are concerned with helping teachers to teach more effectively. One important aspect of such work involves helping teachers develop a better understanding of students' mathematical understanding, as can be revealed by appropriately constructed assessments (see the chapters by Foster and Fisher in this volume). Detailed information from assessments can help PD staff identify content and curricular areas that need attention. The staff can, therefore, make good use of rich and detailed assessment reports (or better, student work), but they, like teachers, will find the kinds of summary information typically available from high-stakes tests to be of limited value.

Students. Assessments should help students figure out what they know and what they don't know; they should be and feel fair. Thus tests that simply return a number — especially tests that return results weeks or months later — are of little use to students. Here is an interesting sidebar on scoring. A study by Butler [1987] indicates that placing grades on test papers can have a negative impact on performance. Butler's study had three groups: (1) students given feedback as grades only; (2) students given feedback as comments but with no grades written on their papers; and (3) students given feedback as comments *and* grades. Not surprisingly, students from Group 2 outperformed students from Group 1, for the straightforward reason that feedback in the form of comments helped the students in the second group to learn more. More interesting, however, is that there were no significant differences in performance between Groups 1 and 3. Apparently the information about their grades focused the attention of students in Group 3 away from the content of the comments on their papers.

In any case, assessments can serve useful proposes for students. The challenge is to make them do so.

Discussion. The preceding narrative shows that many of the "stakeholders" in assessment, especially in standardized testing, have goals for and needs from assessments that can be complementary and even contradictory. If any conclusions should be drawn from this section of this chapter, they are that the goals of and mechanisms for testing are complex; that different testing constituencies have very different needs for the results of assessments; and that no one type of measure, much less any single assessment, will serve the needs of all those constituencies. To compound the problem, these constituencies do not often communicate with each other about such issues. These facts, among others, led MSRI to bring together the various groups identified in this part of this chapter to establish lines of communication between them.

Unintended Consequences of Assessment

Test score inflation and the illusion of competence. If you practice something a lot, you are likely to get good at it. But the question is, what have you gotten good at? Have you learned the underlying ideas, or are you only competent at things that are precisely like the ones you've practiced on? In the latter case, you may give the illusion of competence while actually not possessing the desired skills or understandings.

Suppose, say, that elementary grade students study subtraction. Typical "two-column" subtraction problems found on standardized tests are of the form

1. $\begin{array}{r} 87 \\ -24 \\ \hline \end{array}$

Problem 1 is, of course, equivalent to the following:

2.   Subtract 24 from 87.

A priori one would expect students to do more or less as well on both problems — perhaps slightly worse on Problem 2 because it gives a verbal instruction, or because students who do the problem in their heads might find it slightly harder to keep track of the digits in the two numbers when they are not lined up vertically. But, if the students understand subtraction, Problems 1 and 2 are the same problem, and one would expect students to perform comparably on both.

Now, what would happen in a school district where students spent hour after hour being drilled on problems like Problem 1?

Roberta Flexer [1991] studied the test performance of students in a school district that had a "high-stakes" test at the end of the year. She compared test performance with an equating sample of students from another district that did not have a high-stakes test. She obtained the following statistics.

Sample	Problem 1 $\begin{array}{r} 87 \\ -24 \\ \hline \end{array}$	Problem 2 subtract 24 from 87
High-stakes district	83% correct	66% correct
Low-stakes equating sample	77% correct	73% correct

On problems that looked just like the ones they had practiced, such as Problem 1, students in the high-stakes district did substantially better than students from the low-stakes district. But their scores plummeted on the equivalent problem, and they did far worse than the "control" students. Thus the procedural skill of the students in the high-stakes district came at a significant cost.

Flexer's statistics demonstrated this kind of pattern on a range of tasks. In short, drilling students on tasks just like those known to be on the high-stakes exam resulted in the *illusion of competence*. (My mentor Fred Reif tells the story of his visit to a medical clinic. A technician asked him for the index finger of his left hand, in order to take a blood sample. Reif asked the technician if she could use his right index finger instead — he wanted the left hand untouched because he had a viola recital coming up soon after the blood test and did fingering with his left hand. The technician said no when he first asked. When he asked again, she allowed after some deliberation, and some misgivings, that "it might be OK" to use his right hand instead of the left for the blood sample. This is a case in point. Do we really want students, or professionals, who follow procedures without understanding?)

It is folk knowledge in the administrative community that an easy way for a new superintendent to appear effective is to mandate the use of a new test the first year he or she is in office. Partly because students are unfamiliar with the test format, scores are likely to be very low the first year. Then, because students and teachers become more familiar with the test format, scores go up over the next two years. Whether the students have actually learned more is open to question — but the administrator can take credit for the increase in test scores.

Shepard and Dougherty [1991] highlight a case in point — actually, 50 cases in point. "A national study by Linn [1990] documented that indeed all 50 states and most districts claimed to be above average." (That is, compared to the average score established when the test was instituted.) However, "achievement gains on norm-referenced tests during the 1980s were not corroborated by gains on NAEP."

Curriculum deformation. In the years immediately before California instated high-stakes testing in reading and mathematics, California's students scored, on average, in the middle tier of the NAEP science examinations. In the 2000 administration of the NAEP science examinations, the first since the advent of high-stakes testing in California, California student performance dropped to the very bottom. The reason: teachers stopped teaching science, because their students were not being held accountable for their science knowledge. (NAEP is a "low-stakes" test, which affects neither the teacher nor the student in any way.) They devoted their classroom time to reading and mathematics instead.

This is an example of what has been called the WYTIWYG phenomenon — "What You Test Is What You Get." WYTIWYG can play out in various ways. For example, if the high-stakes assessment in mathematics focuses on procedural skills, teachers may drill their students for procedural fluency — and conceptual understanding and problem solving skills may be left unaddressed as a consequence. (Recall the subtraction example in the previous section.)

These negatives can be contrasted with some significant positives. Colleagues have reported that in some schools where little curriculum time had been devoted to mathematics or literacy prior to the advent of high-stakes tests, as much as two hours per day are now being devoted to each. In both subject areas, a substantial increase in instructional time can be seen as a significant plus. Thus, high-stakes testing can be seen as a powerful but double-edged sword.

Stifling innovation. In the spring of the year before my daughter entered middle school I visited a number of mathematics classrooms. A number of the teachers told me that I should return to their classrooms after the testing period was over. The state mathematics exams were coming up in a few weeks, and the teachers felt they had to focus on skills that were related to items on the test. Hence what they were teaching — in some cases for weeks or months — did not reflect the practices they wished to put in place.

In some cases, curricular innovators have faced the problem that without "proof of concept" (evidence that a non-standard approach will guarantee high enough test scores) school districts are reluctant to let people try new ideas. At the MSRI conference, civil rights leader and mathematical literacy advocate Robert Moses spoke of the testing regime he had to put into place, in order to reassure administrators that his students would do OK. This represented a significant deflection of time and energy from his larger educational goals. (See [Moses and Cobb 2001], for instance.)

Disenfranchisement due to linguistic or other issues. Here is a problem taken from the Arizona high-stakes math exam:

If x is always positive and y is always negative, then xy is always negative. Based on the given information, which of the following conjectures is valid?

A. $x^n y^n$, where n is an odd natural number, will always be negative.

B. $x^n y^n$, where n is an even natural number, will always be negative.

C. $x^n y^m$, where n and m are distinct odd natural numbers, will always be positive.

D. $x^n y^m$, where n and m are distinct even natural numbers, will always be negative.

Imagine yourself a student for whom English is a second language. Just what is this question assessing? Lily Wong Fillmore [2002, p. 3] writes:

What's difficult about it? Nothing, really, if you know about, can interpret and use—

- exponents and multiplying signed numbers;
- the language of logical reasoning;
- the structure of conditional sentences;
- technical terms such as negative, positive, natural, odd, and even for talking about numbers.
- ordinary language words and phrases such as if, always, then, where, based on, given information, the following, conjecture, distinct, and valid.

I would also add that the problem statement is mathematically contorted and ambiguous. It is hardly clear what getting the question right (or wrong) would indicate, even for a native English speaker.

An impact on drop-out rates? California has instituted High School Exit Examinations (CAHSEE) in English in Mathematics. Students will have multiple chances to take the exams, but the bottom line is this: As of June 2006, a senior who has not passed both the CAHSEE mathematics exam and the CAHSEE English language exam will not graduate from high school. Instead, the student will be given a certificate of attendance. The odds are that the certificate of attendance will not have a great deal of value in the job market.

It remains to be seen how this policy will play out. But, imagine yourself to be a somewhat disaffected high school sophomore who has taken the CAHSEE for the first time. Months later your scores arrive, and they are dismal. On the basis of your scores, you judge that the likelihood of your passing the test later on, without Herculean effort, is small. You don't really like school that much. What incentive is there for you to not drop out at this point? Some have argued, using this chain of reasoning, that the net effects of instituting the exams will be a sharp rise in early drop-outs.

Discussion

My purpose in this introductory chapter has been to provide readers a sense of the assessment landscape. Developing an understanding of the mathematics that someone knows (assessing that person's mathematical proficiency) is a complex art. Different stakeholder groups (mathematicians, mathematics education researchers, parents, policy-makers, test manufacturers and publishers, teachers, professional developers, and students) all have some related but some very different needs from assessments. Not only is it the case that one assessment size does not fit all, it may well be the case that one assessment size does not fit anyone's needs very well. Thus the issue of figuring out which kinds of assessments would provide the right kinds of information to the right stakeholders is a non-trivial and important enterprise. Moreover, the set of examples discussed in the second part of this chapter shows that assessment (especially high-stakes assessment) can often be a blunt instrument, with the tests perturbing the system they measure. My hope is that these observations help to open the doors to the insights and claims in the chapters that follow.

References

[Butler 1987] R. Butler, "Task-involving and ego-involving properties of evaluation", *Journal of Educational Psychology* **79** (1987), 474–482.

[Fillmore 2002] L. W. Fillmore, *Linguistic aspects of high-stakes mathematics examinations*, Berkeley, CA: University of California, 2002.

[Flexer 1991] R. Flexer, "Comparisons of student mathematics performance on standardized and alternative measures in high-stakes contexts", paper presented at the annual meeting of the American Educational Research Association, Chicago, April 1991.

[Linn et al. 1990] R. L. Linn, M. E. Graue, and N. M. Sanders, "Comparing state and district test results to national norms: Interpretations of scoring 'above the national average' ", Technical Report 308, Los Angeles: Center for Research on Evaluation, Standards, and Student Testing, 1990.

[Moses and Cobb 2001] R. Moses and C. Cobb, *Radical equations: Math literacy and civil rights*, Boston: Beacon Press, 2001.

[NCTM 2000] National Council of Teachers of Mathematics, *Principles and standards for school mathematics*, Reston, VA: Author, 2000.

[NRC 2001] National Research Council (Mathematics Learning Study: Center for Education, Division of Behavioral and Social Sciences and Education), *Adding it up: Helping children learn mathematics*, edited by J. Kilpatrick et al., Washington, DC: National Academy Press, 2001.

[Shepard and Dougherty 1991] L. A. Shepard and K. C. Dougherty, "Effects of high-stakes testing on instruction", paper presented at the annual meeting of the American Educational Research Association, Chicago, April 1991.

[U.S. Congress 2001] U.S. Congress, H. Res. 1, 107th Congress, 334 Cong. Rec. 9773, 2001. Available at http://frwebgate.access.gpo.gov.

Assessing Mathematical Proficiency
MSRI Publications
Volume **53**, 2007

Chapter 2
Aims of Mathematics Education

JUDITH A. RAMALEY

Education is our means to instruct our youth in the values and accomplish-ments of our civilization and to prepare them for adult life. For centuries, argu-ments have been made about what an education means and how to distinguish an educated person from an uneducated one. Two views have contended for our allegiance since the time of the ancient Greeks [Marrou 1956]. One perspective is the rational and humane vision of the Sophists and later the philosopher-teacher Isocrates, for whom the test of an education was its ability to prepare a citizen to engage in public affairs. The other view is that of Plato and Socrates, who taught that education must guide the student toward an uncovering of the Truth and Beauty that underlie our human experience, the universal themes and natural laws that a well schooled mind can discern beneath the surface confusion of life and the awakening of the spirit within, that allow us to care intensely about life and learning.

We cannot clear up some of the controversies about mathematics education and how to assess learning until we deal with two underlying issues. The first is the mindset that underlies our approach to assessment. The other is to articulate and then discuss our often unspoken assumptions about what it means to be well educated.

First, let us consider what drives our current approaches to assessment. In a recent workshop on assessment [NRC 2003], the point was made that the public accountability movement is driving assessment toward increasingly large-scale tests of what students know. These tests "do not easily conform to curricula devised to match state and national standards" [NRC 2003, p. ix]. A basic problem is that testing has been shaped by psychometric questions (How can we measure this?) and used increasingly for political purposes rather than ed-ucational questions that can support learning (Is this worth measuring? What do students really need to know and can we measure that knowledge?). We must bridge the gap between what the large-scale tests measure and how the test results are interpreted and used, on the one hand, and what students and teachers are trying to accomplish in the classroom, on the other. To do this, we can profit by studying the NRC workshop report on assessment. It recommends that large-scale assessments and classroom assessments (a) share a common model of student learning, (b) focus on what is most highly valued rather than what is easy to measure, (c) signal to teachers and students what is important for them to teach and learn. The report goes on to offer some helpful technical and design elements that can increase the usefulness of both levels of tests.

If we are to assess what is most highly valued, we then must address the second underlying problem, namely, what *do* we value and what do we seek as the goal of education? We cannot talk about assessment until we are clear about our underlying philosophy of education and our goals for all of our young people. As long as we continue to approach the role of mathematics in the curriculum from different perspectives, we will have difficulty agreeing on what students should know and how they should learn. While we seek clarity of purpose, we need to keep in mind that our discussions must have genuine consequences for all students, including those that we do not serve well today. Robert Moses [2001] has made the case that children who are not quantitatively literate may be doomed to second-class economic status in our increasingly technological society. We have compelling evidence that "poor children and children of color are consistently shortchanged when it comes to mathematics" [Schoenfeld 2002]. Schoenfeld argues that we can serve all children well if we attend to four critical conditions in our schools.

- A high quality curriculum.
- A stable, knowledgeable and professional teaching community.
- High quality assessment that is aligned with curricular goals.
- Mechanisms for the continued development of curricula, assessment and professional development of our teachers and administrators.

To put all of these conditions in place, however, we need to develop a consensus about what it means to be mathematically literate. How shall we define "basic skills" and "conceptual understanding and problem solving," the relationship of these things to each other, and the appropriate balance of the two in our curriculum?

In *The Great Curriculum Debate*, Tom Loveless traces our current dilemmas back to John Dewey, who, in 1902 described two "sects." One subdivides every subject into studies, then lessons, and finally specific facts and formulae; the other focuses on the development of the child and active learning. Loveless describes the first sect as the educational-traditionalist mode or *rational* mode, a teacher-centered model that seeks classical explicit goals, expects discipline and order in the classroom where the class is led by the teacher and students are assessed by regular testing.

According to Loveless [2001, p. 2], "Traditionalists are skeptical that children naturally discover knowledge or will come to know much at all if left to their own devices." They are "confident that evidence, analysis and rational thought are greater assets in the quest for knowledge and virtue than human intuition and emotions" [2001, p. 3].

Loveless characterizes the child-centered model as the educational progressive or *romantic* tradition that "reveres nature and natural learning and allows learning to unfold without standards, rules, hierarchies of skill, rote practice and memorization." Critics of the "traditional view" tend to describe it as "drill and kill." Critics of reform-oriented mathematics dismiss it as "fuzzy math" and point out errors in the mathematics itself, arguing that the curriculum does not make mathematical sense. It is clear that the problem we have in deciding how to assess mathematics is that *we do not agree on a philosophy of education* that can offer guidance about what should be taught and how, and most importantly, *for what reasons*.

A recent study by the Mathematics Learning Study Committee of the National Research Council draws on elements of both traditions, "the basics" as well as "conceptual understanding," and links them together into a larger vision of what it means to know and be able to use mathematics. Perhaps this more integrative model can move us toward a shared understanding about what mathematics must be taught, how and to what end. Only then can we really agree on how to go about assessing mathematics learning.

The NRC in its booklet *Helping Children Learn Mathematics* [Kilpatrick and Swafford 2002] summarizes the exploration of mathematics education that appeared in fuller form in *Adding It Up: Helping Children Learn Mathematics* (2001). The "interwoven and interdependent" components of mathematics proficiency advanced by the NRC Committee are:

- Understanding: Comprehending mathematical concepts, operations and relations and knowing what mathematical symbols, diagrams and procedures mean.
- Computing: Carrying out mathematical procedures, such as adding, subtracting, multiplying and dividing numbers flexibly, accurately, efficiently and appropriately.
- Applying: Being able to formulate problems mathematically and to devise strategies for solving them using concepts and procedures appropriately.
- Reasoning: Using logic to explain and justify a solution to a problem or to extend from something known to something unknown.

- Engaging: Seeing mathematics as sensible, useful and doable — if you work at it — and being willing to do the work.

[Kilpatrick and Swafford 2002, p. 9]

This balanced approach is consistent with the work of Jerome Bruner, who argued that "any subject can be taught effectively in some intellectually honest form to any child at any stage of development" [Bruner 1977, p. 33].

Roger Geiger [1993] has pointed out that we academics tend to picture ourselves as "communities of scholars, free and ordered spaces, dedicated to the unfettered pursuit of teaching and learning." According to Geiger, in these intellectual spheres we produce increasingly specialized knowledge, not "wisdom, sagacity, or liberal learning." In recent years, there has been a great deal of exploration about how to link theory and utility in the scholarly pursuits of both faculty and students. The concepts of Donald Stokes are worth considering as a starting point for making peace across intellectual domains and clarifying our expectations about what we expect students to learn about mathematics. Although his approach was directed toward technology transfer, his ideas apply equally well to the design of the curriculum and its goals.

Stokes [1997] sought a "more realistic view of the relationship between basic research and technology policies" (p. 2) and hence between private interests (those of the researcher) and the public good (the advance of technology and its effects on society and the economy). To pave the way toward an effective blending of the imperative of knowledge for its own sake and knowledge that has consequences, Stokes developed the concept of intellectual spaces that he calls *Quadrants*. These are defined by the balance of theoretical interests and practical use pursued. Thomas Alva Edison's work fits nicely into the space framed by high interest in use and low interest in advancing understanding. He was "the applied investigator wholly uninterested in the deeper scientific implication of his discoveries." As Stokes puts it, "Edison gave five years to creating his utility empire, but no time at all to the basic physical phenomena underlying his emerging technology" [Stokes 1997, p. 24].

Niels Bohr represents the classic researcher engaged in a search for pure understanding as he explored the structure of the atom. For him, any possible practical use of his modeling was not even a consideration.

Occupying the quadrant where theory and use reinforce each other is Louis Pasteur, who had a strong commitment to understanding the underlying microbiological processes that he had discovered and, simultaneously, a motivation to use that knowledge to understand and control food spoilage, support the French wine industry and treat microbial-based disease.

It would be helpful if our discussions about mathematics education took place in Pasteur's Quadrant while recognizing that some of us are more comfortable in either Bohr's Quadrant or Edison's Quadrant. Some of our students will be drawn to deeper study of mathematics and become academic mathematicians. Others will want to develop a deeper understanding of mathematics in order to pursue careers in science, technology, or engineering. They will need to have a capacity for problem-solving and quantitative reasoning in their repertoire, but they will not advance our understanding of mathematics or pursue careers that have a rich mathematical base. Although we academics must serve all students well, they will ultimately use mathematics in very different ways. We must keep all of our students in mind and teach them authentically and honestly, being faithful to the discipline of mathematics and mindful of our students and how they are developing.

References

[Bruner 1977] J. Bruner, *The process of education*, revised ed., Cambridge, MA: Harvard University Press, 1977.

[Geiger 1993] R. L. Geiger, *Research and relevant knowledge: American research universities since World War II*, New York: Oxford University Press, 1993.

[Kilpatrick and Swafford 2002] J. Kilpatrick and J. Swafford, *Helping children learn mathematics*, Washington, DC: National Academy Press, 2002.

[Loveless 2001] T. Loveless, *The great curriculum debate: How should we teach reading and math?*, Washington, DC: Brookings Institution Press, 2001.

[Marrou 1956] H. I. Marrou, *A history of education in antiquity*, Madison: University of Wisconsin Press, 1956.

[Moses and Cobb 2001] R. Moses and C. Cobb, *Radical equations: Math literacy and civil rights*, Boston: Beacon Press, 2001.

[NRC 2003] National Research Council, *Assessment in support of instruction and learning: Bridging the gap between large-scale and classroom assessment*, Washington, DC: National Academy Press, 2003.

[Schoenfeld 2002] A. H. Schoenfeld, "Making mathematics work for all children: Issues of standards, testing and equity", *Educational Researcher* **31**:1 (2002), 13–25.

[Stokes 1997] D. E. Stokes, *Pasteur's quadrant: Basic science and technological innovation*, Washington, DC: Brookings Institution Press, 1997.

Chapter 3
The No Child Left Behind Act:
Political Context and National Goals

SUSAN SCLAFANI

This is an important moment, a time when the United States government has articulated the expectation that all U.S. students will learn mathematics. The No Child Left Behind Act (NCLB) is a bipartisan commitment to the children of the United States. It has a fundamental premise: Adults in our schools and communities must take responsibility for ensuring that all students develop the fundamental knowledge and skills in mathematics, as well as in all core subjects, that will enable them to succeed in the twenty-first century. In order to achieve this goal, NCLB draws on four basic principles: Accountability for results, local control and flexibility, choice, and research-based practice.

Accountability. Accountability for results has taken on a new dimension for educators, one which some are not eager to embrace. Not only are teachers, principals, and superintendents responsible for student performance in the aggregate, but, for the first time, they are responsible for the performance of subpopulations of students. The performance of students of color and of different ethnicities must be examined separately to ensure that all are making progress toward the standards their state has set. Achievement in English language acquisition and mathematical knowledge must be measured for students who are not fluent speakers of English. In some states, mathematical knowledge is assessed in the students' home language; in others, it is assessed in English. Students in poverty constitute a distinct group whose achievement must be assessed and improved over time.

Moreover, students with disabilities (SWD) must be assessed and the results shared publicly. Over 52 percent of all SWD are identified as learning disabled, meaning that they have average or above average intelligence and yet have

---

This is a transcript of the talk delivered by the author in her former role as Assistant Secretary in the Office of Vocational and Adult Education of the U.S. Department of Education.

difficulty learning in specific areas. Educators, including teachers and special education staff, are asked to help them develop strategies to accommodate those disabilities so the students can go on successfully in those subjects. Without assessments, it is difficult to know if students are making the progress required by NCLB, and whether the strategies used are effective for those students. Approximately eight percent of SWD, less than one percent of all students, have been identified as having significant cognitive disabilities that preclude their achievement at grade level. Each of these students is assessed at the level of the standards in his or her state that is appropriate.

NCLB is a major step forward for the future of the United States. For the first time in our history, we are taking responsibility for all children; no longer are we just mouthing the slogan "all children can learn," but we are taking responsibility for ensuring all will learn. This is a challenge that has not been taken on before, and there are no silver bullets to ensure that it is accomplished. As a nation we are asking educators to work to bring students who are behind up to grade level, we are asking them to change past practices that have not been effective, and we are asking them to work together to develop practices that are effective.

Local control and flexibility. The second principle of NCLB is local control and flexibility. Given the focus on local control, each state develops an accountability plan and determines the actions to be taken if a school does not meet the established rate of adequate yearly progress. Although the law requires certain approaches to developing an accountability system, for the most part, the policies, standards, and assessments are set by each state. The assessments are measures developed by each state to determine whether students in the state have mastered the knowledge and skills established in that state's standards for subjects and grade levels.

The United States has a decided focus on local control. It does not have a national curriculum or national exams that all students must take. Instead each state determines standards and assessments. The expected levels of achievement are set by each state and differ in difficulty and depth. The United States does have a National Assessment of Educational Progress (NAEP) that is administered to a stratified random sample of students in each state. Participation in NAEP is required by Congress to "confirm" the results of state assessment. A state's participation allows comparisons to be made between the percentage of students classified as proficient by the state exam and the percentage of students in the state sample classified as proficient by NAEP.

Choice. NCLB's third principle is choice. Choice has always been available to middle and upper class parents whose decisions in buying a house are often determined by the quality of schools in that district. Affluent parents can choose between public and private schools, but, for many parents, especially those in the highest levels of poverty, choice has not been an option. NCLB allows parents of students who are enrolled in schools that are in need of improvement the opportunity to request transfer to a school that is successful in providing education to all groups of students. Many parents choose to stay for a variety of reasons, but they have been given the opportunity to make the choice to do so. Parents of a child who stays at a school in need of improvement have a choice of providers of tutorial or supplementary educational services from the public or private sector. In this case also, federal funds are used to give parents in poverty the same choices available to other parents concerned about the quality of education their children are receiving.

Research-based practice. The fourth principle of NCLB is practice based on research. To improve educational practice, we must use evidence as the field of medicine has done for the last forty years. It is not surprising to those at universities that there has been little consistency in what is being taught from classroom to classroom in a single school, much less from school to school and district to district. That is in part because we do not have practices to suggest that are based on research on what works. Thanks to the National Institute of Health, we have gold-standard research, random-assignment studies of how students learn to read. These studies were able to isolate five components of reading that must be included in instruction: phonics, phonemic awareness, vocabulary development, fluency and comprehension. When all five components are effectively taught in grades K–3, 95 percent of third graders read at grade level, in contrast with our current level of 40 percent. The remaining five percent will require additional, more intensive interventions to get to grade level.

Now that teachers have seen the power of this research on reading in their classrooms, they are asking for similar research on mathematics. Unfortunately, we do not have thirty years of high quality research that would enable us to tell them definitively what to teach and how to teach it in mathematics. That is, in part, the purpose of this talk — to convince others that further research is needed on content and pedagogy in order to help teachers more effectively teach their students.

Highly qualified teachers. Another area of NCLB is the issue of highly qualified teachers. One would not consider it controversial to ask that secondary
school teachers have a major in the subject they teach, yet that is so far from
current practice that it had to become part of the law. Currently over 50 percent
of our middle school teachers have neither a major nor a minor in mathematics.
Many of them have K–8 certification that required only one to three courses in
mathematics. It is desirable that they have additional professional development
in mathematics, but some do not. In addition, courses taught by unqualified
teachers are not evenly distributed across all of our schools. One does not find
unqualified teachers teaching mathematics as often in suburban schools as in
urban or rural schools. It occurs in urban high schools or in more advantaged
schools when the master schedule requires another section of algebra and no
mathematics teachers are available to teach it. The solution is often to find
someone else on the faculty who is available that period, regardless of whether
his or her only training was an algebra course taken in high school. It is estimated
that unqualified teachers teach courses in a variety of subjects to over a third of
high school students for this reason.

Assessment.   Finally, I would like to expand my discussion of assessment.
Assessment occurs at many different levels. I have mentioned the state assessments that NCLB requires — at grades three through eight and once in high
school. These are meant to inform policy makers about the progress students
are making. NCLB requires assessments in reading and mathematics in order to
provide an indicator of the health of the educational system, as temperature and
blood pressure are checked in a doctor's office as an indicator of an individual's
health. When large percentages of all student groups in a school are performing
at high levels, the school is making adequate progress, and students can be
expected to succeed at the next grade level. However, these assessments are not
the only measure of what students know and are able to do.

Classroom assessments that use a variety of strategies and forms are critical
to ensuring that students learn at the depth required in mathematics. However,
classroom assessment is also connected to the qualifications of the teachers. If
the teacher is not prepared in mathematics, how can he or she assess the knowledge and skills of students at the depth required? How can he or she develop
valid and reliable measures of what students have learned? It is not likely that
such teachers would teach or assess their students at appropriate levels.

District-level testing can inform district staff about the achievement of students in each school and within each classroom. In Houston, district staff used
test results to know where to intervene with professional development for teachers and with additional classes for their students which were taught by college
and graduate students majoring in mathematics.

The NAEP provides another set of statistics about what our students know and are able to do. Attaining proficiency on NAEP means that students are able to take what they have learned and apply it to solving problems. Disaggregated statistics from the 2003 twelfth grade NAEP tell us that current practices have led to a national disgrace: only twenty percent of our white students, three percent of our African American twelfth graders and only four percent of our Hispanic twelfth graders are proficient in mathematics. And these are the students who have made it successfully to twelfth grade and who are going to graduate from high school within months, unprepared for careers or further education.

Finally, I would like to mention the Mathematics and Science Initiative. In February 2003, Secretary of Education Rod Paige convened a Summit on Mathematics that focused on curriculum, teacher development, assessments, and research in mathematics teaching and learning. Presentations at the Summit made clear that unless future teachers are differently prepared and current teachers are re-educated, the next generation of students will not be prepared for careers in science, technology, engineering or mathematics, or able to pursue the wide variety of other careers that require an understanding and use of mathematics. That is why NCLB is so critical to the future of the United States. Unless we focus on the progress our students are making and change our practices to ensure that all children learn at high levels, we cannot expect our students to succeed or our country to remain at the nexus of power, productivity, and innovation.

Section 2
Perspectives on Mathematical Proficiency

Definitions are important in education, as they are in mathematics. If one is to assess students' mathematical proficiency, then one had better start by defining the term. As the two essays in this section indicate, this is not as straightforward as it might seem. To echo R. Buckminster Fuller, a fundamental question is whether one considers mathematics be a noun or a verb. One's view makes a difference: what one defines mathematics to be has significant implications both for teaching and for assessment.

One way to view mathematics (the "noun" view) is as a wonderful and re-markably structured body of knowledge. From this perspective, the question becomes: How should that body of knowledge be organized so that students can best apprehend it? A second way to view mathematics is to think of it as What Mathematicians Do, with an emphasis on the verb. Even here, there are multiple levels of description. At the action level, for example, there are mathematical activities such as solving problems and proving theorems. At a deeper process level there are the activities of abstracting, generalizing, organizing, and reflect-ing (among others), which are called into service when one solves problems and proves theorems. In this section, R. James Milgram (in Chapter 4) and Alan H. Schoenfeld (in Chapter 5) explore aspects of the topics just discussed: the nature of mathematics and what it means to do mathematics, and implications of these views for both instruction and assessment. These two chapters, in combination with the three chapters in the previous section, establish the mathematical basis vectors for the space that is explored in the rest of this volume.

Chapter 4
What is Mathematical Proficiency?

R. JAMES MILGRAM

In February of 2004 Alan Greenspan told the Senate Banking Committee that the threat to the standard of living in the U.S. isn't from jobs leaving for cheaper Asian countries. Much more important is the drop in U.S. educational standards and outcomes.

> "What will ultimately determine the standard of living of this country is the skill of the people," Greenspan pointed out .... "We do something wrong, which obviously people in Singapore, Hong Kong, Korea and Japan do far better. Teaching in these strange, exotic places seems for some reason to be far better than we can do it." [Mukherjee 2004]

Current estimates by Forrester Research (Cambridge, MA) are that over the next 15 years at least 3.3 million jobs and 136 billion dollars in wages will move to Asia.

Introduction

The first job of our education system is to teach students to read, and the majority of students do learn this. The second thing the system must do is teach students basic mathematics, and it is here that it fails. Before we can even think about fixing this — something we have been trying to do without success for many years — we must answer two basic questions.

- What does it mean for a student to be proficient in mathematics?
- How can we measure proficiency in mathematics?

These are hard questions. The initial question is difficult because mathematics is one of the most seriously misunderstood subjects in our entire K–12 educational system. The second question is hard for two reasons.

The research that underlies this chapter was funded by an FIE grant from the U.S. Department of Education. I would like to thank Deborah Loewenberg Ball and Kristin Umland for invaluable discussions and help with many aspects of this work.

(a) Too often test designers and textbook authors do not have a clear idea of what mathematics is. Indeed, something on the order of 25% of the questions on a typical state mathematics assessment are mathematically incorrect. (This will be expanded on in the last section, p. 55.)

(b) Proper assessment is difficult under the best conditions, but becomes essentially impossible when the people writing the tests do not adequately understand the subject.

The first question — what is mathematical proficiency? — will be discussed in what follows. Part of our theme will be to contrast practices in this country with those in the successful foreign countries that Alan Greenspan mentions. We will see that U.S. practices have relatively little connection with actual mathematics, but the programs in the high achieving foreign countries are closer to the mark.

We can understand this surprising assertion better when we understand that school mathematics instruction has drifted to the point where one simply cannot recognize much of actual mathematics in the subject as it is taught in too many of this country's schools. In fact this drift seems to be accelerating. It is fair to say that there has been more drift over the last twenty to thirty years than there was during the previous eighty.

Some schools in our country teach nothing but arithmetic, some nothing but something they call problem solving and mathematical reasoning. Both call what they teach "mathematics." Both are wrong. At this point and in this country, what is taught as mathematics is only weakly connected with actual mathematics, and typical curricula, whether "reform" or "traditional," tend to be off the point. Before continuing we must clarify this assertion.

What is Mathematics?

We must have some idea of what mathematics is in order to start our discussion. Unfortunately, a serious misconception already occurs here. Some things simply cannot be defined in ordinary language and mathematics is almost certainly one of them. This doesn't mean that we can't describe the subject in general terms. We just can't sharply limit it with a definition.

Over the years, a number of people have tried to define mathematics as "the study of patterns" or "the language of science," but professional mathematicians have avoided trying to define mathematics. As best I can recollect, the nearest that a research mathematician came to attempting a definition in print was Roy Adler in the mid-1960s who suggested the semi-serious "Mathematics is what mathematicians do."

A few years back a serious attempt at a short description of mathematics was given privately by Norman Gottlieb at Purdue. He suggested "Mathematics is the study of precisely defined objects." A number of people participating in this discussion said, in effect, "Yes, that's very close, but let's not publicize it since it would not sound very exciting to the current MTV generation, and would tend to confirm the widely held belief that mathematics is boring and useless."

Realistically, in describing what mathematics is, the best we can do is to discuss the most important characteristics of mathematics. I suggest that these are:

(i) Precision (precise definitions of all terms, operations, and the properties of these operations).
(ii) Stating well-posed problems and solving them. (Well-posed problems are problems where all the terms are precisely defined and refer to a single *universe* where mathematics can be done.)

It would be fair to say that virtually all of mathematics is problem solving in precisely defined environments, and professional mathematicians tend to think it strange that some trends in K–12 mathematics education isolate *mathematical reasoning and problem solving* as separate topics within mathematics instruction.

For Item 1 above, the rules of logic are usually considered to be among the basic operations. However, even here mathematicians explore other universes where the "rules of logic" are different. What is crucial is that the rules and operations being used be precisely defined and understood. Mathematics is a WYSIWYG field. There can be no hidden assumptions — *What you see is what you get*.

The Stages Upon Which Mathematics Plays

What we have talked about above is mathematics proper. However, most people do not talk about mathematics but algebra, geometry, fractions, calculus, etc. when they discuss mathematics, and we have not mentioned any of these topics. So where do numbers, geometry, algebra fit in? Mathematics typically plays out on a limited number of stages,[1] and there is often considerable confusion between the stages and the mathematics on these stages.

Here are some examples of stages. The integers build a stage, the rationals build a stage, and the reals build yet another a stage. These are the most important stages for mathematics by far, but you cannot limit mathematics to just these stages.

---

[1] By "stage" I mean — in mathematical terms — a category having objects, maps, maybe Cartesian products, and sufficient structure to do mathematics. However, stage, in the theatric sense, seems a very good description of these structures.

- Patterns, once one has a proper definition, build a stage. This is the theory of groups and group actions — a very advanced subject in mathematics.
- Geometry plays out on another stage. Geometry as we commonly know it, is the mathematics of the plane or space together with its points, lines, planes, distance, angles and transformations. However, the precise definitions here are even more difficult than is the case with patterns.

In practice, in school mathematics, some of the stages above are systematically but heuristically developed for students over a period of years and mathematics is played out to varying degrees. For example, here is a quick description of what happens during the first three years in a program that does things right — the Russian texts translated by the University of Chicago School Mathematics Project [UCSMP 1992a; 1992b; 1992c].

First grade. The stage on which first grade mathematics plays consists of the counting numbers from 1 to 100, addition, subtraction, and simple two- and three-dimensional geometric figures. There are very few definitions here and, since the stage is so small, the definitions can be quite different from definitions students will see later, though they should be present. For example, some of the definitions are given almost entirely via pictures, as is illustrated by the definition of adding and subtracting 1 on page 9:

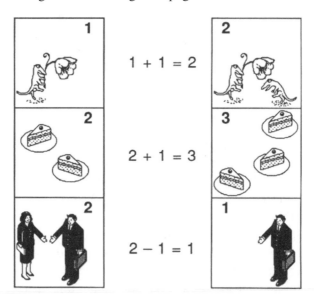

Second grade. The stage is larger, the counting numbers at least from 1 to 1000, all four operations, time, beginning place value, small fractions, and a larger class of geometric figures. There are more definitions and they are more advanced than those in the first-grade universe. Here is the definition of multiplication. Note that it is the first paragraph in the chapter on multiplication and division on page 38:

Multiplication

143.
$$2 + 2 + 2 + 2 + 2 = 10$$
$$2 \cdot 5 = 10$$
The addition of identical addends is called **multiplication**. The sum of the identical addends $2 + 2 + 2 + 2 + 2$ can be written: $2 \cdot 5$.

The multiplication sign is a dot (·). Multiplication problems are read as follows:

$2 \cdot 5 = 10$ **If you take two five times, you get ten, or, two multiplied by five gives ten.**

The vertical bar above appears in these texts to indicate a definition. One can legitimately ask if this definition is sufficient, or if it is possible to do better. But here the main point is that there *is a definition* present, and its position in the exposition indicates that it is an important part of the sequence of instruction.

The definition of division is the first topic in the section on division that begins on page 42, immediately following the section on multiplication:

Division

162.   8 oranges were placed on dishes, 2 oranges to a dish. How many sets of 2 oranges were there? How many dishes were needed?

This problem is solved by the **division** operation. The division sign is two dots (:).

The solution to the problem can be written as follows:
$$8 : 2 = 4$$ Answer: 4 dishes

163.   Grandmother had 10 carrots. She tied them into bunches of 5 carrots each. How many bunches did she get?
$$10 : 5 = 2$$ Answer: 2 bunches

In words, the translation of the definition (which is mostly visual) is that division of b by a is the number of equal groups of a objects making up b. It would likely be the teacher's obligation to see that each student understood the definition. For most, the example and practice would suffice. Some would require the entire verbal definition, and others might need a number of visual examples. This also illustrates an important point. There is no necessity that definitions be entirely verbal. For young learners, visual definitions may be more effective, and these Russian texts show a consistent progression from visual to verbal, as we will see shortly when we give the corresponding third-grade definitions.

As an example of how these definitions are applied, we have the discussion of even and odd numbers (from the section on multiplying and dividing by 2), page 90:

347.   From the series of numbers 1, 2, 3, 4, 5, 6, 7, 8, 9, 10, 11, 12, 13, 14, first give those which are divisible by 2, and then those which are not divisible by 2.

Numbers which are divisible by 2 are called **even**. Numbers which are not divisible by 2 are called **odd**.

348.   From the series of numbers 19, 10, 20, 8, 7, 1, 5, 4, 6, 18, 3, first write out all the odd numbers and then all the even numbers.

It is worth noting that the students have not yet seen fractions so the definition is actually unambiguous. (It has been pointed out to me that children are often ahead of the class instruction, and some children will be familiar with fractions, so that this definition, talking about *numbers* and not *whole numbers* is inappropriate. This is an interesting point, and illustrates one of the points of *tension* between rigorous mathematics and practical teaching. From the mathematical perspective, since the only *numbers* in the universe at the time of the definition are whole numbers, the definition is entirely correct. However, as a practical matter, the teacher must deal with children who want to test the definition on fractions.)

Third grade. The stage is much larger by third grade. Place value and the standard algorithms are fully developed.[2] Area and complex polygonal figures in the plane are present, weights and measures, velocity and the relationship between time and distance traveled at constant velocity have been given, fractions have been developed much further and are now represented on the number line.

---

[2]The details of the algorithms in the Russian texts are different from the details commonly taught in the United States. Nonetheless, the underlying mathematical algorithms are identical, so I use the term *standard algorithm* for the Russian algorithms as well as ours.

Again, there are new definitions appropriate to this new stage, some of which are quite sophisticated. Here is the definition of multiplication — as before the first item in the chapter on multiplication and division on page 89:

Multiplying by a One-Digit Number.

330. To multiply the number 18 by 3 is to take the number 18 as an addend 3 times:
$18 \cdot 3 = 18 + 18 + 18$.
To multiply a number k by 4 is to take it as an addend 4 times.
To multiply a number a by a number b is to take the number a as an addend b times.

331. Are the following equalities true?
$18 \cdot 5 = 18 + 18 + 18 + 18$
$c + c + c + c + c + c = c \cdot 8$
$12 + 12 + 12 + 11 = 12 \cdot 4$
$45 \cdot 6 = 45 + 45 + 45 + 45$

And here is the definition of division. Note the considerable advance on the second-grade definition. This definition will be unchanged in later grades, page 110:

Dividing by a One-Digit Number.

410. Division is related to multiplication; to divide 48 by 4 means to find a number which multiplied by 4 gives 48. This number is 12. That is, $48 : 4 = 12$. What does it mean to divide 72 by 9? 100 by 25?

411. Check by multiplying whether the following divisions have been done correctly:

$95 : 19 = 5$ $\qquad$ $180 : 6 = 30$ $\qquad$ $450 : 3 = 150$

Just for illustration, here are some problems from the end of this chapter on page 137:

27. Solve the following problems by means of equations:

(1) Think of a number. If it is decreased 4 times and 180 is subtracted from the resulting number, 720 is obtained. What number is it?

(2) When the difference of some unknown number and the number 70 is divided by 9, 4 is obtained. Find the unknown number.

(3) 120 people volunteered to landscape a town. Several teams, each with 5 people, worked in a park, and the remaining 40 people planted trees along the streets. How many teams worked in the park?

This is a dramatic demonstration of just how far third-grade students can go with mathematics when the foundations are properly set up. Each concept used in these problems has been defined and discussed in this textbook.

We can compare this program and what is done in other successful foreign programs with what is typically done in this country.[3] For example, the problems above are very similar to those found in fourth-grade Singapore textbooks, so this level of expectation is normal for both countries. In the U.S., where precise definitions are not the norm, problems like these do not appear in textbooks until much later.

To a professional mathematician, one of the most glaring differences between the textbooks in high achieving countries and the United States is the care with which the distinctions above are made for the students and the precision (stage-appropriate, of course, but precision nonetheless) of the definitions in the programs of the successful countries. Moreover, this holds from the earliest grades onwards.

The tools of mathematics. Each stage has its own tools, the rules that we assume are valid there. There are also the overriding tools of mathematics, those tools that tend to have wide applicability over many stages. One of the most important of these tools is abstraction — focusing attention on the most important aspects of a situation or problem and excluding the extraneous.[4]

Problems in mathematics

A problem in mathematics (or a well-posed problem) is a problem where every term is precisely understood in the context of a single stage. It may have no answer, it may have an answer that contains many special subcases. The answer may be complex, as in the case of the problem

Find all quadruples of whole numbers (n, a, b, c) that satisfy the equation $a^n + b^n = c^n$

— this is Fermat's "Last Theorem," of course, and the answer is $\{n = 1$ and $a + b = c$, or $n = 2$ and (a, b, c) forms a Euclidean triple, so $a = 2vw$, $b = v - w$, $c = v + w$ for two integers v, $w\}$; its proof, found by Andrew Wiles some ten years ago, is one of the greatest achievements of modern mathematics.

On the other hand, the answer might be simple, as for the problem "Is the square root of 2 a rational number?", whose answer is "No." The problem can be rephrased as

Find all triples of integers that satisfy the two equations $a^2 + b^2 = c^2$ and $a = b$;

---

[3]To do this properly with any particular program would take much more space than we have available, but there is some discussion of a few of the issues with these programs on the next few pages.

[4]The type of abstraction being discussed here is part of problem solving. Abstraction appears in another form in mathematics, when processes that are common to a number of situations are generalized and given names. We do not discuss this aspect of abstraction here.

rigorously showing that the answer is $a = b = c = 0$, which implies that the square root of two is not rational, was one of the triumphs of Greek mathematics.

As long as all the terms are precisely understood and everything is included in a mathematical stage, we have a problem in mathematics. However, an item like

Find the next term in the sequence $3, 8, 15, 24, \ldots$,

which is extremely similar to items found on almost every state mathematics assessment that I've seen in this country is not a mathematics problem as stated.

Why not? The phrase "next term" has been given no meaning within the context of the question — presumably the stage of whole numbers and their operations. In order to answer this question one would have to *make a guess* as to what the phrase means.[5] At best this is a question in psychology. As such it is somewhat typical of the questions one used to find on IQ tests, or the now discontinued SAT analogies section, where a cultural or taught predisposition to understand "next term" in the same way as the person asking the question biases the results. In short, there most definitely tend to be hidden assumptions in problems of the type described.

It might be helpful to explain in more detail what is wrong with the above problem. Presumably what is wanted is to recognize that the n-th term is given by the rule $(n + 1)^2 - 1 = n(n + 2)$, but this only makes sense if you are told that the rule should be a polynomial in n of degree no more than 2. If you are not given this information then there is no reason that the following sequence is not equally correct (or equally incorrect):

$$3, 8, 15, 24, 3, 8, 15, 24, 3, 8, 15, 24 \ldots$$

with general term

$$\begin{cases} 3 & n \equiv 1 \bmod (4) \\ 8 & n \equiv 2 \bmod (4) \\ 15 & n \equiv 3 \bmod (4) \\ 24 & n \equiv 4 \bmod (4) \end{cases}$$

for $n \geq 1$. But that's only one possibility among an infinity of others. One could, for example, check that the polynomial

$$g(n) = n^4 - 10n^3 + 36n^2 - 48n + 24$$

has 3, 8, 15, 24 as its values at $n = 1, 2, 3, 4$ respectively, but $g(5) = 59$, and

---

[5] Guessing might be an appropriate strategy at some point when *solving* a problem, but not when *understanding* a problem. This is a real distinction between mathematics and some other disciplines. For example, it is perfectly sensible to make guesses about what it is you are trying to understand in science, but in mathematics, if the basic terms have not been defined, there can be no problem.

$g(6) = 168$. So the rule $g(n)$ would give an entirely different fifth term than would $n(n + 2)$ though both give the same values at 1, 2, 3, and 4. Likewise, one could assume the rule is another repeating form, for example

$$3, 8, 15, 24, 15, 8, 3, 8, 15, 24, 15, 8, \ldots$$

where the value at $n + 6$ is the same as the value at n with the values at $n = 1, 2, 3, 4, 5, 6$ given above. In fact the way in which the original sequence could be continued is limited only by your imagination, and who is to say one of these *answers* is more correct than another?

Put yourself in the position of a student who has just been told that the correct answer to the "problem" $\{3, 8, 15, 24\}$ is 35. What will such a student think? There is no *mathematical* reason for such a claim, but the teacher is an authority figure, so this student will tend to accept the teacher's statement and revise any idea he or she might have about what mathematics is. The student will begin to arrive at the understanding that mathematics is, in fact, whatever the instructor wants it to be! Moreover, the student will, as a corollary, learn that answering a mathematical problem amounts to guessing what the person stating the problem wants. Once that happens the student is lost to mathematics. I cannot tell you the number of times that colleagues and even nonprofessionals from Russia, Europe, Japan, and China have mentioned these ubiquitous next-term problems on our tests to me and wondered how we managed to teach mathematics at all.

I wish I could say that these next-term questions are the only problem with K–12 mathematics in this country. But we've only skimmed the surface. For example, *the one thing that is most trumpeted by advocates of so-called reform math instruction in the United States is problem solving. We will see later that the handling of this subject in K–12 is every bit as bad as the next term questions.* To prepare for our discussion of problem solving we need some preliminaries.

School Mathematics as Lists

Perhaps the major reason for the pervasive collapse of instruction in the subject lies in the common view of many educators that learning mathematics consists of memorizing long lists of responses to various kinds of triggers — mathematics as lists.

Over the past eight years I have read a large number of K–8 programs both from this country and others and a number of math methods textbooks. None of the underlying foundations for the subject are ever discussed in this country's textbooks. Instead, it seems as if there is a checklist of disconnected topics. For example, there might be a chapter in a seventh- or eighth-grade textbook with the following sections:

- two sections on *solving one-step equations*,
- two sections on *solving two-step equations*,
- one section on *solving three-step equations*,

but no discussion of the general process of simplifying linear equations and solving them. Sometimes these check lists are guided by *state standards*, but the same structure is evident in older textbooks.

My impression is that too many K–8 teachers in this country, and consequently too many students as well, see mathematics as lists of disjoint, disconnected factoids, to be memorized, regurgitated on the proximal test, and forgotten, much like the dates on a historical time-line. Moreover, and more disturbing, since the isolated items are not seen as coherent and connected, there does not seem to be any good reason that other facts cannot be substituted for the ones that are out of favor.

A good example is long division. Many people no longer see the long-division algorithm as useful since readily available hand calculators and computers will do division far more quickly and accurately than we can via hand calculations. Moreover, it takes considerable class time to teach long division. Therefore many textbooks do not cover it. Instead, discussion of what is called *data analysis* replaces it. As far as I am able to tell, there was no consideration of the mathematical issues involved in this change, indeed no awareness that there are mathematical issues.[6]

The other concern that one has with the "mathematics as lists" textbooks is that it's like reading a laundry list. Each section tends to be two pages long. No section is given more weight than any other, though their actual importance may vary widely. Moreover, each section tends to begin with an example of a trivial application of this day's topic in an area that young students are directly experiencing. Cooking is common. Bicycles are common. But the deeper and more basic contributions of mathematics to our society are typically absent. How can students avoid seeing the subject as boring and useless?

---

[6]To mention but one, the algorithm for long division is quite different from any algorithm students have seen to this point. It involves a process of successive approximation, at each step decreasing the difference between the *estimate* and the exact answer by a factor of approximately 10. This is the first time that students will have been exposed to a convergent process that is not *exact*. Such processes become ever more important the further one goes in the subject. Additionally, this is the first time that estimation plays a major role in something students are doing. Aside from this, the sophistication of the algorithm itself should be very helpful in expanding students' horizons, and help prepare them for the ever more sophisticated algorithms they will see as they continue in mathematics. The algorithm will be seen by students again in polynomial long division, and from there becomes a basic support for the study of rational functions, with all their applications in virtually every technical area. In short, if students do not begin to learn these things using the long-division algorithm, they will have to get these understandings in some other way. These are aspects of mathematics that should not be ignored.

The Applications of Mathematics

The usual reasons given in school mathematics for studying mathematics are because it is "beautiful," for "mental discipline," or "a subject needed by an educated person." These reasons are naive. It doesn't matter if students find the subject beautiful or even like it. Doing mathematics isn't like reading Shakespeare — something that every educated person should do, but that seldom has direct relevance to an adult's everyday life in our society. The main reason for studying mathematics is that our society could not even function without the applications of a very high level of mathematical knowledge. Consequently, without a real understanding of mathematics one can only participate in our society in a somewhat peripheral way. Every student should have choices when he or she enters the adult world. Not learning real mathematics closes an inordinate number of doors.

The applications of mathematics are all around us. In fact, they are the underpinnings of our entire civilization, and this has been the case for quite a long time. Let us look at just a few of these applications. First there are buildings, aqueducts, roads. The mathematics used here is generally available to most people, but includes Euclidean geometry and the full arithmetic of the rationals or the reals.[7] Then there are machines, from the most primitive steam engines of three centuries back to the extremely sophisticated engines and mechanisms we routinely use today.

Sophisticated engines could not even be made until Maxwell's use of differential equations in order to stop the engines of that time from flying apart, stopping, or oscillating wildly, so the mathematics here starts with advanced calculus. Today's engines are far more sophisticated. Their designs require the solutions of complex nonlinear partial differential equations and very advanced work with linear algebra.

Today a major focus is on autonomous machines, machines that can do routine and even nonroutine tasks without human control. They will do the most repetitive jobs, for example automating the assembly line and the most dangerous jobs.

Such jobs would then be gone, to be replaced by jobs requiring much more sophisticated mathematical training. The mathematics needed for these machines, as was case with engines, has been the main impediment to actual widescale implementation of such robotic mechanisms. Recently, it has become clear that the key mathematics is available — mathematics of algebraic and geometric topology, developed over the last century — and we have begun to make dramatic progress in creating the programs needed to make such machines work.

---

[7]The need to build structures resistant to natural disasters like earthquakes requires much more advanced mathematics.

Because of this, we have to anticipate that later generations of students will not have the options of such jobs, and we will have to prepare them for jobs that require proportionately more mathematical education.

But this only touches the surface. Computers are a physical implementation of the rules of (mathematical) computation as described by Alan Turing and others from the mid-1930s through the early 1940s. Working with a computer at any level but the most superficial requires that you understand algorithms, how they work, how to show they are correct, and that you are able to construct new algorithms. The only way to get to this point is to *study* basic algorithms, understand why they work, and even why these algorithms are better (or worse) than others. The highly sophisticated "standard algorithms" of arithmetic are among the best examples to start. But one needs to know other algorithms, such as Newton's Method, as well. What is essential is real knowledge of and proficiency with *algorithms in general*, not just a few specific algorithms.

And we've still only touched the surface. Students have to be prepared to live effective lives in this world, not the world of five hundred years back. That world is gone, and it is only those who long for what never was who regret its passing. Without a serious background in mathematics one's options in our present society are limited and become more so each year. Robert Reich described the situation very clearly [2003]:

> The problem isn't the number of jobs in America; it's the quality of jobs. Look closely at the economy today and you find two growing categories of work — but only the first is commanding better pay and benefits. This category involves identifying and solving new problems.... This kind of work usually requires a college degree....
>
> The second growing category of work in America involves personal services.... Some personal-service workers need education beyond high school — nurses, physical therapists and medical technicians, for example. But most don't.

Mathematical Topics and Stages in School Mathematics

Historically, the choices of the mathematics played out on particular mathematical stages that is taught in K–12 have been tightly tied to the needs of our society. Thus, my own education in upstate New York and Minnesota, with learning to use logarithms and interpolation in fifth and sixth grade, exponentials and compound interest in seventh grade, and culminating in solid geometry and trigonometry was designed to prepare for the areas of finance, architecture, medicine, civil, and mechanical engineering. For example, exponentials are essential for figuring dosages of medicine and for dealing with instability, errors and vibrations in mechanisms.

In the countries most successful in mathematics education, these considerations routinely go into their construction of mathematics standards. People from all concerned walks of life set the *criteria* for the desired outcomes of the education system, and professional mathematicians then write the standards. In the U.S. the notion of overriding criteria and focused outcomes seems to virtually never play a role in writing mathematics standards, and the outcomes are generally chaotic.

Let us now return to our main theme — mathematical proficiency.

Definitions

We have talked about what mathematics is in general terms. The word that was most frequently used was precision. The first key component of mathematical proficiency is the ability to understand, use, and as necessary, create definitions.

A definition selects a subset of the universe under discussion — the elements that satisfy the definition.

Once one has a definition one must understand it. This does not simply mean that one memorizes it and can repeat it verbatim on command. Rather, a student should understand why it is stated the way it is. It is necessary to apply at least the following three questions to every definition:

- What does the statement include?
- What does the statement exclude?
- What would happen if the definition were changed and why is the changed definition not used?

This is so basic that, once it became clear that my father could not dissuade me from becoming a mathematician, he gave me one key piece of advice. He said "Whenever you read a book or a paper in mathematics and you come to a definition, stop. Ask why the definition was given in the way it was, try variations and see what happens. You will not understand what is going on unless you do."

The lack of definitions in U.S. mathematics instruction. Definitions are the most problematic area in K–12 instruction in this country. First, they hardly ever appear in the early grades, and later, when people attempt to use definitions they get things wrong. Consider the following problem from a recent sample of released eighth-grade state assessment questions [Kentucky 2004, p. 5]:

4. Which diagram below best shows a rotation of the pre-image to the image?

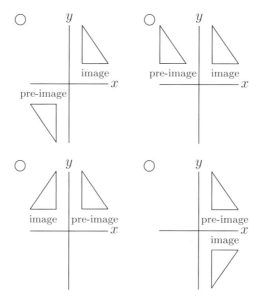

The solution sheet tells us that the upper left choice is "the answer," but is it? Let us ignore the imprecision of the phrase "best shows," and assume that what is being asked is "Which diagram shows the effect of a rotation from the pre-image to the image?"[8] In fact, each of the answers is correct, depending on how well one knows the definition and properties of rotations.

A rotation in space always has an axis, a straight line fixed under the rotation. To see such a rotation take an orange, support it only at the top and bottom and spin the orange. When it stops you will have a rotation through the total angle of the spin. We obtain all but the picture at the upper right in this way, depending on the angle of the line connecting the top and bottom and where we put the triangle in the orange. But even the picture at the upper right can be obtained by a rotation from a more advanced mathematical perspective — that of projective geometry. This is a subject that was extensively studied about a century back, and is usually part of standard undergraduate geometry or applied linear algebra courses at the college level. It is the essential tool in computer graphics. One would hope that the people charged with writing a state math assessment would know mathematics at this level, but they should certainly be aware that rotations in space are easier to understand than rotations in the plane. Also, note the gratuitous use of the terms "image" and especially "pre-image" in this problem.

---

[8] As stated this problem rests on the undefined notion of "best." So, as discussed above, it is not a question in mathematics. But we might guess the intent of the question, and that is what I tried to do in rephrasing it.

It seems to be the norm in K–8 mathematics textbooks in this country that there are no definitions in the early grades, and this seems to also be the practice in virtually all our K–5 classrooms. For example, in the NCTM's *Principles and Standards* [NCTM 2000] we find

- develop understanding of fractions as parts of unit wholes, as parts of a collection, as locations on number lines, and as divisions of whole numbers;
- use models, benchmarks, and equivalent forms to judge the size of fractions.

Note the strict use of models (or representations, in fact *multiple* representations), instead of definitions. We had a difficult time with this standard in California. If you are going to compare fractions you have to have a definition of fraction, and then a definition telling what it means for a fraction to be greater than another fraction.[9] So we changed it to

- Explain different interpretations of fractions, for example, parts of a whole, parts of a set, and division of whole number by whole numbers.

Even this was not entirely satisfactory since there was no explicit definition, but implicitly, the use of the term *interpretations* should give a hint that there is a single notion underlying all these different *representations*. At least that was what we hoped.[10]

Who gets hurt when definitions are not present? The emphasis on precision of language and definitions matters most for exactly the most vulnerable of our students. It is these students who must be given the most careful and precise foundations. The strongest students often seem able to fill in definitions for themselves with minimal guidance. On the other hand, foreign outcomes clearly show that with proper support along these lines, all students can get remarkably far in the subject.

Mathematical Problem Solving

We have seen that there are three basic components to mathematics: stages, definitions, and problem solving. We have discussed the first two. It is now time to discuss problem solving.

---

[9]Educators tend to look at one blankly when we say something like this. They typically respond that, intuitively, there is only one possible ordering. But this is not true. Orderings have been studied in advanced mathematics and it turns out that there are infinitely many *different* consistent meanings for less than or greater than for fractions and integers, each of which has its uses. The actual situation is very far from being intuitive.

[10]Underlying the lacuna in school math around definitions sometimes appears to be a belief, or perhaps a hope, that mathematics is innate, and that students, playing with manipulatives, will find all of mathematics already hiding in their memories. Mathematicians who teach mathematics for pre-service elementary school teachers often have to deal with such claims when students from the education schools come to them explaining that they do not have to learn what is currently being covered in the course since they will automatically know it when they need it.

To start, I think everyone needs to be aware of this basic truth:

PROBLEM SOLVING IS CURRENTLY AN ARCANE ART.

We do not know how to reliably teach problem solving. The most effective method I know is to have a mathematician stand in front of a class and solve problems. Many students seem to be able to learn something of this multi-faceted area in this way, but, as we will see, the stage has to be carefully set before students can take advantage of this kind of experience.

What I will discuss now is what virtually all serious research mathematicians believe, and, as far as I've been able to ascertain, most research scientists. This is not what will be found in a typical math methods textbook. Other theories about mathematical problem solving are current there. It could be that the focus of the views on problem solving in these texts is concerned with routine problems where the biggest effort might be in understanding what the problem is asking. This can be a difficult step, but *here we are talking about solving a problem where the answer is not immediate and requires a novel idea from the student.* It is exactly this level of problem solving that should be the objective for every student, because, at a minimum, this is what virtually all nonroutine jobs will require.

For example, when I was young, dock work was brutal — lifting and carrying. Today, the vast majority of this work is done by huge robotic mechanisms, and the dock worker of today spends most of his or her time controlling a very expensive and complex machine or smaller forklifts. The usual requirement is two years of college to handle the big machines, because running these big machines entails extensive nonroutine problem solving.

The hidden part of problem solving. There is a hidden aspect to problem solving: something that happens behind the scenes, something that we currently do not know how to measure or explain. It is remarkable, when you read the biographies of great mathematicians and scientists that they keep saying of their greatest achievements, "I was doing something else and the answer to my problem just came to me."[11] This is not only true for the greatest, it is true for every serious research mathematician or scientist that I've ever talked to about this kind of issue.

Answers and ideas just seem to come out of the blue. But they don't! There are verbal and nonverbal aspects to problem solving. *Successful researchers have learned how to involve the nonverbal mechanisms in their brains in analyzing and resolving their problems, and it is very clear that these nonverbal regions are much more effective at problem solving than the verbal regions.* (My

---

[11] H.-H. Wu points out that the first example of this that he is aware of in print is due to H. Poincaré.

usual experience was to wake up suddenly at 2:00 a.m. or so, and the answer to a problem that I had been working on without success maybe two weeks back would be evident.)

In order to engage the nonverbal areas of the brain in problem solving, extensive training seems to be needed. This is probably not unlike the processes that one uses to learn to play a musical instrument.[12] Students must practice! One of the effects, and a clear demonstration that the process is working, is when students become fluent with the basic operations and don't have to think about each separate step.

For the common stages of school mathematics, students must practice with numbers. They must add them until basic addition is automatic. The same for subtraction and multiplication. They must practice until these operations are automatic. This is *not* so that they can amaze parents and friends with mathematical parlor tricks, but to facilitate the nonverbal processes of problem solving. At this time we know of no other way to do this, and I can tell you, from personal experience with students, that it is a grim thing to watch otherwise very bright undergraduates struggle with more advanced courses because they have to figure everything out at a basic verbal level. What happens with such students, since they do not have total fluency with basic concepts, is that — though they can often do the work — they simply take far too long working through the most basic material, and soon find themselves too far behind to catch up.

Skill and automaticity with numbers is only part of the story. Students must also bring abstraction into play. This is also very commonly an unconscious process. There are huge numbers of choices for what to emphasize and what to exclude in real problems so as to focus on the core of what matters. Indeed, it is often far from clear what the core actually is. As was the case before, one has to practice to facilitate abstraction. How?

One explores the situation, focusing on one area, then another, and accumulates sufficient data so that nonverbal tools in the brain can sort things out and focus on what matters. But in order to do this, the groundwork has to be laid. That is what algebra does (or is supposed to do). That is why students should practice with abstract problems and symbolic manipulation. Moreover, as we know, Algebra I and more particularly Algebra II are the gate keepers for college [Adelman 1999, p. 17]. When we think of problem solving in this way, that is not so surprising.

The need for further study. Our knowledge here is fragmentary and anecdotal. What I was saying above is highly plausible, and all the research mathematicians that I've discussed it with agree that it fits their experiences. However, it is not

---

[12]It is probably not a coincidence that an inordinate number of professional mathematicians are also skilled musicians.

yet possible to assert this knowledge as fact. Basic research needs to be done, much as was done for reading. The medical and psychological sciences almost certainly have the tools to begin such research now. But so far, such work is only in the earliest stages of development. In the meantime, I would suggest that the observations above not be ignored. It is clear that current approaches to problem solving in K–12 are not working as well as we would like.

We have not discussed the verbal aspects of problem solving. We will turn to them shortly, but first let us discuss one final aspect of mathematics, the interface between mathematics and the real world.

The Art of Creating Well-Posed Problems

This is another thing that we do not know how to teach. Rather, this is one of the most important things that our best Ph.D. students in mathematics actually are learning when they write a thesis. They are initially given small, reasonably well-posed problems to get their feet wet, and, if they survive this, then they are given a real problem, roughly posed, and some guidance.

What the students are then asked to do is to create sensible, appropriate, well-posed problems, that, when taken together, will give a satisfactory answer to the original question. And, of course, the students are expected to be able to resolve the questions they come up with.

It should be realized that not all real-world problems are amenable to mathematical analysis in this way — including those that talk about numbers. For example we have the following problem taken from the *California Mathematics Framework* [California 1992]:

> The 20 percent of California families with the lowest annual earnings pay an average of 14.1 percent in state and local taxes, and the middle 20 percent pay only 8.8 percent. What does that difference mean? Do you think it is fair? What additional questions do you have?

One can apply the processes we discussed earlier to create any number of well-posed questions, but it will be very difficult to find any that are highly relevant. A huge problem is how to give a precise but reasonable definition of "fair." The idea of fairness is subject to much debate among social scientists, politicians, economists, and others. Then, when one attempts to see what the 14.1% and 8.8% might actually mean, further questions arise, including questions about the amounts spent by these two groups in other areas, and what the impact of these amounts might be. In fact, applying rigorous analysis of the type being discussed here with the objective of creating proper questions in mathematics shows just how poorly the question was actually phrased and prevents an educated person from taking such a question at face value. Moreover, it shows that one essentially cannot produce well-posed questions in mathematics that accurately reflect the objectives of this question. This example, even though numbers appear in it, is not a question that can be directly converted into questions in mathematics.

Let us look at one real-life example. A few years back I was asked to help the engineering community solve the basic problem of constructing algorithms that would enable a robotic mechanism to figure out motions without human intervention in order to do useful work in a region filled with obstacles.

The discussion below illustrates the key issues in creating well-posed problems from real-world problems, and the way in which mathematicians, scientists, engineers, and workers in other areas approach such problems.

The first step was to break the problem into smaller parts and replace each part by a precise question in mathematics. The physical mechanism was abstracted to something precise that modeled what we viewed as the most important features of the mechanism. Then the obstacles were replaced by idealized obstacles that could be described by relatively simple equations.

The initial problem was now replaced by an idealized problem that could be formulated precisely and was realistic in the sense that solutions of this new problem would almost always produce usable motions of the actual mechanism.

The second step was to devise a method to solve the mathematical problem. We tested the problem and soon realized that this first approximation was too big a step. A computer had to be able to plan motions when there were no obstacles before it could handle the idealized problem in a region with obstacles. The mathematics of even the problem with no obstacles had been a stumbling block for engineers, and the methods that are currently used for both regions with and without obstacles are quite crude — basically, create a large number of paths and see if any of them work! The difficulty with this approach is that it takes hours to compute relatively simple paths.

The plan was refined and revised. It turned out that the engineering community was not aware of a core body of mathematics that had been developed over the last hundred years. Within this work were basic techniques that could be exploited to completely resolve the problem of motion planning when no obstacles were present. Everything that was needed could be found in the literature. Of course, one needed extensive knowledge of mathematics to be able to read the literature and know where to look for what was known.

We could now resolve the simplified problem, but could we solve the original mathematical problem? With the solution of the first problem as background, we studied the problem with obstacles. The new techniques were applicable to this problem as well. But here new and very focused problem solving was needed, since the required results were not in the literature. It turned out that this, too, could be done. What was needed was real understanding and fluency with the mathematics being used.

Having solved the mathematical problem, could we apply the solution to the original real-world problem? Once the mathematical problem was solved, the solution and its meaning had to be communicated to the engineering community. Translating the mathematics into practical algorithms was the final step. It is currently being finished but already, programs have been written that do path planning for simple mechanisms in regions with lots of obstacles in fractions of a second.

What do we learn from this example? The first step is the key. One abstracts the problem, replacing it by problems that can be precisely stated within a common stage — hence are problems in mathematics — and that have a realistic chance of being solved. This requires real knowledge of the subject, and is a key reason why students have to learn a great deal of mathematics.[13] When solving or creating problems, knowledge of similar or related situations is essential.

But one also has to be sure that the resulting answers will be of use in the original problem. For this, one must be cognizant of what has been left out in the abstracted problem, and how the missing pieces will affect the actual results. *This is why understanding approximation and error analysis are so important.* When one leaves the precise arena of mathematics and does actual work with actual measurements, one has to know that virtually all measurements have errors and that error build up that has not been accounted for can make all one's work useless.

There is no one *correct problem* in this process, in the sense that there can be many different *families of mathematical problems* that one can usefully associate with the original real-world problem. However, it must be realized that each "mathematical problem" will have *only one answer*. This is another point where there has been confusion in school mathematics. There has been much discussion of real-world problems in current K–12 mathematics curricula, but then it is sometimes stated that these problems have many correct mathematical answers, a confounding of two separate issues.[14]

Here is a very elementary example. This is a popular third-grade problem.

Two friends are in different classrooms. How do they decide which classroom is bigger?

---

[13] More precisely, I should add "carefully selected mathematics," where the selection criteria include the liklihood that the mathematics taught will be needed in targeted occupations and classes of problems.

[14] Perhaps this should be recognized as an example of what happens if one is not sufficiently precise in doing mathematics. Confounding *problems* with *answers* confused an entire generation of students and was one of the precipitating factors in the California math wars.

This is a real-world problem but not a problem in mathematics. The issue is that the word *bigger* does not have a precise meaning. Before the question above can be *associated to a problem in mathematics* this word must be defined. Each different definition gives rise to a different problem.

The next step is to solve these new problems. That's another story.

Mathematical Problem Solving II

We have discussed the general issues involved in solving problems in mathematics, the distinction between verbal processes and nonverbal processes. It is now time to talk about the verbal processes involved in problem solving.

Many people regard my former colleague, George Pólya, as the person who codified the verbal processes in problem solving.[15] He wrote five books on this subject, starting with *How to Solve It* [Pólya 1945], which was recently reprinted. We will briefly discuss his work on this subject.

One needs the context in which his books were written in order to understand what they are about. Pólya together with his colleague and long-time collaborator, G. Szegö, believed that their main mathematical achievement was the two-volume *Aufgaben und Lehrsätze aus der Analysis* [1971], recently translated as *Problems and Theorems in Analysis* and published by Springer-Verlag. These volumes of problems were meant to help develop the art of problem solving for graduate students in mathematics, and they were remarkably effective.[16] I understand that Pólya's main motivation in writing his problem-solving books was to facilitate and illuminate the processes he and Szegö hoped to see developed by students who worked through their two volumes of problems. (Pólya indicates in the two-volume *Mathematics and Plausible Reasoning* [1954] that graduate students were his main concern. But he also ran a special junior/senior seminar on problem solving at Stanford for years, so at many points he does discuss less advanced problems in these books.)

Another thing that should be realized is that the audience for these books was modeled on the only students Pólya really knew, the students at the Eidgenössische Technische Hochschule (Swiss Federal Institute of Technology) in Zürich, and the graduate and undergraduate mathematics majors at Stanford. Thus, when Pólya put forth his summary of the core verbal steps in problem solving:

---

[15] Other people, particularly Alan Schoenfeld, have studied and written on mathematical problem solving since, and I take this opportunity to acknowledge their work. However, the discussions in most mathematics methods books concentrate on Pólya's contributions, so these will be the focus of the current discussion.

[16] For example, my father helped me work through a significant part of the first volume when I was 18, an experience that completely changed my understanding of mathematics.

1. Understand the problem
2. Devise a plan
3. Carry out the plan
4. Look back

he was writing for very advanced students and he left out many critical aspects of problem solving like "check that the problem is well-posed," since he felt safe in assuming that his intended audience would not neglect that step. But, as we've seen in the discussion above, this first step cannot be left out for a more general audience. Indeed, for today's wider audience we have to think very carefully about what should be discussed here.

The other thing that appears to have been left out of Pólya's discussion is the fact that problem solving divides into its verbal and nonverbal aspects. This is actually not the case. Pólya was well aware of the distinction. Here is a core quote from [Pólya 1945, p. 9], where he talks about getting ideas, one of the key nonverbal aspects of problem solving:

> We know, of course, that it is hard to have a good idea if we have little knowledge of the subject, and impossible to have it if we have no knowledge. Good ideas are based on past experience and formerly acquired knowledge. Mere remembering is not enough for a good idea, but we cannot have any good idea without recollecting some pertinent facts; materials alone are not enough for constructing a house but we cannot construct a house without collecting the necessary materials. The materials necessary for solving a mathematical problem are certain relevant items of our formerly acquired mathematical knowledge, as formerly solved problems, or formerly proved theorems. Thus, it is often appropriate to start the work with a question: Do you know a related problem?

It is worth noting how these context difficulties have affected the importation of Pólya's ideas into the K–12 arena. Keep in mind that one of the key aspects of problem solving — illustrated by all the previous remarks — is the degree of *flexibility* that is needed in approaching a new problem. By contrast, today's math methods texts apply the rigidity of list-making even to Pólya's work. Thus we have the following expansion of Pólya's four steps taken from a widely used math methods book[17] that pre-service teachers are expected to learn as "problem solving."

1. Understanding the problem

    (i) Can you state the problem in your own words?

---

[17] The book's title and authors are not mentioned here because this represents a general failure and is not specific to this book.

(ii) What are you trying to find or do?

(iii) What are the unknowns?

(iv) What information do you obtain from the problem?

(v) What information, if any, is missing or not needed?

2. Devising a plan

(i) Look for a pattern

(ii) Examine related problems and determine if the same technique applied to them can be applied here

(iii) Examine a simpler or special case of the problem to gain insight into the solution of the original problem.

(iv) Make a table.

(v) Make a diagram.

(vi) Write an equation.

(vii) Use guess and check.

(viii) Work backward.

(ix) Identify a subgoal.

(x) Use indirect reasoning.

3. Carrying out the plan

(i) Implement the strategy or strategies in step 2 and perform any necessary actions or computations.

(ii) Check each step of the plan as you proceed. This may be intuitive checking or a formal proof of each step.

(iii) Keep an accurate record of your work.

4. Looking back

(i) Check the results in the original problem. (In some cases, this will require a proof.)

(ii) Interpret the solution in terms of the original problem. Does your answer make sense? Is it reasonable? Does it answer the question that was asked?

(iii) Determine whether there is another method of finding the solution.

(iv) If possible, determine other related or more general problems for which the technique will work.

Here are very similar expansions from another widely used math methods text, to reinforce the fact that the rigid expansion of Pólya's four steps above is the norm, rather than the exception:

Understand the problem

- Read the information
- Identify what to find or pose the problem
- Identify key conditions; find important data
- Examine assumptions.

Develop a plan

- Choose Problem-Solving Strategies

    (i) Make a model.

    (ii) Act it out.

    (iii) Choose an operation.

    (iv) Write an equation.

    (v) Draw a diagram.

    (vi) Guess-check-revise.

    (vii) Simplify the problem.

    (viii) Make a list.

    (ix) Look for a pattern.

    (x) Make a table.

    (xi) Use a specific case.

    (xii) Work backward.

    (xiii) Use reasoning.
- Identify subproblems.
- Decide whether estimation, calculation, or neither is needed.

Implement the plan

- If calculation is needed, choose a calculation method.
- Use Problem-Solving Strategies to carry out the plan.

Look back

- Check problem interpretation and calculations.
- Decide whether the answer is reasonable.
- Look for alternate solutions.
- Generalize ways to solve similar problems.

The consequences of the misunderstanding of problem solving in today's textbooks and tests. I mentioned at the beginning of this essay that state assessments in mathematics average 25% mathematically incorrect problems each. Indeed, with Richard Askey's help, I was responsible for guiding much of the development of the evaluation criteria for the mathematics portion of the recently released report on state assessments by Accountability Works [Cross et al. 2004]. In this role we had to read a number of state assessments, and 25% was consistent. These errors were not trivial typos, but basic misunderstandings usually centered around problem solving and problem construction.

Sadly, Pólya was fully aware of these risks, but could do nothing to prevent them even though he tried. It is not common knowledge, but when the School Mathematics Study Group (SMSG)[18] decided to transport Pólya's discussion to K–12, Pólya strenuously objected. Paul Cohen told me that at one of the annual summer meetings of SMSG at Stanford in the late 1960s, Pólya was asked to give a lecture, and in this lecture he explained why the introduction of "problem solving" as a key component of the SMSG program was a very bad mistake. Afterwards, Cohen told me that Pólya was well aware that his audience had applauded politely, but had no intention of following his advice. So Pólya asked Cohen, who had just won the Fields medal, if he would help. But it was not possible to deflect them.

In hindsight we can see just how accurate Pólya was in his concerns.

Summary

Mathematics involves three things: precision, stages, and problem solving. The awareness of these components and the ways in which they interact for basic stages such as the real numbers or the spaces of Euclidean geometry and the stages where algebra plays out are the essential components of mathematical proficiency. Perhaps the biggest changes in K–12 instruction that should be made to bring this to the forefront are in the use of definitions from the earliest grades onwards. Students must learn precision because if they do not, they will fail to develop mathematical competency. There is simply no middle ground here.

It is well known that early grade teachers are very concerned with making mathematics *accessible* to students, and believe that it is essential to make it fun. However, while many educators may believe that precision and accessibility are in direct opposition to each other, a study of the mathematics texts used in the programs of the successful foreign countries shows that this is not the case. Problems can be interesting and exciting for young students, and yet be precise.

Problem solving is a very complex process involving both verbal and nonverbal mental processes. There are traps around every corner when we attempt to codify problem solving, and current approaches in this country have not

---

[18]The School Mathematics Study Group was the new-math project that met in the summer at Stanford during the 1960s.

been generally successful. The concentration has been almost entirely on verbal aspects. But verbal problem solving skills, by themselves, simply do not get students very far. On the other hand, the only way mathematicians currently know to develop the nonverbal part involves hard work — practice, practice, more practice, and opportunities to see people who are able to work in this way actually solve problems.

Over a period of generations, teachers in the high-performing countries have learned many of these skills, and consequently I assume that students there are exposed to all the necessary ingredients. It seems to be remarkable how successful the results are. The percentages of students who come out of those systems with a real facility with mathematics is amazing.

There is no reason that our students cannot reach the same levels, but there is absolutely no chance of this happening overnight. Our entire system has to be rebuilt. The current generation of pre-service teachers must be trained in actual mathematics.

It is likely to take some time to rebuild our education system, and we cannot be misled by "false positives" into prematurely thinking that we've reached the goal. For example, right now, in California, the general euphoria over the dramatic rise in test outcomes through the middle grades puts us in the most dangerous of times, especially with limited resources and the fact that we have been unable to change the educations that our K–8 teachers receive.

Our current teachers in California have done a remarkable job of rebuilding their own knowledge. I'm in awe of them. I had assumed, when we wrote the current *California Mathematics Standards*, that the result would be a disaster since the new standards represented such a large jump from previous expectations and consequent teacher knowledge, but these teachers proved themselves to be far more resilient and dedicated than I had ever imagined.

As remarkable as our teachers have been, they could only go so far. But given the demonstrated quality of the people that go into teaching, if *mathematicians and mathematics educators* manage to do things right, I've every confidence that *they* will be able to go the rest of the way.

As Alan Greenspan's remarks at the beginning of this essay show, the stakes are simply too high for failure to be an option.

References

[Adelman 1999] C. Adelman, *Answers in the tool box: academic intensity, attendance patterns, and bachelor's degree attainment*, Washington, DC: U.S. Department of Education, 1999.

[California 1992] *California mathematics framework*, Sacramento: California Department of Education, 1992.

[Cross et al. 2004] R. W. Cross, J. T. Rebarber, J. Torres, and J. Finn, C. E., *Grading the systems: The guide to state standards, tests and accountability policies*, Washington, DC: Thomas B. Fordham Foundation and Accountability Works, January 2004. Available at http://edexcellence.net/institute/publication/publication.cfm?id=3D328. Retrieved 22 Feb 2006.

[Kentucky 2004] *Grade 8 sample released questions*, Frankfort, KY: Kentucky Department of Education, January 2004.

[Mukherjee 2004] A. Mukherjee, "Greenspan worries about math gap, not wage gap", Bloomburg.com (February 17, 2004). Available at http://www.bloomberg.com/apps/news?pid=10000039&refer=columnist_mukherjee&sid=angrr7YVphls. Retrieved 9 Jan 2007.

[NCTM 2000] National Council of Teachers of Mathematics, *Principles and standards for school mathematics*, Reston, VA: Author, 2000.

[Pólya 1945] G. Pólya, *How to solve it: A new aspect of mathematical method*, Princeton University Press, 1945.

[Pólya 1954] G. Pólya, *Mathematics and plausible reasoning* (2 vols.), Princeton University Press, 1954.

[Pólya and Szegö 1971] G. Pólya and G. Szegö, *Aufgaben und Lehrsätze aus der Analysis*, 4th ed., Heidelberger Taschenbücher **73, 74**, 1971.

[Reich 2003] R. Reich, "Nice work if you can get it", *Wall Street Journal* (December 26, 2003). Available at www.robertreich.org/reich/20031226.asp.

[UCSMP 1992a] M. I. Moro, M. A. Bantova, and G. V. Beltyukova, *Russian grade 1 mathematics*, Chicago: UCSMP, 1992. Translated from the Russian 9th edition, 1980.

[UCSMP 1992b] M. I. Moro and M. A. Bantova, *Russian grade 2 mathematics*, Chicago: UCSMP, 1992. Translated from the Russian 12th edition, 1980.

[UCSMP 1992c] A. S. Pchoiko, M. A. Bantova, and M. I. Moro, *Russian grade 3 mathematics*, Chicago: UCSMP, 1992. Translated from the Russian 9th edition, 1978.

Assessing Mathematical Proficiency
MSRI Publications
Volume **53**, 2007

Chapter 5
What is Mathematical Proficiency and How Can It Be Assessed?

ALAN H. SCHOENFELD

To establish a common point of departure with Jim Milgram's chapter, this chapter is framed around the two basic questions with which his chapter began:

- What does it mean for a student to be proficient in mathematics? (What should students be learning?)
- How can we measure proficiency in mathematics? (How can we tell if we are succeeding?)

My main emphasis is on the first question, because much of the rest of this volume addresses the second.

In the introduction to this volume and in the first chapter, I pointed to the fact that the "cognitive revolution" (see [Gardner 1985], for instance) produced a significant reconceptualization of what it means to understand subject matter in different domains (see also [NRC 2000]). There was a fundamental shift from an exclusive emphasis on knowledge — what does the student *know*? — to a focus on what students know and can do with their knowledge. The idea was not that knowledge is unimportant. Clearly, the more one knows, the greater the potential for that knowledge to be used. Rather, the idea was that having the knowledge was not enough; being able to use it in the appropriate circumstances is an essential component of proficiency.

Some examples outside of mathematics serve to make the point. Many years ago foreign language instruction focused largely on grammar, vocabulary, and literacy. Students of French, German, or Spanish learned to read literature in those languages — but when they visited France, Germany, or Spain, they found themselves unable to communicate effectively in the languages they had studied. Similarly, years of instruction in English classes that focused on grammar instruction resulted in students who could analyze sentence structure but who were not necessarily skilled at expressing themselves effectively in writing. Over the

past few decades, English and foreign language instruction have focused increasingly on communication skills — on mastering the basics, of course (e.g.,
conjugating verbs, acquiring a solid vocabulary, mastering grammar) *and* learning the additional skills that enable them to use what they have learned.

A similar evolution took place in mathematics. The knowledge base remains
important; it goes without saying that anyone who lacks a solid grasp of facts,
procedures, definitions, and concepts is significantly handicapped in mathematics. But there is much more to mathematical proficiency than being able to
reproduce standard content on demand. A mathematician's job consists of at
least one of: extending known results; finding new results; and applying known
mathematical results in new contexts. The problems mathematicians work on,
in academia or in industry, are not the kind of exercises that get solved in a few
minutes or hours; they are problems that may take days, weeks, months, or years
to solve. Thus, in addition to possessing a substantial amount of specialized
knowledge, mathematicians possess other things as well. Good problem solvers
are flexible and resourceful. The have many ways to think about problems —
alternative approaches if they get stuck, ways of making progress when they
hit roadblocks, of being efficient with (and making use of) what they know.
They also have a certain kind of mathematical disposition — a willingness to pit
themselves against difficult mathematical challenges under the assumption that
they will be able to make progress on them, and the tenacity to keep at the task
when others have given up. As will be seen below, all of these are aspects of
mathematical proficiency; all of them can be learned (or not) in school; all of
them can help explain why some attempts at problem solving are successful and
some not.

Proficiency, Part A: Knowledge Base

There is a long history of attempts to prescribe the mathematical content
that students should know. Many of those efforts have involved having groups
of scholars working together for a number of years. It would be foolish for
me to try to supplant their work, especially given the small amount of space
available. Hence I will defer to the judgments made in volumes such as the
National Council of Teachers of Mathematics' two major standards documents
[NCTM 1989; 2000], especially the latter, *Principles and Standards for School
Mathematics*; and the National Research Council's [NRC 2001] volume *Adding
It Up*. (Note: a summary of the dimensions of mathematical proficiency found
in those volumes was given in Chapter 1 of this volume, "Issues and Tensions
in the Assessment of Mathematical Proficiency.")

Instead, I will discuss different interpretations of what it means to know that
content. A major source of controversy over the past decade has involved not

only the level of procedural skills expected of students, but also what is meant by "understanding." For example, what does it mean for an elementary school student to understand base-ten subtraction?

For some people — notably, those who wrote the current California mathematics standards, understanding a concept means being able to compute the answers to exercises that employ that concept. For example, here are the California mathematics standards related to arithmetic in grade 3 [CSBE 1997]:

2.0 Students calculate and solve problems involving addition, subtraction, multiplication, and division:

2.1 Find the sum or difference of two whole numbers between 0 and 10,000.

2.2 Memorize to automaticity the multiplication table for numbers between 1 and 10.

2.3 Use the inverse relationship of multiplication and division to compute and check results.

2.4 Solve simple problems involving multiplication of multidigit numbers by one-digit numbers ($3,671 \times 3 = $ ___).

2.5 Solve division problems in which a multidigit number is evenly divided by a one-digit number ($135 \div 5 = $ ___).

2.6 Understand the special properties of 0 and 1 in multiplication and division.

2.7 Determine the unit cost when given the total cost and number of units.

2.8 Solve problems that require two or more of the skills mentioned above.

and the California Grade 5 number sense standards:

2.0 Students perform calculations and solve problems involving addition, subtraction, and simple multiplication and division of fractions and decimals:

2.1 Add, subtract, multiply, and divide with decimals; add with negative integers; subtract positive integers from negative integers; and verify the reasonableness of the results.

2.2 Demonstrate proficiency with division, including division with positive decimals and long division with multidigit divisors.

2.3 Solve simple problems, including ones arising in concrete situations, involving the addition and subtraction of fractions and mixed numbers (like and unlike denominators of 20 or less), and express answers in the simplest form.

2.4 Understand the concept of multiplication and division of fractions.

2.5 Compute and perform simple multiplication and division of fractions and apply these procedures to solving problems.

Although the term "problem solving" is used in these *standards,* it refers to computational proficiency: For example, in the third grade standard 2.4, we are told that students will be able to "solve" the "problem" $(3,671 \times 3 = \underline{\hphantom{xx}})$.

This approach stands in stark contrast to the one taken in the National Council of Teachers of Mathematics volume *Principles and Standards for School Mathematics* [NCTM 2000]. Consider, for example, the language used in describing the Number and Operations Standard for grades 3–5, from p. 148:

Instructional programs from prekindergarten through grade 12 should enable all students to:

Understand numbers, ways of representing numbers, relationships among numbers, and number systems.

In grades 3–5 all students should

• understand the place-value structure of the base-ten number system and be able to represent and compare whole numbers and decimals;

• recognize equivalent representations for the same number and generate them by decomposing and composing numbers;

• develop understanding of fractions as parts of unit wholes, as parts of a collection, as locations on number lines, and as divisions of whole numbers;

• use models, benchmarks, and equivalent forms to judge the size of fractions;

• recognize and generate equivalent forms of commonly used fractions, decimals, and percents;

• explore numbers less than 0 by extending the number line and through familiar applications;

• describe classes of numbers according to characteristics such as the nature of their factors.

Understand meanings of operations and how they relate to one another

(detail omitted)

Compute fluently and make reasonable estimates

(detail omitted)

The differences in terminology, meaning, and intended competencies are clear, as is the exemplification in *Principles and Standards.* Now the question is, does such a difference in setting standards make a difference? And does it make a difference in assessment?

The simple answer is that it can make a great deal of difference. For example, consider the following statistics from [Ridgway et al. 2000].

In 2000, the Silicon Valley Mathematics Assessment Collaborative gave two tests to a total of 16,420 third, fifth, and seventh graders. One was the SAT-9, a skills-oriented test consistent with the California mathematics standards. The other was the Balanced Assessment test, a much broader test (including questions that focus on skills, concepts and problem solving) that is aligned with NCTM's (1989) *Curriculum and Evaluation Standards for School Mathematics*, the precursor to *Principles and Standards*. The simplest analysis assigns each student a score of "proficient" or "not proficient" on each examination. Needless to say, student scores on any two mathematics assessments are likely to be highly correlated — but, the differences between student scores on the SAT-9 and Balanced Assessment tests are very informative. Consider Table 1.

Balanced Assessment	SAT-9	
	Not proficient	Proficient
Not proficient	29%	22%
Proficient	4%	45%

Table 1. Aggregated scores for 16,420 Students on the SAT-9 and Balanced Assessment tests (grades 3, 5, and 7). From [Ridgway et al. 2000].

As expected, 74% of the students at grades 3, 5, and 7 were given the same "proficient" or "not proficient" ratings on the two tests. But, compare the second row of data to the second column. More than 90% of the students who were declared proficient on the Balanced Assessment test were declared proficient on the SAT-9 — that is, doing well on the Balanced Assessment tests is a reasonably good guarantee of doing well on the SAT-9. The converse is not true: approximately one third of the students declared proficient on the SAT-9 exam were declared to be not proficient on the Balanced Assessment test.

This is critically important, and entirely consistent with a small but growing body of literature comparing "reform" and "traditional" curricula in mathematics. (See, for example, [Senk and Thompson 2003], which examines the performance of students who studied curricula that had been developed with support from the National Science Foundation with more traditional, skills-oriented curricula.) In brief, the findings in that literature are as follows. Students who experience skills-focused instruction tend to master the relevant skills, but do not do well on tests of problem solving and conceptual understanding. Students who study more broad-based curricula tend to do reasonably well on tests of skills (that is, their performance on skills-oriented tests is not statistically different from the performance of students in skills-oriented courses), and they do much better than those students on assessments of conceptual understanding and problem solving.

In short, one's concept of what counts as mathematics matters a great deal — and, what you assess counts a great deal. First, students are not likely to learn what they are not taught. Hence teaching a narrow curriculum has consequences. Second, one only finds out about what students don't know if one assesses for that knowledge. Thus, for example, the SAT-9 does not reveal the problem with a narrow curriculum: 22% of the total population is declared proficient although a more broad-based test calls that proficiency into question.

In what follows I shall briefly delineate additional aspects of mathematical proficiency.

Proficiency, Part B: Strategies

One of the strands of mathematical proficiency described in *Adding It Up* is "*strategic competence* — ability to formulate, represent, and solve mathematical problems" [NRC 2001, p. 5]. It goes without saying that "knowing" mathematics, in the sense of being able to produce facts and definitions, and execute procedures on command, is not enough. Students should be able to use the mathematical knowledge they have.

The starting place for any discussion of problem solving strategies is the work of George Pólya. In 1945, with the pioneering first edition of *How to Solve It*, Pólya opened up the study of problem solving strategies. The core of the book was devoted to a "short dictionary of heuristic." To quote from Pólya, "The aim of heuristic is to study the methods and rules of discovery and invention. The present book is an attempt to revive heuristic in a modern and modest form" [Pólya 1945, pp. 112–113]. "Modern Heuristic endeavors to understand the process of solving problems, especially the *mental operations typically useful* in this process" [Pólya 1945, pp. 129–130]. In *How to Solve It*, Pólya described powerful problem solving strategies such as making use of analogy, making generalizations, re-stating or re-formulating a problem, exploiting the solution of related problems, exploiting symmetry, and working backwards. Polya's subsequent volumes, *Mathematics and Plausible Reasoning* [Pólya 1954] and *Mathematical Discovery* [Pólya 1981], elaborated substantially on the ideas in *How to Solve It*, showing how one could marshal one's mathematical knowledge in the service of solving problems.

To cut a long story short (see [Schoenfeld 1985; 1992] for detail), the heuristic strategies Pólya describes are more complex than they appear. Consider, for example, a strategy such as "if you cannot solve the proposed problem ... could you imagine a more accessible related problem?" [Pólya 1945, p. 114]. The idea is that although the problem you are trying to solve may be too difficult for now, you might be able to solve a simpler version of it. You might then use the result, or the idea that led to the solution of the simpler problem, to solve the original problem.

At this level of generality, the strategy sounds straightforward — but the devil is in the details. For example, Pólya [1945, p. 23] discusses the solution of this problem:

> Using straightedge and compass, and following the traditional rules for geometric constructions, inscribe a square in a given triangle. Two vertices of the square should be on the base of the triangle, the other two vertices of the square on the other two sides of the triangle, one on each.

That is, you are given a triangle such as in the left diagram; you wish to produce a square such as seen on the right, using only straightedge and compass.

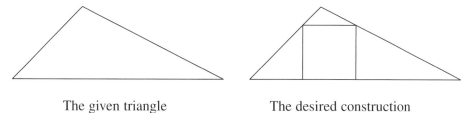

The given triangle   The desired construction

In an idealized discussion, Pólya shows how the right kind of questioning can lead a student to consider an easier related problem (demanding only that three of the corners of the desired square lie on the given triangle), noting that there are infinitely many squares that meet this condition, and that the locus of the fourth vertices of such squares can be used to find the one such square that has its fourth (and hence all four) vertices on the sides of the triangle. The discussion is logical and straightforward — and, once one sees the solution, it is wonderfully elegant. But the question is, what will non-ideal students do?

I have told undergraduates that the problem can be solved by exploiting the solution to an easier related problem, and asked them what easier related problem they might want to try to solve. Typically, they will suggest inscribing the square in a "special" triangle (such as an equilateral or isosceles triangle) or inscribing the square in a circle. Unfortunately, both of these approaches lead to dead ends. The former problem is no easier to solve than the original; the latter can be solved (it is a standard construction) but I know of no way to exploit it to solve the given problem. With further prodding, students will suggest inscribing a rectangle in the given triangle. They recognize that there are infinitely many rectangles that can be inscribed, and that one of them must be a square — but this is an existence proof, and it does not generate a construction. Then, students may suggest either (a) requiring only that three vertices of the square lie on the sides of the triangle, or (b) trying to circumscribe a square with a triangle similar to the given triangle. Each of these approaches does yield a solution, but not without some work.

In short, to use the strategy Pólya describes, the problem solver needs to

a. think to use the strategy,
b. generate a relevant and appropriate easier related problem,
c. solve the related problem, and
d. figure out how to exploit the solution or method to solve the original problem.

All of these can be nontrivial.

The evidence, however, is that accomplished mathematicians use such strategies (see not only Pólya's books but [DeFranco 1991]) and that high school and college students can learn to master such strategies; see, for example, [Lester 1994; Schoenfeld 1985; 1992].

Proficiency, Part C: Metacognition (Using What You Know Effectively)

Picture someone stepping into a bog, and beginning to sink in — then taking a second step forward and then a third, rather than turning back. By the time the person realizes that he or she is in quicksand, all is lost. One has to wonder, why didn't that person take stock before it was too late?

Now consider the following time-line graph (Figure 1) of two students working a problem.

In the specific problem session represented here (see [Schoenfeld 1985] for detail), the students read the problem statement and hastily made a decision about what to do next. Despite some clear evidence that this approach was not productive for them, they persevered at it until they ran out of time. When they did, I asked them how the approach they had taken — they had chosen to calculate a particular area — was going to help them. They were unable to say.

As it happens I knew these students well, and knew that each of them had enough mathematical knowledge to solve the given problem (which was similar to a problem on a final examination the students had taken just a week or two earlier). The point is that there is just so much that a student can be doing at any one time. Because the students were focused on doing a particular computation, and they had never stopped to consider how wise it was to invest their time in doing so, they never reconsidered — and thus never got to use the knowledge they had.

Reflecting on progress while engaged in problem solving, and acting accordingly ("monitoring and self-regulation") is one aspect of what is known as metacognition — broadly, taking one's thinking as an object of inquiry. As the

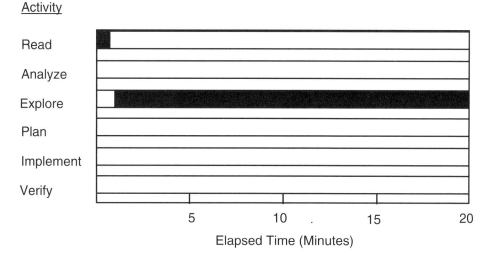

Figure 1. A time-line representation of a problem solving attempt. Reprinted from [Schoenfeld 1992], with permission.

graph above indicates, failing to do so can guarantee failure at problem solving: if one is fully occupied doing things that do not help to solve the problem, one may never get to use the "right" knowledge in the right ways.

Although I have told this story as an anecdote, there are lots of data to back it up. Over more than 25 years, more than half of the hundreds of problem sessions that my research assistants and I have videotaped have been of the type represented in Figure 1. This kind of finding has been widely replicated; for a general discussion see [Lester 1994] or [Schoenfeld 1987].

As was the case with strategies, the story here is that (a) effective problem solvers behave differently; and (b) students can learn to be much more efficient at monitoring and self-regulation, and become more successful problem solvers thereby.

Figure 2 shows the time-line graph of a mathematician working a two-part problem in a content area he had not studied for years. Each of the triangles in the figure represents a time that the mathematician assessed the state of his solution attempt and then acted on the basis of that assessment (sometimes deciding to change direction, sometimes deciding to stay the course — but always with decent reason). There is no question that he solved the problem because of his efficiency. He did not choose the right approaches at first, but, by virtue of not spending too much time on unproductive approaches, managed to find productive ones.

Figure 3 shows the time-line graph of a pair of students working a problem after having taken my problem solving course. During the course, I focused

Activity

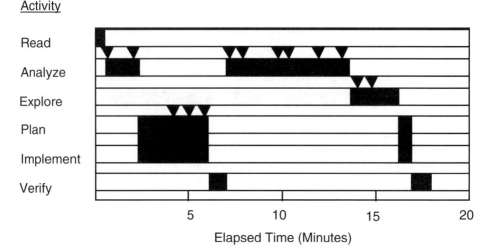

Figure 2. A time-line representation of a mathematician's problem solving attempt. Reprinted from [Schoenfeld 1992], with permission.

a great deal on issues of metacognition, acting as a "coach" while groups of students worked problems. (That is, I would regularly intervene to ask students if they could justify the solution paths they had chosen. After a while, the students began to discuss the rationales for their problem solving choices as they worked the problems.)

As Figure 3 indicates, the students hardly became "ideal" problem solvers. In the particular solution represented in Figure 3, the students jumped into a solution with little consideration after reading the problem. However, they reconsidered about four minutes into the solution, and chose a plausible solution direction. As it happens, that direction turned out not to be fruitful; about eight minutes later they took stock, changed directions, and went on to solve the problem. What made them effective in this case was not simply that they had the knowledge that enabled them to solve the problem. It is the fact that they gave themselves the opportunity to use that knowledge, by truncating attempts that turned out not to be profitable.

Proficiency, Part D: Beliefs and Dispositions

I begin this section, as above, with a description of one of the strands of mathematical proficiency described in *Adding It Up*: "productive disposition" — an "habitual inclination to see mathematics as sensible, useful and worthwhile, coupled with a belief in diligence and one's own efficacy" [NRC 2001, p. 5].

Readers who know the research literature will find the idea that "beliefs and dispositions" are aspects of mathematical proficiency is familiar — but those

Activity

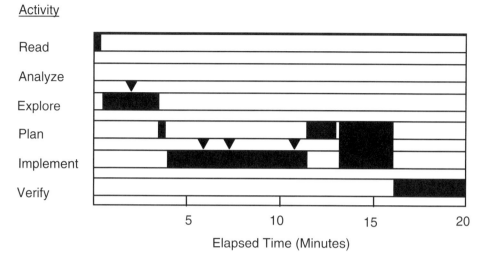

Figure 3. A time-line representation of two students' problem solving attempt, following the completion of a course on problem solving. Reprinted from [Schoenfeld 1992], with permission.

who do not are likely to find it rather strange. After all, what do beliefs have to do with doing mathematics? The simple answer is, "a great deal." Also as above, I shall indicate that this category of behavior matters (a) for students, and (b) for professional mathematicians.

What follows is a straightforward arithmetic problem taken from the 1983 National Assessment of Educational Progress, or NAEP [Carpenter et al. 1983]:

> An army bus holds 36 soldiers. If 1128 soldiers are being bussed to their training site, how many buses are needed?

The solution is simple. If you divide 1128 by 36, you get a quotient of 31 and a remainder of 12. Hence you need 32 buses, under the (tacit) assumptions that each bus will be filled to capacity if possible and that no buses will be allowed to carry more 36 soldiers.

NAEP is a nationwide survey of U.S. students' mathematical performance, with a carefully undertaken sampling structure. Some 45,000 students worked this problem. Here is how the responses were distributed:

29% gave the answer "31 remainder 12"
18% gave the answer "31"
23% gave the correct answer, "32"
30% did the computation incorrectly.

A full 70% of the students did the computation correctly, but only 23% of the students rounded up correctly. How could this be? How could it be possible

that 29% of the students answered that the *number of buses needed* involves
a remainder? Imagine asking these same students to call a bus company to
arrange for buses to take their school on an outing. Would any of them mention
remainders?

The brief explanation for this astounding behavior (see [Schoenfeld 1985;
1992] for more detail) is that these students had learned their counterproductive
behavior in their mathematics classes. In the late 1970s and early 1980s, there
was an increased emphasis on "problem solving" in mathematics classrooms in
the U.S. The reality (and the reason for the quotation marks) is that the focus was
almost entirely superficial: instead of being given pages of rote computational
problems such as

$$7 - 4 = \underline{\hspace{1cm}}$$

students were given pages of comparably rote problems of the type

> John had 7 apples. He gave 4 apples to Mary.
> How many apples does he have left?

The students soon figured out that the word problems were "cover stories" that
had little or nothing to do with the real world. The most efficient way to solve the
problems was to read the text, ignoring the "real world" context. The students
learned to pick out the numbers, identify the operation to perform on them,
do the computation, and write the answer. That is, a quick scan of the word
problem produces the following: "7 and 4 are the numbers. Subtract, and write
the answer down." This procedure got students the right answer, almost all the
time. It became a habit.

Now apply it to the NAEP busing problem. A quick scan of the problem
statement produces "1128 and 36 are the numbers. Divide, and write the answer
down." If you do — and 29% of the students taking the exam did — you write
that the answer is "31 remainder 12."

In short, if you believe that mathematics is not supposed to make sense, and
that working mathematics problems involves rather meaningless operations on
symbols, you will produce nonsensical responses such as these. Hence beliefs
are important — and, students pick up their beliefs about the nature of mathematics from their experiences in the mathematics classroom. Other typical student
beliefs, documented over many years (see [Lampert 1990; Schoenfeld 1992]),
include:

- Mathematics problems have one and only one right answer.
- There is only one correct way to solve any mathematics problem — usually
  the rule the teacher has most recently demonstrated to the class.

- Ordinary students cannot expect to understand mathematics; they expect simply to memorize it, and apply what they have learned mechanically and without understanding.
- Mathematics is a solitary activity, done by individuals in isolation.
- Students who have understood the mathematics they have studied will be able to solve any assigned problem in five minutes or less.
- The mathematics learned in school has little or nothing to do with the real world.
- Formal proof is irrelevant to processes of discovery or invention.

(Reprinted with permission from [Schoenfeld 1992, p. 359])

One might think that professional mathematicians, especially those who have earned the Ph.D. in mathematics, would have productive beliefs about themselves and their engagement with mathematics. Interestingly, that is not the case. De Franco [1991] compared the problem solving performance of eight mathematicians who had achieved national or international recognition in the mathematics community with that of eight published mathematicians (the number of publications ranged from 3 to 52) who had not achieved such clear recognition. He also had the mathematicians fill out questionnaires regarding their beliefs about and mathematics and their problem solving practices (e.g., whether they tried alternative ways to approach problems if their initial methods did not pan out). The conclusion [DeFranco 1991, p. 208]:

> The responses to the questionnaire indicate that the beliefs about mathematics and problem solving held by subjects in group A [the prominent mathematicians] are dissimilar to those held by the subjects in group B. To the extent that beliefs impact problem solving performance, it would appear that the beliefs acquired by group A (group B) would positively (negatively) influence their performance on the problems.

Implications for Assessment

As highlighted in this chapter, a person's mathematical knowledge is far from the whole story. If you are interested in someone's mathematical *proficiency* — that is, what someone knows, can do, and is disposed to do mathematically — then it is essential to consider all four aspects of mathematical proficiency discussed in this chapter. Knowledge plays a central role, as it must. But, an individual's ability to employ problem solving strategies, the individual's ability to make good use of what he or she knows, and his or her beliefs and dispositions, are also critically important. As DeFranco's research showed, this is not just the case for students as they learn mathematics. It is the case for professional mathematicians as well.

With regard to assessments aimed at capturing students' mathematical proficiency, the key operative phrase was coined by Hugh Burkhardt some years ago: the nature of assessment (or testing) is critically important because "What You Test Is What You Get" (WYTIWYG).

The WYTIWYG principle operates both at the curriculum level and at the individual student level. Given the "high stakes" pressures of testing under the No Child Left Behind law [U.S. Congress 2001], teachers feel pressured to teach to the test — and if the test focuses on skills, other aspects of mathematical proficiency tend to be given short shrift. (This is known as curriculum deformation; see Chapter 1 for an example.) Similarly, students take tests as models of what they are to know. Thus, assessment shapes what students attend to, and what they learn.

As noted in Chapter 1, skills are easy to test for, and tests of skills are easy to defend legally (they have the "right" psychometric properties). However, there are still significant issues with regard to problem solving and conceptual understanding. In this chapter, the comparison of students' SAT-9 and Balanced Assessment scores in my discussion of the knowledge base makes that point dramatically. Aspects of strategy, metacognition, and beliefs are much more subtle and difficult to assess. Yet, doing so is essential. Some of the chapters in the balance of this volume will describe assessments that attempt to capture some of the aspects of mathematical proficiency discussed here. To the degree that these assessments succeed in doing so, this represents real progress.

References

[Carpenter et al. 1983] T. P. Carpenter, M. M. Lindquist, W. Matthews, and E. A. Silver, "Results of the third NAEP mathematics assessment: Secondary school", *Mathematics Teacher* **76**:9 (1983), 652–659.

[CSBE 1997] California State Board of Education, *Mathematics: Academic content standards for kindergarten through grade twelve*, Sacramento, 1997. Available at http://www.cde.ca.gov/be/st/ss/mthmain.asp. Retrieved 1 Nov 2005.

[DeFranco 1991] T. DeFranco, "A perspective on mathematical problem-solving expertise based on the performances of male Ph.D. mathematicians", pp. 195–213 in *Research in collegiate mathematics education, II*, edited by J. Kaput et al., Washington, DC: Conference Board of the Mathematical Sciences, 1991.

[Gardner 1985] H. Gardner, *The mind's new science: A history of the cognitive revolution*, New York: Basic Books, 1985.

[Lampert 1990] M. Lampert, "When the problem is not the question and the solution is not the answer", *American Educational Research Journal* **27** (1990), 29–63.

[Lester 1994] F. Lester, "Musings about mathematical problem solving research: 1970–1994", *Journal for Research in Mathematics Education* **25**:6 (1994), 660–675.

[NCTM 1989] National Council of Teachers of Mathematics, *Curriculum and evaluation standards for school mathematics*, Reston, VA: Author, 1989.

[NCTM 2000] National Council of Teachers of Mathematics, *Principles and standards for school mathematics*, Reston, VA: Author, 2000.

[NRC 2000] National Research Council (Committee on Developments in the Science of Learning, Commission on Behavioral and Social Sciences and Education), *How people learn: Brain, mind, experience, and school*, expanded ed., edited by J. D. Bransford et al., Washington, DC: National Academy Press, 2000.

[NRC 2001] National Research Council (Mathematics Learning Study: Center for Education, Division of Behavioral and Social Sciences and Education), *Adding it up: Helping children learn mathematics*, edited by J. Kilpatrick et al., Washington, DC: National Academy Press, 2001.

[Pólya 1945] G. Pólya, *How to solve it: A new aspect of mathematical method*, Princeton University Press, 1945.

[Pólya 1954] G. Pólya, *Mathematics and plausible reasoning* (2 vols.), Princeton University Press, 1954.

[Pólya 1981] G. Pólya, *Mathematical discovery*, combined ed., New York: Wiley, 1981.

[Ridgway et al. 2000] J. Ridgway, R. Crust, H. Burkhardt, S. Wilcox, L. Fisher, and D. Foster, *MARS report on the 2000 tests*, San Jose, CA: Mathematics Assessment Collaborative, 2000.

[Schoenfeld 1985] A. H. Schoenfeld, *Mathematical problem solving*, Orlando, FL: Academic Press, 1985.

[Schoenfeld 1987] A. H. Schoenfeld, "What's all the fuss about metacognition?", pp. 189–215 in *Cognitive science and mathematics education*, edited by A. H. Schoenfeld, Hillsdale, NJ: Erlbaum, 1987.

[Schoenfeld 1992] A. H. Schoenfeld, "Learning to think mathematically: Problem solving, metacognition, and sense-making in mathematics", pp. 334–370 in *Handbook of research on mathematics teaching and learning*, edited by D. Grouws, New York: Macmillan, 1992.

[Senk and Thompson 2003] S. L. Senk and D. R. Thompson (editors), *Standards-based school mathematics curricula: What are they? What do students learn?*, Mahwah, NJ: Erlbaum, 2003.

[U.S. Congress 2001] U.S. Congress, H. Res. 1, 107th Congress, 334 Cong. Rec. 9773, 2001. Available at http://frwebgate.access.gpo.gov.

Section 3
What Does Assessment Assess?
Issues and Examples

It is said that a picture is worth a thousand words. When it comes to assessment, an example is worth that and more. Someone may claim to assess student understanding of X, but what does that mean? The meaning becomes clear when one sees what is actually being tested. Examples help for many reasons. They make authors' intentions clear, and they teach. In working through good examples of assessments, one learns how to think about student understanding. This section of the book offers a wide range of examples and much to think about.

In Chapter 6, Hugh Burkhardt takes readers on a tour of the assessment space. He asks a series of questions related to the creation and adoption of assessments, among them the following: Who is a particular assessment intended to inform? What purpose will it serve? (To monitor progress? To guide instruction? To aid or justify selection? To provide system accountability?) Which aspects of mathematical proficiency are valued? How often should assessment take place to achieve the desired goals? What will the consequences of assessment be, for students, teachers, schools, parents, politicians? What will it cost, and is the necessary amount an appropriate use of resources? Burkhardt lays out a set of design principles, and illustrates these principles with a broad range of challenging tasks. The tasks, in turn, represent the mathematical values Burkhardt considers central: specifically, that the processes of *mathematizing* and mathematical modeling are centrally important, as is the need for students to explain themselves clearly using mathematical language.

In Chapter 7, Jan de Lange continues the tour of mathematical assessments. Like Burkhardt, he believes that assessment development is an art form, and that like any art form, it follows certain principles, in the service of particular goals. He introduces the framework for the development of the Program for International Student Assessment. PISA assessments, like those of TIMSS (which formerly stood for Third International Mathematics and Science Study, and now for Trends in International...), are international assessments of mathematical competence. PISA differs substantially from TIMSS in that it focuses much more on students' ability to use mathematics in applied contexts. Hence, a design characteristic of PISA problems is that they must be realistic: the mathematics in the problems must correspond, in a meaningful way, to the phenomena they characterize. De Lange also argues that valuable assessments should highlight not only what students can do, but what they find difficult, sometimes pinpointing significant (and remediable) omissions in curricula.

In Chapter 8, Bernard Madison makes a somewhat parallel argument regarding the need for sense-making in an increasingly quantified world. For the most part, he notes, students exposed to the traditional U.S. curriculum have the formal mathematical tools they need in order to make sense of problems in context; what they lack is experience in framing problems in ways that make sense. This is increasingly important, as consumers and voters are bombarded with graphs and data that support contradictory or pre-determined positions. Full participation in a democratic society will call for being able to sort through the symbols to the underlying assumptions, and to see if they really make sense.

One virtue of cross-national studies is that is they raise questions about fundamental assumptions. People tend to make assumptions about what is and is not possible on the basis of their experience in particular contexts, which are often regional or national. Cross-national comparisons can reveal that something thought to be impossible is not only possible, but has been achieved in another culture. What needs to be done here to achieve it? In Chapter 9, Richard Askey uses a range of mathematics assessments to take readers on a tour of the possible. Some of these assessments are cross-national; others, which play the same role, are historical. It is quite interesting, for example, to compare the mathematical skills required of California teachers in 1875, and 125 years later!

In Chapter 10, David Foster, Pendred Noyce, and Sara Spiegel point to yet another use of assessment the way in which the systematic examination of student work can lead to teachers' deeper understanding of mathematics, of the strengths and weaknesses of the curricula they are using, and of student thinking. Foster, Noyce, and Spiegel describe the work of the Silicon Valley Mathematics Initiative (SVMI), which orchestrates an annual mathematics assessment given to more than 70,000 students. SVMI uses the information gleaned from the student work to produce a document called *Tools for Teachers,* which is the basis of professional development workshops with teachers. As Chapter 10 shows, such attention to student thinking pays off.

Readers of a certain age may remember the warnings that accompanied trial runs of the National Emergency Broadcast System: This is a test. This is only a test! The chapters in this section show that, properly constructed and used, assessments are anything but "only" tests. They are reflections of our values, and vital sources of information about students, curricula, and educational systems.

Assessing Mathematical Proficiency
MSRI Publications
Volume **53**, 2007

Chapter 6
Mathematical Proficiency:
What Is Important?
How Can It Be Measured?

HUGH BURKHARDT

This chapter examines important aspects of mathematical performance, and illustrates how they may be measured by assessments of K–12 students, both by high-stakes external examinations and in the classroom. We address the following questions:

- *Who does assessment inform?* Students? Teachers? Employers? Universities? Governments?
- *What is assessment for?* To monitor progress? To guide instruction? To aid or justify selection? To provide system accountability?
- *What aspects of mathematical proficiency are important and should be assessed?* Quick calculation? The ability to use knowledge in a new situation? The ability to communicate precisely?
- *When should assessment occur to achieve these goals?* Daily? Monthly? Yearly? Once?
- *What will the consequences of assessment be?* For students? For teachers? For schools? For parents? For politicians?
- *What will it cost, and is the necessary amount an appropriate use of resources?*

There are, of course, multiple answers to each of these interrelated questions. Each collection of answers creates a collection of constraints whose satisfaction may require a mix of different kinds of assessment: summative tests, assessment embedded in the curriculum, and daily informal observation and feedback in the

---

Malcolm Swan and Rita Crust led the design of many tasks in this chapter. The tasks were developed and refined in classrooms in the U.K. and the U.S. by the Mathematics Assessment Resource Service team. I have been fortunate to work with them all.

classroom. Rather than discuss each type of assessment, this chapter describes principles that should guide the choice of a system of assessment tasks created with these questions in mind, particularly the third: *What aspects of mathematical proficiency are important and should be assessed?* Every assessment is based on a system of values, often implicit, where choices have to be made (see [NRC 2001], for example); here I seek to unpack relationships between aspects of mathematical proficiency and types of assessment tasks, so the choices can be considered and explicit.

The discussion will mix analysis with illustrative examples. Specific assessment tasks are, perhaps surprisingly, a clear way of showing what is intended — a short item cannot be confused with a long, open investigation, whereas "show a knowledge of natural numbers and their operations" can be assessed by either type of task, although each requires very different kinds of mathematical proficiency.

Assessment Design Principles

Measure what is important, not just what is easy to measure. This is a key principle — and one that is widely ignored. Nobody who knows mathematics thinks that short multiple-choice items really represent mathematical performance. Rather, many believe it makes little difference what kinds of performance are assessed, provided the appropriate mathematical topics are included. The wish for cheap tests that can be scored by machines is then decisive, along with the belief that "Math tests have always been like this."[1] This approach is widely shared in all the key constituencies, but for very different reasons. Administrators want to keep costs down. Psychometricians are much more interested in the statistical properties of items than what is assessed. Moreover, the assumptions underlying their procedures are less-obviously flawed for short items. Teachers dislike all tests and want to minimize the time spent on them as a distraction from "real teaching" — ignoring the huge amounts of time they now spend on test preparation that is not useful for learning to do mathematics. Parents think "objectively scored" multiple-choice tests are "fairer" than those scored by other methods, ignoring the values and biases associated with multiple-choice tasks. None of these groups seems to be aware that assessment may affect students' learning of, view of, and attitude to mathematics. This chapter describes tasks that assess aspects of mathematical proficiency that may be difficult or impossible to assess by multiple-choice tasks.

---

[1] Only in the U.S., particularly in mainstream K–12 education. Other countries use much more substantial tasks, reliably scored by people using carefully developed scoring schemes.

Assess valued aspects of mathematical proficiency, not just its separate components. Measuring the latter tells you little about the former — because, in most worthwhile performances, the whole is much more than the sum of the parts. Is a basketball player assessed only through "shooting baskets" from various parts of the court and dribbling and blocking exercises? Of course not — scouts and sports commentators watch the player in a game. Are pianists assessed only through listening to scales, chords and arpeggios (though all music is made of these)? Of course not — though these may be part of the assessment, the main assessment is on the playing of substantial pieces of music. Mathematical performance is as interesting and complex as music or basketball, and should by the same token be assessed holistically as well as analytically. When we don't assess in this way (which, for U.S. school mathematics, is much of the time), is it any surprise that so many students aren't interested? No intelligent music student would choose a course on scales and arpeggios.

What do these principles imply for assessment in K–12 mathematics? Consider the following simple task:

A triangle has angles $2x$, $3x$ and $4x$.

(a) Write an expression in terms of x for the sum of the angles.

(b) By forming an equation, find the value of x.

If a 16-year-old cannot find x without being led through the task by (a) and (b), is this worthwhile mathematics? For the student who can do the task without the aid of (a) and (b), this already-simple problem is further trivialized by fragmentation. Compare the triangle task to the following task, modified from the Balanced Assessment for the Mathematics Curriculum Project *Middle Grades Assessment Package 1* [BAMC 1999, p. 40], for students of the same age:

Consecutive Addends

Some numbers equal the sum of consecutive natural numbers:

$$5 = 2 + 3$$
$$9 = 4 + 5$$
$$ = 2 + 3 + 4$$

• Find out all you can about sums of consecutive addends.

This is an *open investigation* of a surprisingly rich pure mathematical microcosm, where students have to formulate questions as well as answer them. It is a truly an *open-ended* task, i.e., one where diverse (and incomplete) solutions are expected, and can be used and assessed at various grade levels. (Note the crucial difference between an open-ended task and a *constructed response*.)

Scaffolding can be added to give students easier access, and a well-engineered ramp of difficulty, as illustrated by the following version.

- Find a property of sums of two consecutive natural numbers.
- Find a property of sums of three consecutive natural numbers.
- Find a property of sums of n consecutive natural numbers.
- Which numbers are not a sum of consecutive addends?

In each case, explain why your results are true.

The proof in the last part is challenging for most people. However, the scaffolding means students only have to *answer* questions, not to *pose* them — an essential part of doing mathematics. Is this the kind of task 16-year-old students should be able to tackle effectively? What about the following task, modified from the *Be a Paper Engineer* module of [Swan et al. 1987–1989]? Is it worthwhile, and does it involve worthwhile mathematics?

Will It Fold Flat?

Diagram A is a side view of a pop-up card.

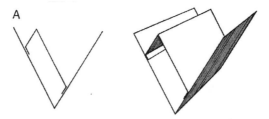

- Look at the diagrams below.
- Which cards can be closed without creasing in the wrong place?
- Which can be opened flat without tearing?
- Make up some rules for answering such questions.

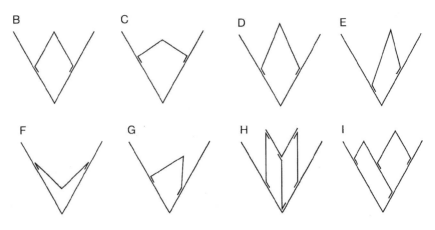

What about the following, more practical, task, adapted from the high-school level materials from the Balanced Assessment for the Mathematics Curriculum Project [BAMC 2000, p. 78]?

Design a Tent

Your task is to design a tent like the one in the picture. It must be big enough for two adults to sleep in (with baggage). It should also be big enough so that someone can move around while kneeling down. Two vertical poles will hold the tent up.

Would the following more scaffolded version of the prompt in Design a Tent be a more suitable performance goal, or does it lead them by the hand too much? (Feedback in development of tasks with students guides such design decisions.) One might ask:

- Estimate the relevant dimensions of a typical adult.

- Estimate the dimensions of the base of your tent.

- Estimate the length of the vertical tent poles you will need.

- Show how you can make the top and sides of the tent from a single piece of material. Show all the measurements clearly.

    Calculate any lengths or angles you don't know.
    Explain how you figured these out.

This version is a typical fairly closed *design task*, requiring sensible estimation of quantities, geometric analysis, and numerical calculations (and even the Pythagorean Theorem).

These tasks (particularly Will It Fold Flat? and Design a Tent) are also seen as worthwhile by people who are *not* mathematicians or mathematics teachers. (Most people will not become either — but they *all* have to take high school mathematics.) The choice of performance targets, illustrated by the exemplars above, is at the heart of determining the content and nature of the K–12 mathematics curriculum. All sectors of society have an interest in these choices; mathematicians and mathematics educators need their views, and their informed consent. This requires the kind of well-informed debate that remains rare — and, too often, is obfuscated by the emotional over-simplifications of partisans on both sides of the "math wars."

Correlation Is Not Enough

It is often argued that, though tests only measure a small part of the range of performances we are interested in, the results correlate well with richer measures. Even if that were true (it depends on the meaning of "correlate well"), it is *not* a justification for narrow tests. Why? Because assessment plays *three* major roles:

- A. to measure performance — i.e. "to enable students to show what they know, understand and can do;"

but also, with assessment that has high stakes for students and teachers, *inevitably*

- B. to exemplify the performance goals. Assessment tasks communicate vividly to teachers, students and their parents what is valued by society.

Thus

- C. to drive classroom learning activities via the WYTIWYG principle: What You Test Is What You Get.

The roles played by assessment have implications for test designers. Correlation is never enough, because it only recognizes A. The effects through C of cheap and simple tests of short multiple-choice items can be seen in classrooms — the fragmentation of mathematics, the absence of substantial chains of reasoning, the emphasis on procedure over assumptions and meaning, the absence of explanation and mathematical discourse...The list goes on.

Balanced assessment takes A, B, and C into account. The roles played by assessment suggest that a system of assessment tasks should be designed to have two properties:

- *Curriculum balance*, such that teachers who "teach to the test" are led to provide a rich and balanced curriculum covering *all* the learning and performance goals embodied by state, national, or international standards.
- *Learning value* — because such high-quality assessment takes time, the assessment tasks should be worthwhile learning experiences in themselves.

Assessment with these as prime design goals will, through B and C above, support rather than undermine teaching and learning high-quality mathematics. This is well recognized in some countries, where assessment is used to actively encourage improvement. In the U.S., a start has been made. The Mathematics Assessment Resource Service [Crust 2001–2004; NSMRE 1998] have developed better-balanced assessment, as have some states. However, cost considerations too often lead school systems to choose cheap multiple-choice tests that assess only a few aspects of mathematical performance, and that drive teaching and learning in the wrong directions. This is despite the fact that only a tiny fraction of educational spending is allocated to assessment.

It follows from B and C that choosing the range of task types to use in an assessment system together with lists of mathematical content is a rather clear way to determine a curriculum. (Lists of mathematics content alone, while essential, do not answer many key questions about the aspects of mathematical performance that are valued, so do not specify the types and frequency of assessment tasks. For example: What should be the balance of short items, 15-minute tasks, or three-week projects?) This issue and its relationship with a more analytic approach, are discussed in a later section.

Some common myths about assessment are worth noting:

Myth 1: Tests are precision instruments. They are not, as test-producers' fine print usually makes clear. Testing and then retesting the same student on parallel forms, "equated" to the same standard, can produce significantly different scores. This is ignored by most test-buyers who know that measurement uncertainty is not politically palatable, when life-changing decisions are made on the basis of test scores. The drive for precision leads to narrow assessment objectives and simplistic tests. (This line of reasoning suggests that we should test by measuring each student's height, a measure which is well-correlated with mathematics performance for students from ages 5 to 18.)

Myth 2: Each test should cover all the important mathematics in a unit or grade. It does not and cannot, even when the range of mathematics is narrowed to short content-focussed items; testing is always a sampling exercise. This does not matter as long as the samples in different tests range across all the goals — but some object: "We taught (or learned) X but it wasn't tested this time." (Such sampling is accepted as the inevitable norm in other subjects. History examinations, year-by-year, ask for essays on different aspects of the history curriculum; final examinations in literature or poetry courses do not necessarily expect students to write about every book or poem studied.)

Myth 3: "We don't test that but, of course, all good teachers teach it." If so, then there are few "good teachers;" the rest take very seriously the measures by which society chooses to judge them and, for their own and their students' futures, concentrate on these.

Myth 4: Testing takes too much time. This is true if testing is a distraction from the curriculum. It need not be, if the assessment tasks are also good learning (i.e., curriculum) tasks. Feedback is important in every system; in a later section we shall look at the cost-effectiveness of assessment time.

What Should We Care About?

We now take a further look at this core question. Is "Will these students be prepared for our traditional undergraduate mathematics courses?" still a sound criterion for judging K–12 curricula and assessment? What other criteria should be considered? (Personal viewpoint: the traditional imitative algebra–calculus route was a fine professional preparation for my career as a theoretical physicist[2]; however, for most people it is not well-matched to their future needs — except for its "gatekeeping" function which could be met in various ways. (Latin was required for entrance to both Oxford and Cambridge Universities when I was an undergraduate. All now agree that this is an inappropriate gatekeeper.)

In seeking a principled approach to goal-setting, it is useful to start with a look at societal goals — what capabilities people want kids to have when they leave school. Interviews with widely differing groups produce surprisingly consistent answers, and their priorities are not well-served by the current mathematics curriculum. I have space to discuss just a few key aspects.

Automata or thinkers? Which are we trying to develop? Society's demands are changing, and will continue to change, decade by decade — thus students need to develop flexibility and adaptability in using skills and concepts, and in self-propelled learning of new ones. American economic prosperity is said to depend on developing *thinkers* at all levels of technical skill, whether homebuilder, construction-site worker, research scientist, or engineer. Equally, it is absurd economics to spend the approximately $10,000 required for a K–12 mathematics education to develop the skills of machines that can be purchased for between $5 and $200. *Thinkers* appear to have more fun than drones, which is important for motivation. So, how do we assess *thinkers*? We give them problems that make them *think*, strategically, tactically, and technically — as will many of the problems student will face after they leave K–12 education, where mathematics can help.

Mathematics: Inward- or outward-looking? Mathematicians and many good mathematics teachers are primarily interested in mathematics itself. For them, its many uses in the world outside mathematics are a spin-off. Mathematics and mathematics teaching are two admirable and important professions — but their practitioners are a tiny minority of the population, in school and in society as a whole. They rightly have great influence on the design of the K–12 mathematics curriculum, but should the design priorities be theirs, or more outward-looking ones that reflect society's goals? The large amount of curriculum time devoted

---

[2]Not surprising, since it was essentially designed by Isaac Newton — and not much changed in content since.

to mathematics arose historically because of its utility in the outside world.[3] That priority, which now implies that the mathematics curriculum must change, should continue to be respected.

Mathematical Literacy

Mathematical literacy is an increasing focus of attention, internationally (see, e.g., [PISA 2003]) and in the U.S. (see, e.g., [Steen 2002]). The Organisation for Economic Co-operation and Development Programme of International Student Assessment, seeks to assess mathematical literacy, complementing the mathematically inward-looking student assessments of the Third International Mathematics and Science Study (see de Lange's chapter for more discussion of the design of these tests). Various terms[4] are used for mathematical literacy. In the U.S. "quantitative literacy" is common; in the U.K., where the term "numeracy" was coined [Crowther 1959], it is now being called "functional mathematics" [UK 2004b]. Each of these terms has an inherent ambiguity. Is it literacy *about* or *in using* mathematics? Is it functionality *inside* or *with* mathematics? The latter is the focus:

> *Functional mathematics* is mathematics that most *nonspecialist adults will benefit from using in their everyday lives* to better understand and operate in the world they live in, and to make better decisions.

Secondary school mathematics is not functional mathematics for most people. (If you doubt this, ask nonspecialist adults, such as English teachers or administrators, when they last used some mathematics they first learned in secondary school.) Functional mathematics is distinct from the "specialized mathematics" important for various professions.

The current U.S. curriculum is justifiable as specialized mathematics for some professions. However, as a gatekeeper subject, which is a key part of everyone's education, should mathematics education not have a large component of functional mathematics that every educated adult will actually use?

I shall outline what is needed to make the present U.S. high school mathematics functional, the core of which is the teaching of modeling. Modeling also reinforces the learning of mathematical concepts and skills (see [Burkhardt and Muller 2006]. This is not a zero-sum game.

---

[3] The argument that mathematics is an important part of human culture is clearly also valid — but does it justify more curriculum time than, say, music? Music currently gives much more satisfaction to more people.

[4] Each of these terms each has an inherent ambiguity. Is it literacy *about* or *using* mathematics? Is it functionality *inside* or *with* mathematics? It is the latter that is the focus of those concerned with mathematical literacy.

Modeling

Skill in modeling is a key component in "doing mathematics." The figure
below shows a standard outline of its key phases; see, for example, [Burkhardt
1981].[5] In current mathematics assessment and teaching, only the SOLVE phase
gets much attention. (The situation is sometimes better in statistics curricula.)

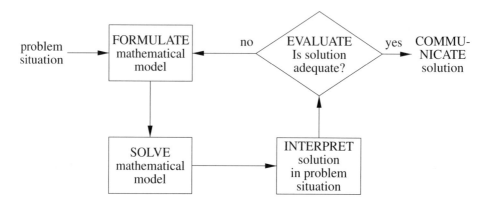

Key phases in modeling

Mathematical modeling is not an everyday term in school mathematics; indeed, it is often thought of as an advanced and sophisticated process, used only
by professionals. That is far from the truth; we do it whenever we *mathematize*
a problem. The following tasks illustrate this:

- Joe buys a six-pack of coke for $5.00 to share among his friends. How much
  should he charge for each bottle?
- If it takes 40 minutes to bake 5 potatoes in the oven, how long will it take to
  bake one potato?
- If King Henry the Eighth had 6 wives, how many wives did King Henry the
  Fourth have?

The difference between these tasks is in the appropriate choice of mathematical
model. The first is a standard proportion task. However, *all the tasks in most
units on proportion need proportional models, so skill in choosing an appropriate model is not developed.* In the second task, the answer depends on the type
of oven (what remains constant: for traditional constant temperature ovens the
answer is about 40 minutes, and for constant power microwave ovens there is
rough proportionality, so an approximate answer would be $40 \div 5 = 8$ minutes).

---

[5]The phases of pure mathematical problem solving are similar.

For each problem, as usual, more refined models could also be discussed. For the third task, if students laugh they pass.

Mathematics teachers sometimes argue that choosing the model is "not mathematics" — but it is essential for mathematics to be functional. Of course, the situations to be modeled in mathematics classrooms should not involve specialist knowledge of another school subject but should be, as in the examples above, situations that children encounter or know about from everyday life. Teachers of English reap great benefits from making instruction relevant to students' lives; where mathematics teachers have done the same (see, e.g., [Swan et al. 1987–1989]), motivation is improved, particularly but not only with weaker students. Relationships in their classrooms are also transformed.[6] Mathematics acquires human interest. Curriculum design is not a zero-sum game; the use of "math time" in this way enhances students' learning of mathematics itself [Burkhardt and Muller 2006].

What Content Should We Include?

There will always be diverse views on content. This is not the place to enter into a detailed discussion of what mathematical topics should have what priority (for such a discussion see, [NRC 2001], for instance). Here I shall only discuss a few aspects of U.S. curricula that, from an international perspective, seem questionable. Is a year of Euclidean geometry a reasonable, cost-effective use of every high school graduate's limited time with mathematics, or should Euclidean geometry be considered specialized mathematics — an extra option for enthusiasts? Should not the algorithmic and functional aspects of algebra, including its computer implementation in spreadsheets and programming, now play a more central role in high school algebra? (Mathematics everywhere is now done with computer technology — except in the school classroom.) Should calculus be a mainstream college course, to the exclusion of discrete mathematics and its many applications, or one for those whose future lies in the physical sciences and traditional engineering?

In the U.K., policy changes [UK 2004a; 2004b] have addressed such issues by introducing "double mathematics" from age 14, with a challenging functional "mathematics for life" course for all and additional specialized courses with a science and engineering, or business and information technology focus. It will be interesting to see how this develops. (The U.K. curriculum already has separate English language and English literature courses. All students take the first; about half take both.)

---

[6]"The Three R's for education in the 21st century are Rigor, Relevance and Relationships," Bill Gates, U.S. National Governor's Conference, 2005. Functional mathematics develops them all.

A Framework for Balance

Mathematical Content Dimension

- *Mathematical content* will include some of:

  Number and quantity including: concepts and representation; computation; estimation and measurement; number theory and general number properties.

  Algebra, patterns and function including: patterns and generalization; functional relationships (including ratio and proportion); graphical and tabular representation; symbolic representation; forming and solving relationships.

  Geometry, shape, and space including: shape, properties of shapes, relationships; spatial representation, visualization and construction; location and movement; transformation and symmetry; trigonometry.

  Handling data, statistics. and probability including: collecting, representing, interpreting data; probability models — experimental and theoretical; simulation.

  Other mathematics including: discrete mathematics, including combinatorics; underpinnings of calculus; mathematical structures.

Mathematical Process Dimension

- *Phases* of problem solving, reasoning and communication will include, as broad categories, some or all of:

  Modeling and formulating;
  Transforming and manipulating;
  Inferring and drawing conclusions;
  Checking and evaluating;
  Reporting.

Task Type Dimensions

- *Task type*: open investigation; nonroutine problem; design; plan; evaluation and recommendation; review and critique; re-presentation of information; technical exercise; definition of concepts.
- *Nonroutineness*: context; mathematical aspects or results; mathematical connections.
- *Openness*: open end with open questions; open middle.
- *Type of goal*: pure mathematics; illustrative application of the mathematics; applied power over the practical situation.
- *Reasoning length:* expected time for the longest section of the task. (An indication of the amount of scaffolding).

Circumstances of Performance Dimensions

- *Task length*: short tasks (5–15 minutes), long tasks (15–60 minutes), extended tasks (several days to several weeks).
- *Modes of presentation*: written; oral; video; computer.
- *Modes of working*: individual; group; mixed.
- *Modes of response*: written; built; spoken; programmed; performed.

Dimensions of Mathematical Performance

Whenever curriculum and assessment choices are to be made, discussion should focus on performance as a whole, not just the range of mathematical topics to be included. To support such an analysis, the Mathematics Assessment Resource Service has developed a *Framework for Balance*, summarized on the facing page. The *Framework* includes, as well as the familiar *content* dimension, the *phases of problem solving* from the figure on page 86, and various others including one holistic dimension, *task type*. This multidimensional analytic framework (it is dense, and takes time to absorb) is a way to examine how the major dimensions of performance are balanced in a particular test or array of assessment tasks. In most current tests, balance is sought only across the content dimension, and the only task type is short exercises that require only transforming and manipulating (the SOLVE phase).[7] The ability to formulate a problem is trivialized, and interpretation, critical evaluation and communication of results and reasoning are rarely assessed.

Task types. I will briefly illustrate the holistic dimension of the otherwise analytic *Framework for Balance* with tasks of each type. I chose to illustrate the holistic dimension because it brings out something of the variety of challenges that mathematics education and assessment should aim to sample (as in literature, science, social studies, music, etc.). Tasks are mostly given here in their core form rather in a form engineered for any specific grade. The tasks are designed to enable *all* students who have worked hard in a good program to make significant progress, while offering challenges to the most able. This can be achieved in various ways by including "open tasks" or "exponential ramps" to greater generality, complexity, and/or abstraction. We start the examples with two *planning tasks* — the second being more open, giving less specific guidance.

Ice Cream Van

You are considering driving an ice cream van during the summer break. Your friend, who knows everything, says that "it's easy money." You make a few enquiries and find that the van costs $100 per week to lease. Typical selling data is that one can sell an average of 30 ice creams per hour, each costing 50 cents to make and each selling for $1.50.

How hard will you have to work in order to make this "easy money"?

---

[7]The common argument that "You need a solid basis of mathematics before you can do these things" is simply untrue. However small or large your base of concepts and skills, you can deploy it in solving worthwhile problems — as young children regularly show, using counting. Deferring these practices to graduate school excludes most people, and stultifies everyone's natural abilities in real problem solving. It is also an equity issue — such deferred gratification increases the achievement gap, probably because middle class homes have time and resources to encourage their children to persist in school activities that lack any obvious relevance to their current lives.

Timing Traffic Lights

A new set of traffic lights has been installed at an intersection formed by the crossing of two roads. Left turns are *not* permitted at this intersection. For how long should each road be shown the green light?

Treilibs et al. [1980] analyzed responses to these tasks from 120 very high-achieving grade 11 mathematics students and found that *none* used algebra for the modeling involved. (The students used numbers and graphs, more or less successfully.) These students all had five years of successful experience with algebra but, with no education in real problem solving, their algebra was non-functional. Modeling skill is important and, as many studies (see Swan et al. 1987–1989, for example) have shown, teachable.

The next task [Crust 2001–2004] is typical of a genre of *nonroutine problems* in pure mathematics, often based on pattern generalization, in which students develop more powerful solutions as they mature.

Square Chocolate Boxes

Chris designs chocolate boxes.
The boxes are in different sizes.
The chocolates are always arranged in the same kind of *square* pattern.
The shaded ovals are dark chocolates and the white ovals are milk chocolates.

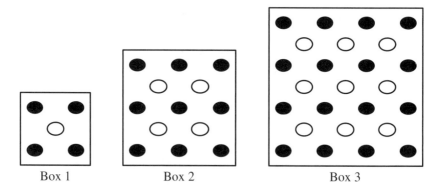

Box 1   Box 2   Box 3

Chris makes a table to show how many chocolates are in each size of box.

Box number	1	2	3	4	5
number of dark chocolates	4	9			
number of milk chocolates	1	4			
total number of chocolates	5	13			

- Fill in the missing numbers in Chris's table.
- How many chocolates are there in Box 9? Show how you figured it out.
- Write a rule or formula for finding the total number of chocolates in Box *n*. Explain how you got your rule.
- The total number of chocolates in a box is 265. What is the box number? Show your calculations.

The scaffolding shown for this task fits the current range of performance in good middle school classrooms. One would hope that, as problem solving strategies and tactics become more central to the curriculum, part 3 alone would be a sufficient prompt. The following is an *evaluate and recommend* task — an important type in life decisions, where mathematics can play a major role.

Who's For The Long Jump?

Our school has to select a girl for the long jump at the regional championship. Three girls are in contention. We have a school jump-off. These are their results, in meters.

Elsa	Ilse	Olga
3.25	3.55	3.67
3.95	3.88	3.78
4.28	3.61	3.92
2.95	3.97	3.62
3.66	3.75	3.85
3.81	3.59	3.73

Hans says "Olga has the longest average. She should go to the championship." Do you think Hans is right? Explain your reasoning.

This task provides great opportunities for discussing the merits and weaknesses of alternative measures. Ironically, in the TIMSS video lesson (from Germany, but it could be in the U.S.) on which this task is based, the students calculate the mean length of jump for each girl and use that for selection. Olga wins, despite having shorter longest jumps than either of the others. The teacher moves on without comment! A splendid opportunity is missed — to discuss other measures, their strengths and weaknesses, the effect of a "no jump," or any other situational factors. (Bob Beamon — who barely qualified for the Olympics after two fouls in qualifying jumps — would have been excluded. He set a world record.) Is this good mathematics? I have found research mathematicians who defend it as "not wrong." What does this divorce from reality do for students' image of mathematics?

Magazine Cover [Crust 2001–2004] is a *re-presentation of information* task (for grade 3, but adults find it nontrivial). It assesses geometry and mathematical communication.

Magazine Cover

This pattern is to appear on the front cover of the school magazine.

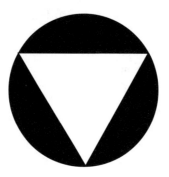

You need to call the magazine editor and describe the pattern as clearly as possible in words so that she can draw it.

Write down what you will say on the phone.

The rubric for Magazine Cover illustrates how complex tasks can, with some scorer training, be *reliably* assessed — as is the practice in most countries and, in the United States, in some of the problems in the Advanced Placement exams.

Magazine Cover: Grade 3	Points
Core element of performance: describe a geometric pattern	
Based on these, credit for specific aspects of performance should be assigned as follows:	
A circle.	1
A triangle.	1
All corners of triangle on (circumference of) circle.	1
Triangle is equilateral. Accept: All sides are equal/the same.	1
Triangle is standing on one corner. Accept: Upside/going down.	1
Describes measurements of circle/triangle.	1
Describes color: black/white.	1
Allow 1 point for each feature up to a maximum of 6 points.	
Total possible points:	6

For our last example, we return to a type that, perhaps, best represents "doing" both mathematics and science — *investigation*. Consecutive Addends (page 79) is an open investigation in pure mathematics. Equally, there are many important situations in everyday life that merit such investigation. One important area, where many children's quality of life is being curtailed by their parents' (and society's) innumeracy, is tackled in:

Being Realistic About Risk

Use the Web to find the chance of death each year for an average person of the same age and gender as

- you
- your parents
- your grandparents

List some of the things that people fear (or dream of), such as being

- struck by lightning
- murdered
- abducted by a stranger
- killed in a road accident
- a winner of the lottery

For each, find out the proportion of people it happens to each year.

Compare real and perceived risks and, using this information, write advice to parents on taking appropriate care of children.

There will need to be more emphasis on *open investigations*, pure and real-world, if the quality of mathematics education, and students' independent reasoning, is to improve.

The tasks above, and the *Framework for Balance*, provide the basis for a response to our question, "What mathematics values should assessment reflect?" Taken together, they give a glimpse of the diversity of assessment tasks that enable students to show how well they can do mathematics — "making music" not just "practicing scales." There is a place in the *Framework for Balance* for *technical exercises* too — but even these don't have to be boring:

Square Peg

Lee has heard of an old English proverb used when someone is doing a job that they are not suited to. The proverb describes the person as "fitting like a square peg in a round hole."

Lee wondered how much space was left if a square peg was fitted into a round hole.

Lee constructed a square that just fit inside a circle of radius 10 cm.

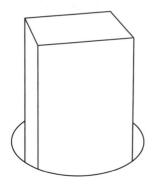
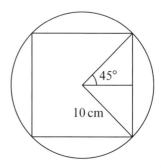

- What percentage of the area of the circle is filled by the area of the square?

Explain your work and show all your calculations.

Another part of this task asks for the same calculation for a circle inside a 10 cm square hole.

Published examples of tasks of these various types include: a set of annual tests for grades 3 through 10 [Crust 2001–2004]; the *New Standards* exam-related tasks [NSMRE 1998], and classroom materials for assessment and teaching (*Balanced Assessment for the Mathematics Curriculum*: see, for instance, [BAMC 1999; 2000]). The *World Class Tests* [MARS 2002–2004] provides a more challenging range of tasks, aimed at high-achieving students.

Improving quality in assessment design

Designing and developing good assessment tasks, which have meaning to students and demand mathematics that is important for them, is among the most difficult educational design challenges. The tasks must enable students *to show what they know, understand and can do* without the help from teachers that classroom activities can provide. Task design is usually subject to too-tight constraints of time and form. Starting with a good mathematics problem is necessary, but far from sufficient. As in all design: *Good design principles are not enough; the details matter.*[8]

---

[8]The difference between Mozart, Salieri, and the many other composers of that time we have never heard of was not in the principles (the rules of melody, harmony, counterpoint, and musical form). Students deserve tasks with some imaginative flair, in mathematics as well as in music and literature.

Thus it is important to recognize high quality in assessment tasks, and to identify and encourage the designers who regularly produce outstanding work. The latter are few and hard to find. [Swan 1986] contains some well-known benchmark examples.

The emphasis in this chapter on the task *exemplars* is no accident, but it is unconventional. However, without them the analytic discussion lacks meaning. In a misguided attempt to present assessment as more "scientific" and accurate than it is, most tests are designed to assess elements in a model of the domain, which is often just a list of topics. All models of performance in mathematics are weak, usually taking no account of how the different elements interact.

Our experience with assessment design shows that it is much better to start with the tasks. Get excellent task designers to design and develop a wide range of good mathematics tasks, classify them with a domain model, then fill any major gaps needed to balance each test.

Interestingly and usefully, when people look at specific tasks, sharply differing views about mathematics education tend to soften into broad agreement as to whether a task is worthwhile, and the consensus is, "Yeah, our kids should be able to do that."

Having looked in some depth at tasks that measure mathematical performance, we now have the basis for answering the other questions with which I began. I shall be brief and simplistic.

Who is assessment for? What is it for? Governments, and some parents, want it for accountability. Universities and employers for selection. They all want just one reliable number. Teachers and students, on the other hand, can use a lot of rich and detailed feedback to help diagnose strengths and weaknesses, and to guide further instruction. Some parents are interested in that too.

When should it happen to achieve these goals? For teachers and students in the classroom, day-by-day — but, to do this well,[9] they need much better tools. For accountability, tests should be as rare as society will tolerate; the idea that frequent testing will drive more improvement is flawed. Good tests, that will drive improvements in curriculum, need only happen every few years.

What will the consequences be? Because effective support for better teaching is complex and costs money while pressure through test scores is simple and cheap, test-score-based sanctions seem destined to get more frequent and more severe. The consequences for mathematics education depend on the quality of the tests. Traditional tests will continue to narrow the focus of teaching, so learning, which relies on building rich connections for each new element, will suffer. Balanced assessment will, with some support for teachers, drive

---

[9] The classroom *assessment for learning* movement is relatively new. There is much to do.

continuing improvement. Currently, with the air full of unfunded mandates, the chances of improved large-scale assessment do not look good.

Cost and cost-effectiveness

Finally: *What will assessment cost, and would this expenditure be an appropriate use of resources?*

Feedback is crucial for any complex interactive system. Systems that work well typically spend approximately 10% turnover on its "instrumentation." In U.S. education, total expenditure is approximately $10,000 per student-year, which suggests that approximately $1,000 per student-year should be spent on assessment across all subjects. Most of spending should be for assessment for learning in the classroom,[10] with about 10%, or approximately $100 a year, on summative assessment linked to outside standards. This is an order of magnitude more than at present but still only 1% of expenditure. Increases will be opposed on all sides for different reasons: budget shortage for administrators and dislike of assessment for teachers. Yet while "a dollar a student" remains the norm for mathematics assessment, students' education will be blighted by the influence of narrow tests. If, for reasons of economy and simplicity, you judge the decathlon by running only the 100 meters, you may expect a distortion of the training program!

References

[BAMC 1999] *Balanced Assessment for the Mathematics Curriculum: Middle grades Package 1, Grade 6–9*, White Plains, NY: Dale Seymour Publications, 1999.

[BAMC 2000] *Balanced Assessment for the Mathematics Curriculum: High school Package 2, Grade 9–11*, White Plains, NY: Dale Seymour Publications, 2000.

[Burkhardt 1981] H. Burkhardt, *The real world and mathematics*, Glasgow: Blackie, 1981.

[Burkhardt and Muller 2006] H. Burkhardt and E. Muller, "Applications and modelling for mathematics", in *Applications and modelling in mathematics education*, edited by W. Blum et al., New ICMI Studies Series **10**, New York: Springer, 2006.

[Crowther 1959] Central Advisory Council for Education, *15–18: A report of the Central Advisory Council for Education* [Crowther Report], London: Her Majesty's Stationery Office, 1959.

[Crust 2001–2004] R. Crust and the Mathematics Assessment Resource Service Team, *Balanced assessment in mathematics* [annual tests for grades 3 through 10], Monterey, CA: CTB/McGraw-Hill, 2001–2004.

---

[10]Professor, on seeing abysmal student scores a third of the way through his analysis course: "We've gone so far in the semester, I don't know what to do except to go on — even though it's hopeless."

[MARS 2002–2004] Mathematics Assessment Resource Service, *World class tests of problem solving in mathematics, science, and technology*, London: Nelson, 2002–2004. Shell Centre Team: D. Pead, M. Swan, R. Crust, J. Ridgway, & H. Burkhardt for the Qualifications and Curriculum Authority.

[NRC 2001] National Research Council (Mathematics Learning Study: Center for Education, Division of Behavioral and Social Sciences and Education), *Adding it up: Helping children learn mathematics*, edited by J. Kilpatrick et al., Washington, DC: National Academy Press, 2001.

[NSMRE 1998] *New standards mathematics reference examination*, San Antonio, TX: Harcourt Assessment, 1998.

[PISA 2003] Programme for International Student Assessment, *The PISA 2003 assessment framework: Mathematics, reading, science and problem solving knowledge and skills*, Paris: Organisation for Economic Co-operation and Development, 2003. Available at http://www.pisa.oecd.org/dataoecd/46/14/33694881.pdf. Retrieved 13 Jan 2007.

[Steen 2002] L. A. Steen (editor), *Mathematics and democracy: The case for quantitative literacy*, Washington, DC: National Council on Education and the Disciplines, 2002. Available at http://www.maa.org/ql/mathanddemocracy.html. Retrieved 28 Feb 2006.

[Swan 1986] M. Swan and the Shell Centre Team, *The language of functions and graphs*, Manchester, UK: Joint Matriculation Board, 1986. Reissued 2000, Nottingham, U.K: Shell Centre Publications.

[Swan et al. 1987–1989] M. Swan, J. Gillespie, B. Binns, H. Burkhardt, and the Shell Centre Team, *Numeracy through problem solving* [Curriculum modules], Nottingham, UK: Shell Centre Publications, 1987–1989. Reissued 2000, Harlow, UK: Longman.

[Treilibs et al. 1980] V. Treilibs, H. Burkhardt, and B. Low, *Formulation processes in mathematical modelling*, Nottingham: Shell Centre Publications, 1980.

[UK 2004a] Department for Education and Skills: Post-14 Mathematics Inquiry Steering Group, *Making mathematics count*, London: Her Majesty's Stationery Office, 2004. Available at http://www.mathsinquiry.org.uk/report/index.html. Retrieved 28 Feb 2006.

[UK 2004b] Department for Education and Skills: Working Group on 14–19 Reform, *14–19 curriculum and qualifications reform*, London: Her Majesty's Stationery Office, 2004. Available at http://publications.teachernet.gov.uk/eOrderingDownload/DfE-0976-2004.pdf. Retrieved 1 Mar 2006.

Assessing Mathematical Proficiency
MSRI Publications
Volume **53**, 2007

Chapter 7
Aspects of the Art of Assessment Design

JAN DE LANGE

Educational design in general is a largely underestimated and unexplored area of design, and its relationship with educational research can be characterized as somewhat less than satisfying. The design of assessments is often seen as an afterthought. And it shows.

Of course, there is a wealth of publications on assessment, but quite often these focus on psychometric concerns or preparation for high-stakes tests (a very profitable industry). What is lacking is a tight linkage between research findings and the creation of mathematically rich and revealing tasks for productive classroom use. The report "Inside the black box" [Black and Wiliam 1998], which looks in depth at current research, shows clearly that we should not only invest more in classroom assessment in mathematics, but also that the rewards will be high if we do so.

Linking Items with Framework

To provoke some discussion, and to invite the reader to reflect on the items and make a judgment, I start the examples with an item from the Third International Mathematics and Science Study (TIMSS). According to the TIMSS web site http://www.timss.org, TIMSS was "the largest and most ambitious international study of student achievement ever conducted. In 1994–95, it was conducted at five grade levels in more than 40 countries (the third, fourth, seventh, and eighth grades, and the final year of secondary school)." Mathematics coverage on TIMSS was, in essence, international consensus coverage of the traditional curriculum.

The item in question, given to students about 14 years old in the mid-1990s, involved a simplified portion of a regional road map, replicated on the next page. The scale is reinforced by a sentence above the map:

> One centimeter on the map represents 8 kilometers on the land.

Distance on a Map

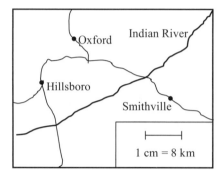

About how far apart are Oxford and Smithville on the land?

A. 4 km   B. 16 km   C. 25 km   D. 35 km

The problem seems simple, actually too simple, for the intended age group. The multiple-choice format makes it even easier for the students to find the correct answer, 35 km. There is nothing wrong with easy questions as such; the surprise lies in the "intended performance category," which is "using complex procedures." The consequence of this classification is that students who have chosen answer C are considered able to use complex mathematical procedures. This seems somewhat far-fetched.

This example shows clearly that we need to make sure that there is a reasonable relationship between "expected performances" (or "competencies" or "learning goals") and assessment items. This may seem straightforward but is at the heart of the design of any coherent and consistent assessment design.

Framework and Competencies

Discussions of mathematical competencies have really evolved during the last decade, not least because of large-scale assessments like TIMSS, the Program for the International Student Assessment, known as PISA (see http://nces.ed.gov/surveys/pisa/AboutPISA.asp), and the U.S. National Assessment of Educational Progress, known as NAEP (see http://nces.ed.gov/nationsreportcard/). Items from TIMSS and PISA are discussed in this chapter. (NAEP is discussed in Chapters 1 and 3.)

The context used in the TIMSS item above is excellent. The task is an authentic problem: estimate distances using a map and a scale. But simply placing a problem in a context can create difficulties as well. This has been discussed in the literature, although not widely. That it is necessary to pay more attention to the connection between context and content is shown by the following example, taken from a popular U.S. textbook series:

One day a sales person drove 300 miles in $x^2 - 4$ hours.
The following day, she drove 325 miles in $x + 2$ hours.

- Write and simplify a ratio comparing the average rate the first day with the average the second day.

This example shows clearly what we would classify as a non-authentic problem in the sense that the mathematics is sound, but is unlikely to be needed in the context in which the problem has been set. To call this "problem solving using algebra," as the textbook suggests, clearly shows that we need clearer understandings of and guidelines for the design of authentic problem solving items.

Figure 1, adapted from [PISA 2003, p. 30], suggests concerns illustrated by the examples. In the center of the figure are the problem as posed, and the desired solution. The ovals at the top of Figure 1 indicate that special attention is given to mathematical situations and contexts, and that problems on the assessment will focus on overarching mathematical concepts. These are the mathematical arenas in which students will be examined. The bottom oval in Figure 1 refers to competencies — the specific mathematical understandings we want to measure. The question is, do the students have these competencies? Especially in classroom assessment, this question is often overlooked: the teacher poses a question, the students give an answer, and the issue considered is whether or

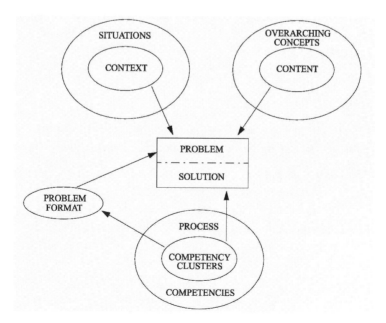

Figure 1. Problems, competencies, context, and content. Adapted from [PISA 2003, p. 30].

not the answer is correct. More often than not both teacher and students fail to reflect on what competencies were used in order to solve the "problem."

Figure 1 by itself does not shed any light on how these relationships function in good task design, it just indicates that these relationships need to function properly. But we need much more research in order to come close to answers that are scientifically based. At this moment many high-quality tasks are the result of creative impulses, and often seem more the result of an art than the result of scientific principles.

As we look at different frameworks we see some correlation between the competencies identified, albeit with slight differences. In an earlier *Framework for Classroom Assessment* [de Lange 1999] the following clusters of competencies were identified:

- *Reproduction*: simple or routine computations, definitions, and (one-step or familiar) problems that need almost no mathematization.

- *Connections*: somewhat more complex problem solving that involves making connections (between different mathematical domains, between the mathematics and the context).

- *Reflection*: mathematical thinking, generalization, abstraction and reflection, and complex mathematical problem solving.

These clusters of competencies are at the heart of the PISA framework. Each of the clusters can be illustrated with examples. (Note: These are not meant as examples of psychometrically valid test items, merely as illustrations of the competencies needed.)

Examples of reproduction

What is the average of 7, 12, 8, 14, 15, 9?

1000 zed are put in a savings account at a bank, at an interest rate of 4%. How many zed will there be in the account after one year?

The savings account problem above is classified as reproduction, because it will not take most students beyond the simple application of a routine procedure. The next savings account problem goes beyond that: it requires the application of a chain of reasoning and a sequence of computational steps that are not characteristic of the reproduction cluster competencies, hence is classified as in the connection cluster.

Examples of connections

1000 zed are put in a savings account at a bank. There are two choices: one can get a rate of 4%, or one can get a 10 zed bonus from the bank and a 3% rate. Which option is better after one year? And after two years?

Mary lives two kilometers from school. Martin five. What can you say about how far Mary and Martin live from each other?

Example of reflection

In Zedland the national defense budget is 30 million zed for 1980. The total budget for that year is 500 million zed. The following year the defense budget is 35 million zed, while the total budget is 605 million zed. Inflation during the period covered is 10%.

(a) You are invited to give a lecture for a pacifist society. You intend to explain that the defense budget decreases over this period. Explain how you would do this.

(b) You are invited to give a lecture for a military academy. You intend to explain that the defense budget increases over this period. Explain how you would do this.

Item Difficulty and Other Considerations

I observed, when discussing the TIMSS example, that items may be "simple," not in the sense that many students succeed in doing the task, but in the sense that students seem to have the mathematical skills required to do so. The story of how the following example stopped being a potential PISA item raises questions about the kinds of tasks that should be included in large-scale assessments.

Heartbeat

For health reasons people should limit their efforts, for instance during sports, in order not to exceed a certain heartbeat frequency.

For years the relationship between a person's recommended maximum heart rate and the person's age was described by the following formula:

$$Recommended\ maximum\ heart\ rate = 220 - age$$

Recent research showed that this formula should be modified slightly. The new formula is as follows:

$$Recommended\ maximum\ heart\ rate = 208 - (0.7 \times age)$$

Question 1

A newspaper article stated: "A result of using the new formula instead of the old one is that the recommended maximum number of heartbeats per minute for young people decreases slightly and for old people it increases slightly."

From which age onwards does the recommended maximum heart rate increase as a result of the introduction of the new formula? Show your work.

Question 2

The formula *Recommended maximum heart rate* = $208 - (0.7 \times age)$ also helps determine when physical training is most effective: this is assumed to be when the heartbeat is at 80% of the recommended maximum heart rate.

Write down a formula for calculating the heart rate for most effective physical training, expressed in terms of age.

This problem seems to meet most standards for a good PISA item. It is authentic; it has real mathematics; there is a good connection between context and content; and the problem is nontrivial. Therefore it was no great surprise that when this item was field tested, it did well as a test of the relevant student knowledge. There was only one small difficulty: fewer than 10% of the students were successful. This was a reason to abandon the problem.

Abandoning such problems is a problem! We should look into why this item is difficult for so many 15-year-old students! Why are we excluding items of this kind, with a rather low success rate, from these kinds of studies? We are probably missing a valuable source of information, especially if we consider the longitudinal concept underlying the whole PISA study.

Knowing How to Think

TIMSS underscores this point—at least in my opinion. Let us look at an item for 18-year-olds from the 1990s. This item was the subject of a *New York Times* article by E. Rothstein, titled "It's not just numbers or advanced science, it's also knowing how to think" (9 Mar 1998, p. D3). The title is not a quote from the PISA framework regarding the clusters, but a very good observation that the journalist made when looking at this item.

Rod

A string is wound symmetrically around a circular rod. The string goes exactly four times around the rod. The circumference of the rod is 4 cm and its length is 12 cm.

Find the length of the string. Show all your work.

The *Times* article included a solution that has all the essentials in the first sentence: "Imagine that you unwrap the cylinder and then flatten it." If you do that, you see that the cylinder becomes a rectangle that is 4 cm (the circumference of the cylinder) in height. The string unrolls as a straight line, and it first hits the bottom of the rectangle at a distance 1/4 of the length of the cylinder, or 3 cm. The next part of the string starts again at the top, hits the bottom 3 cm further along, and so on.

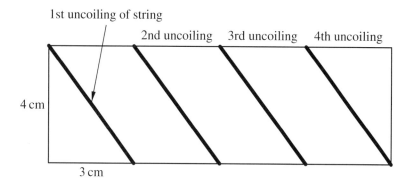

Once the unwrapping has been visualized, an application of the Pythagorean theorem is all that remains to be done. However imagining the unwrapping is something that caused a major problem for almost all students (about 10% got the problem correct) because "knowing how to think" in this creative, applied sense is not part of many curricula. Despite the low success rate it is very useful to have an item like the Rod Problem in the study, because it can reveal significant problems with the curriculum.

This last example might suggest that we reconsider the aims of mathematics education. *If* we consider that the TIMSS Distance on a Map Problem shows that kids are able to carry out complex mathematical procedures; *if* we accept that the Heartbeat Problem should not be in the item collection because it is too difficult; and *if* we similarly reject the Rod Problem, then there seems to be reason to reflect on mathematics education as a whole, and on assessments in particular. Are our students really so "dumb," or as a German magazine asked its readers in reaction to PISA: "Are German kids *doof*?"

This is an interesting question, worth serious reflection, especially in light of some more or less recent outcomes of educational research about mathematics learning. An alternative is to consider the possibility that poor student performance on such tasks is a result of the lack of curricular opportunities, both in terms of specific mathematical experiences (such as in the case of the rod problem) and because, in general those designing curricula have seriously underestimated what students might be able to do if appropriately challenged and supported.

Enhancing Mathematics Learning

Neuroscientists have entered the arena that was considered the domain of cognitive psychologists, mathematics educators, and others. Barth et al. [2006] stated: "Primitive (innate) number sense can be used as a tool for enhancing early mathematics." This is a point well made (about young children) and once again mentioned by an Organisation for Economic Co-operation and Development report [OECD 2003, p. 25]: "Children can do much better than was expected." (The question here of course is: who expected what? But we will not go into that discussion here.) The German mathematics educator Wittmann expressed similar sentiments, probably shared by many who study young children's learning of mathematics: "Children are much smarter than we tend to think" (quoted in *Der Spiegel*, 12 Jun 2004, p. 190).

If all this is true (and we think it is), this has great implications, not only for curriculum design but for assessment as well. It means that assessment should assess as many learning outcomes as possible, at the edge of what children can solve. We need challenging tasks that give us real and valuable information about students' thinking.

Here are examples, taken from regular schools and regular kids, that not only show that there seems to be much truth in these quotes, but also that good assessment design can deliver us little miracles.

This is a test designed for 5- to 6-year-olds. The images shown here (continued on the next page) contain written responses from a student, including the lines from the camera to the shoes in the upper left rectangle:

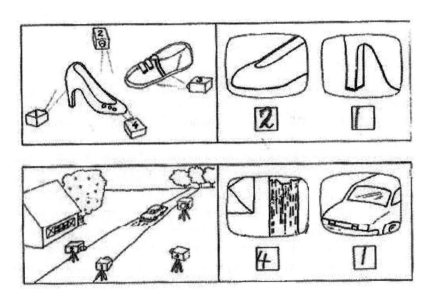

The only question the children were asked is: "Who makes the photo?" The teacher explained orally that this means that the kids had to find the camera that made the pictures shown on the right.

Here is a student's explanation for his response to the first problem (see preceding page):

Look, you have camera 1, 2, 3 and camera 4. And you see two shoes. Camera 2 makes photo 1. This is because camera 2 points towards the nose of the shoe [points to the two lines he has drawn]. And camera 1 films the heel of the shoe.

Camera 4 films the other side; otherwise you would have seen the three dots. And camera 3 films the other shoe!

And here are his comments on the last of the four problems shown:

This is a difficult one!

Camera 3 films the left side of the girl and in the picture you face the other side. Camera 1 films that side, and camera 2 films the back of that girl. Camera 4 shows no lens. This [the lens] sits in front of the camera [points to the front of camera 4] because they are filming towards the playground. And so [puts finger on line of vision] the camera films the shovel.

Yes, this problem does involve complex geometric reasoning. But as we see, many students are capable of doing it, if given the experience.

Reasoning Is an Art

Another example that was trialed both at middle school level and with graduate mathematics students will give us some insight in the possibilities of assessment, and how to challenge students to become mathematical reasoners.

The problem starts with a little strip that shows how the following figure was created.

The first figure of the strip is a "cross." The horizontal and vertical segments are extended to form the second figure. The second figure is rotated 22.5 degrees about its center and four new segments are drawn to create the third figure. (Thus the angle marked in the third figure is 22.5 degrees.)

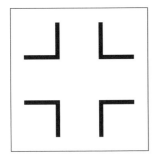 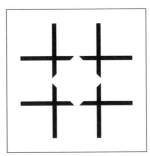 

The task: Use transparencies that show the shape to make a Moorish star pattern. An example of such a Moorish star pattern is shown at the top of the next page.

After a while most students are able to make the star. The real question follows: if you know that a key angle is 22.5 degrees, what are the measurements of all the other angles in the star pattern? (See middle picture on the next page.)

And how did we get the following star with the same transparencies? What do you know about the number of star points, and about all angles?

It goes almost without saying that both the 14-year-olds and 24-year-olds were indeed challenged and surprised. But above all they encountered the beauty of mathematics in a very unexpected way.

Here is a 22-year-old student's work:

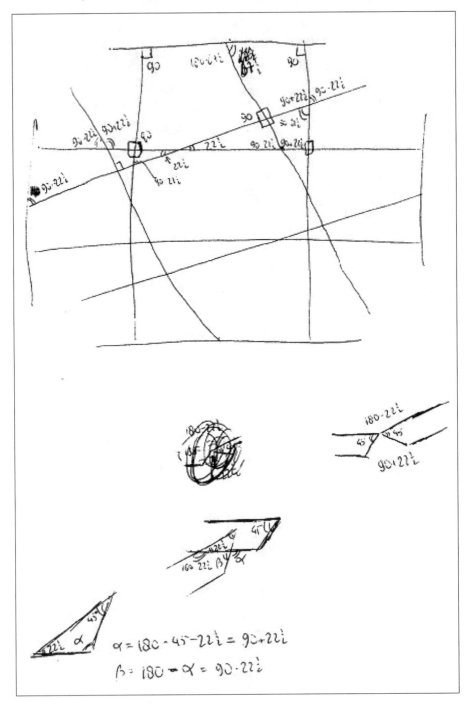

In sum: Mathematics meets art; mathematics assessment needs to be an art.

References

[Barth et al. 2006] H. Barth, K. La Mont, J. Lipton, S. Dehaene, N. Kanwisher, and E. S. Spelke, "Non-symbolic arithmetic in adults and young children", *Cognition* **98**:3 (2006), 199–222.

[Black and Wiliam 1998] P. Black and D. Wiliam, *Inside the Black Box: Raising standards through classroom assessment*, London: King's College, 1998.

[de Lange 1999] J. de Lange, *A framework for classroom assessment in mathematics*, Madison, WI: National Center for Improving Student Learning and Achievement in Mathematics and Science, 1999. Available at http:/www.fi.uu.nl/catch.

[OECD 2003] Organisation for Economic Co-operation and Development, *A report on the literacy network and numeracy network deliberations*, Paris: Author, 2003.

[PISA 2003] Programme for International Student Assessment, *The PISA 2003 assessment framework: Mathematics, reading, science and problem solving knowledge and skills*, Paris: Organisation for Economic Co-operation and Development, 2003. Available at http://www.pisa.oecd.org/dataoecd/46/14/33694881.pdf. Retrieved 13 Jan 2007.

Assessing Mathematical Proficiency
MSRI Publications
Volume **53**, 2007

Chapter 8
Mathematical Proficiency for Citizenship

BERNARD L. MADISON

Number and U.S. Democracy

"Numbers have immeasurably altered the character of American society," wrote Patricia Cohen in 1982 in the concluding pages of her book *A Calculating People*. Cohen continues, "Our modern reliance on numbers and quantification was born and nurtured in the scientific and commercial worlds of the seventeenth century and grew in scope in the early nineteenth century, under the twin impacts of republican ideology and economic development." In the two decades since Cohen's book, reliance on numbers and quantification has increased well beyond what could have been imagined, and no end is in sight.

Quantification emerged in seventeenth-century Western Europe as an alternative to the classical Aristotelian systems of classification as a way to make sense of the world [Crosby 1997]. England was becoming a commercial capital and numbers were needed to describe the economy. Arabic numerals finally had been adopted and the first books on arithmetic had appeared. Education in arithmetic was beginning but was an arcane subject, cut into small bits, and not easy to learn. Arithmetic as a commercial subject was disconnected from formal mathematics [Cohen 1982]. To a larger extent than I believe healthy, both these circumstances still exist today.

Inherent demands of a democracy, a large diverse country, and persistently aggressive economic development have fed the historical inclination of U.S. culture toward quantification to its enormity of today. Characteristics of the U.S. democracy that increase need for quantitative reasoning include:

- Constitutional mandate for a census.
- Protection of diverse and minority interests.
- Informing political debate.
- Confirming national identity.

- Free market system with minimally regulated labor markets.
- Recent deregulations of markets and services.
- Environmental protection and occupational safety.
- Emphasis on individual wealth accumulation.

Winston Churchill observed that first we shape our buildings and then they shape us. Driven by the characteristics above, our higher education system has permeated U.S. society with vast storehouses of knowledge and technologies based largely on computers. Consequently, floods of data and numbers are daily fare in the U.S. In turn, as with Churchill's buildings, number, quantification, data, and analysis shape U.S. culture and are more powerful than most realize.

What mathematics is critically important for informed participation in this highly quantitative U.S. society? This question has been considered since colonial times but has never been as difficult to answer as it is now, at the beginning of the twenty-first century. A more difficult question is: How we can design a developmental approach to achieve mastery of this mathematics and assessments to measure progress in this development?

I do not pretend to have complete answers to these questions, but I am quite sure that our traditional introductory college mathematics courses and traditional assessments are inadequate responses.

The Emergence of U.S. Higher Education

A century ago, only about 10% of the 14- to 17-year-olds were enrolled in high school. Half a century ago, only a small fraction of high school graduates went on to higher education. Both have changed. Indeed, college enrollments have grown from 10% of the 18- to 24-year-old cohort in 1945 to just over 50% of that cohort in 1990. Thus, today as many students enroll in postsecondary education as in secondary school. For many reasons, the U.S. public has made college the locus of expectations for leading a decent life — the need for sophisticated quantitative literacy is surely among those reasons [Steen 2004].

As enrollments in higher education have grown nearly four-fold during the last forty years (from approximately 4 to 15 million students), the curriculum has remained relatively stagnant. Our system of higher education was conceived at a time when 2% of the U.S. population went to college and its purpose then was to educate students to become a class apart. But life in the twenty-first century is more complex, so college has replaced high school as the educational standard for a democratic society in the twenty-first century. However, the task of reconceiving the structures of higher education for a democratic purpose — especially in fields that rely on quantitative and mathematical methods — has hardly begun. Indeed, much the reverse has taken place. The curriculum in mathematics has gradually narrowed, forcing almost everyone through the bottleneck of calculus. It may well be that with regard to a democratic conception of higher education it is undergraduate mathematics that is most out of date [Steen 2004].

In the development of undergraduate education in the U.S., among all the subjects or disciplines, mathematics has some unique characteristics. The present-day model of undergraduate education — that of majors and electives with a "general education" core — can be traced to Charles Eliot's reform of Harvard's undergraduate curriculum at the end of the nineteenth century. This new "general education" — actually a reaction to Eliot — became the conveyance for the democratic ideal of higher education — to ensure that graduates of all majors left college equipped with the breadth of general knowledge and skills necessary for full participation in democratic society. Majors were developed in standard college disciplines and all but mathematics, the one survivor of the classical curriculum, developed general education courses. Mathematics served up its standard courses — algebra, trigonometry, solid and analytic geometry — for general education. Unlike professors in other disciplines, a college mathematics professor from 1850 would not find today's college algebra course at all unfamiliar. In short, the bulk of U.S. collegiate mathematics has not changed in a century in spite of the college population and the U.S. society and culture being drastically different [Steen 2004].

Transfer to Contexts

Several suggestions for college mathematics courses for general education are documented in twentieth-century literature. Courses with titles such as finite mathematics, liberal arts mathematics, and college mathematics have entered the college curriculum, but the bulk of general education mathematics enrollments remain in the traditional courses of algebra, trigonometry, analytic geometry and calculus. None of these courses, as traditionally taught, is effective in educating for the quantitative literacy required in twenty-first-century United States. However, except for probability and statistics, almost all the mathematics needed for quantitative literacy is in these traditional courses or their prerequisites. The difficulty is with transfer of the mathematical knowledge and skills to the multitude of contexts where citizens confront quantitative situations. It is the ability to make that transfer that I believe to be essential for being proficient in mathematics, and this ability is apparently incredibly difficult to assess [Wiggins 2003]. The following examples illustrate what I mean.

Example 1. The following article appeared in the *Lincoln (NE) Journal Star* on August 21, 2001.

Forcing fuel efficiency on consumers doesn't work
By Jerry Taylor

Although the late, great energy crisis seems to have come and gone, the political fight over yesterday's panic rages on. The big dust-up this fall will be over SUVs, light trucks and minivans. Should the government order Detroit to make them get more miles per gallon? Conservationists say "yes." Economics 101 says "no."

Let's start with a simple question: Why should the government mandate conservation? When fuel becomes scarce, fuel prices go up. When fuel prices go up, people buy less fuel. Economists have discovered over the long run, a 20 percent increase in gasoline costs, for instance, will result in a 20 percent decline in gasoline consumption. No federal tax, mandate or regulatory order is necessary.

Notice the phrase "over the long run." Energy markets are volatile because consumers do not change their buying habits much in the short run.

This has led some critics to conclude that people don't conserve enough when left to their own devices. They do, but consumers need to be convinced that the price increases are real and likely to linger before they'll invest in energy-efficient products or adopt lifestyle changes. But even in the short run, people respond. Last summer was a perfect example: For the first time in a non-recession year, gasoline sales declined in absolute terms in response to the $2 per gallon that sold throughout much of the nation.

Mandated increases in the fuel efficiency of light trucks, moreover, won't save consumers money. A recent report from the National Academy of Sciences, for instance, notes that the fuel efficiency of a large pickup could be increased from 18.1 miles per gallon to 26.7 miles per gallon at a cost to automakers of $1,466. But do the math: It would take the typical driver 14 years before he would save enough in gasoline costs to pay for the mandated up-front expenditure. A similar calculation for getting a large SUV up to 25.1 miles per gallon leads to a $1,348 expenditure and, similarly, more than a decade before buyers would break even.

"Fine with me," you say? But it's one thing to waste your own money on a poor investment; it's entirely another to force your neighbor to do so. You could take that $1,466, for instance, put it in a checking account yielding 5 percent interest, and make a heck of a lot more money than you could by investing it in automobile fuel efficiency.

Even if government promotion of conservation were a worthwhile idea, a fuel efficiency mandate would be wrong. That's because increasing the mileage a vehicle gets from a gallon of gasoline reduces the cost of driving. The result? People drive more.

Energy economists who've studied the relationship between automobile fuel efficiency standards and driving habits conclude that such mandates are offset by increases in vehicle miles traveled.

If we're determined to reduce gasoline consumption dramatically, the right way to go would be to increase the marginal costs of driving by increasing the tax on gasoline. Now, truth be told, I don't support this idea much either. A recent study by Harvard economist Kip Viscusi demonstrates that the massive fuel taxes already levied on drivers (about 40 cents per gallon) fully "internalize" the environmental damages caused by driving. But conservationists reject this approach for a different reason: Consumers hate gasoline taxes and no Congress or state legislature could possibly increase them.

Look, it's a free country. If you want to buy a fuel-efficient car, knock yourself out. But using the brute force of the government to punish consumers who don't share your taste in automobiles serves no economic or environmental purpose.[1]

What's involved in analyzing the argument?

- Glean the relevant information.
- Have confidence to take up the challenge.
- Estimate to see if assertions are reasonable.
- Do the mathematics — linear and exponential functions.
- Generalize the situation.
- Reflect on the results.

Some relevant information

- The cost of increasing the fuel efficiency of large pickup trucks from 18.1 miles per gallon to 26.7 miles per gallon is $1466 per vehicle.
- Claim 1: It would take the typical driver 14 years to make up the $1466.
- Claim 2: You can make a "heck of a lot" more money by putting the $1466 in an account yielding 5% interest.

Ambiguities

- Typical driver's annual mileage — 10,0000 to 20,000 is reasonable.
- Price of gallon of fuel — about $1.40 at the time the article was written.

---

[1] *Lincoln (NE) Journal Star*, 21 Aug 2001, p. 5B. © 2001, Jerry Taylor and Cato Institute. Used with permission.

- Frequency of compounding of interest — annually to continuously.
- How long is the $1466 left to accumulate interest?
- Are the savings on fuel invested similarly?

Estimating time required to save $1466

Assumptions: Annual mileage is 10,000, cost of gasoline is $1.50 per gallon, fuel efficiency is 20 or 30 miles per gallon. We use $1.50 instead of $1.40 to simplify calculation of the estimate.

Estimated cost of fuel for a large pickup truck for one year:

$$\frac{10{,}000 \text{ miles per year}}{20 \text{ miles per gallon}} = 500 \text{ gallons per year at } \$1.50 \text{ per gallon}$$

$$= \$750 \text{ per year}$$

Estimated time to save $1466 by using a more fuel-efficient truck:

If the truck's fuel efficiency increased from 20 miles per gallon to 30 miles per gallon, then the price of gasoline would be multiplied by 2/3, so savings is 1/3 of $750 or $250 per year.

If $250 per year is saved on fuel, then the amount of time needed to save enough in gasoline costs to make up the $1466 in Claim 1 is less than 6 years:

$$\text{Number of years to save } \$1466 = \frac{\$1500}{\$250} = 6.$$

Estimated time to save $1466 by investing $1466 in a checking account:

5% per year on $1500 yields $75 per year at simple interest. So it takes approximately 1500/75 or 20 years to earn $1466.

Calculating time required to save $1466

Assumptions: Annual mileage is 10,000, price of gasoline is $1.40 per gallon, fuel efficiency is 18.1 or 26.7 miles per gallon. The number of years required to save $1466 is

$$\frac{1466}{\left(\dfrac{10000}{18.1} - \dfrac{10000}{26.7}\right) \times 1.40} = 5.88$$

This answer of 5.88 years is quite different from the 14 years claimed in the article.

Calculating comparison of costs over time

Let M be miles driven per year, G the price of gasoline per gallon, t the time in years.

Let C_1 be the price of gasoline for a large pickup truck with fuel efficiency of 18.1 miles per gallon that is driven M miles per year for t years.

Let C_2 be the price of gasoline for a large pickup truck with increased fuel efficiency (26.7 miles per gallon) that is driven M miles per year for t years, plus $1466.

Then, setting M equal to 10,000 and setting G equal to 1.4:

$$C_1(t) = \frac{MG}{18.1}t = 773.48t$$

$$C_2(t) = \frac{MG}{26.7}t + 1466 = 524.34t + 1466$$

At year t the earnings from $1466 invested at a rate of 5% are

$$1466e^{.05t} - 1466 \text{ if compounded continuously,}$$

$$1466(1.05^t) - 1466 \text{ if compounded annually.}$$

The annual difference in the amount spent on gasoline for the truck with fuel efficiency 18.1 miles per gallon and the truck with fuel efficiency 26.7 miles per gallons is:

$$773.48 - 524.34 = 249.14.$$

If the savings of $249.14 per year due to increased fuel efficiency are invested in an account at 5% compounded continuously, then after t years the value of the account will be:

$$\sum_{n=1}^{t} 249.14e^{.05(t-n)}.$$

Some computations using calculators yield the following information:

Year	Investment earnings on $1466	Gas savings	Investment earnings on gas savings
10	951	2491	3152
20	2519	4982	8350
30	5104	7474	16918
40	9366	9965	31046
50	16339	12457	54339

These data, based on the stated assumptions, show that it would take approximately 40 years for the $1466 invested at 5% to return as much as savings on gasoline. If these savings are invested annually, the income far exceeds the returns on the $1466.

Generalize the situation

One direction for generalizing this is analyzing the linear functions C_1 and C_2 with either the price of gasoline or the miles driven per year as parameters. Another is to introduce more complex investment issues such as present value.

Reflection on the results

One such reflection is that the article does not reveal some information that is needed to check to see if the author's conclusions are valid or reasonable. Based on the calculations above with assumptions that seem reasonable, the author's conclusions are not valid. Further analysis might reveal possible assumptions on the cost of gasoline and miles driven per year by the "typical driver" that would support the author's conclusions. For example, using $1.40 per gallon of gasoline and 4200 miles per year, the author's claim of 14 years to save the $1466 would be valid.

Another reflection is that the author's claim that the 5% return on the checking account would exceed the savings on gasoline will be correct if one is willing to wait long enough. The exponential growth will eventually exceed the linear growth.

Example 2. This letter to the editor appeared in the *Arkansas Democrat-Gazette* on April 9, 2002 (p. 7B).

> My children asked me how many ancestors and how many acts of these ancestors they are responsible for after reading and listening to the Razorbacks' coaching dilemma.
>
> They have been taught that they are responsible for their own actions and sometimes the actions of their friends or even their parents. They just want to know how far this goes back.
>
> My daughter had visited the slave ship exhibit at one of our downtown museums and recognized a family name as being a builder of slave ships back in the 1500s in Britain. She also knew that another relative brought six slaves over to Jamestown in the 1600s. How much was she going to have to pay in retribution? Was she the only one responsible or were there others?
>
> Before this got even more out of hand, we decided to do the math. Assuming four generations per century and only one child per family, that would be 19 generations. Two to the power of 19 would be 524,288 people who shared the responsibility.
>
> Then we started laughing at the total absurdity of the idea of one person today paying for the sins of another when there had been 524,288 people in between.
>
> And that wasn't even counting brothers and sisters.

Conclusion: Get a life. Forgive and forget all 524,288 of them.[2]

Analysis of the argument

Accepting the author's assumption that there are 19 generations between the sixteenth-century ancestor and the daughter, one explores how the author arrives at $2^{19} = 524,288$ "people in between" the ancestor and the daughter.[3]

One way to arrive at 2^{19} is to assume that the number of descendents of the ancestor doubles every generation. However, this model of exponential growth of population would require two children of the ancestor and each of these children (along with a spouse) would have two children. This doubling-each-generation yields 2^{19} descendents of the ancestor at the daughter's generation. However, this violates the author's assumption of "only one child per family," and most of the descendents of the ancestor would not be considered in between the ancestor and the daughter.

A second way to arrive at 2^{19} is to begin with the daughter and consider her parents, then her grandparents, then great-grandparents, etc. Continuing, one would arrive at 2^{19} ancestors of the daughter in the sixteenth century. Although this can be consistent with the assumption of one child per family, the 2^{19} ancestors are not in between the daughter and the sixteenth-century ancestor. There is a possibility that one would sum up this chain of ancestors of the daughter, two parents plus four grandparents, etc. This would yield a sum as follows:

$$2 + 2^2 + 2^3 + \cdots + 2^{19} = 2(2^{19} - 1)$$

or

$$2 + 2^2 + 2^3 + \cdots + 2^{18} = 2(2^{18} - 1).$$

However, neither of these is 2^{19}.

Two reasonable ways to count the number of people in between the ancestor and the daughter would be by counting the child of the ancestor, then the grandchild, then the great grandchild, etc. This would yield 18 or 19 (depending on where one begins counting) people in between. Since at each generation, the child takes a spouse, one could count the children and the spouses, giving 36 or 38 people in between.

Extensions of this discussion can move to models of population growth, showing why the simple exponential growth (doubling each generation) has limitations and moving to bounded exponential growth and logistic growth.

---

[2] © 2002 *Arkansas Democrat-Gazette*. Used with permission.

[3] I have used this letter with more than 130 students in four different classes. When presented with the situation as described in the letter, some students have believed that the counting of generations is flawed, and only a few students independently have detected flaws in the counting and language. Because of the serious nature of the issue discussed in the letter, I have insisted that we accept only that the author is making an argument for a position and focus on the validity of the argument that is given credibility by "doing the math." Because of the nature of the issue and the diverse opinions about it, I only use this item in class or homework rather than on a quiz or examination.

As in the previous example, we see that the author's assumptions about what constitutes a relevant mathematical model shape the conclusions that can be drawn from it. One draws very different conclusions from the same basic data when different models are used. The best defense against being misled by inappropriate arguments is to be able to figure out the right arguments oneself.

Example 3. The False Positive Paradox (conditional probability).

Use the following information to answer the question.

- The incidence of breast cancer in women over 50 is approximately 380 per 100,000.
- If cancer is present, a mammogram will detect it approximately 85% of the time (15% false negatives).
- If cancer is not present, the mammogram will indicate cancer approximately 7% of the time (7% false positives).

If a mammogram is positive, which of the following is nearest to the odds that cancer is present? Assume that all women over 50 get tested.

A. 9 out of 10

B. 8 out of 10

C. 7 out of 10

D. 1 out of 6

E. 1 out of 14

F. 1 out of 22

The correct choice of F is surprising to many; hence, the term paradox. In fact, the 380 cancer cases would generate roughly 325 positive tests, while the 99,620 noncancer cases would generate nearly 7000 false positives, so the odds that a positive test indicates cancer is about 1 in 22.

Example 4. The graphs in Figure 1 on the next page appeared in the *New York Times* on August 11, 2002 (p. 25).

Sample tasks

1. Explain why these graphs are titled "The Rise in Spending."

2. From 1990 to 1995, how did the cost of health insurance premiums change?

3. Explain why inflation is included in each of the graphs.

4. In what year was the rate of increase in prescription drug costs the greatest?

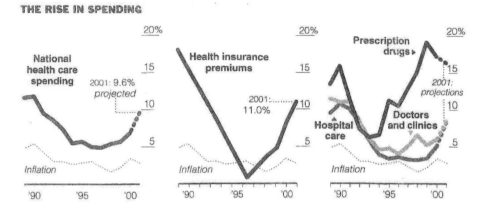

Figure 1. Used with permission, © 2002, *New York Times.*

Example 5. The graphs in Figure 2 appeared in the *New York Times* on April 7, 1995 (p. A29).

Sample tasks

1. Can both of these views be correct? Explain.
2. In each bar graph there is a "bar" over $20,000 to $30,000. Do these two bars represent the same quantity? Explain.

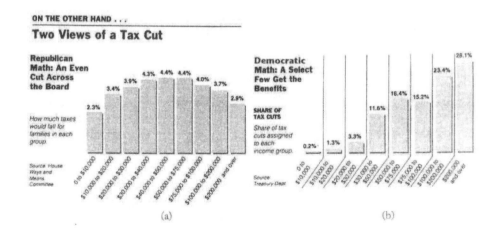

Figure 2. Used with permission. © 1995, *New York Times*

What the examples imply. What do the above examples imply about critical proficiencies in mathematics needed for citizenship? The following seem clear:

- Calculate and estimate with decimals and fractions.
- Recognize and articulate mathematics.
- Generalize and abstract specific mathematical situations.
- Understand and use functions as process, especially to describe linear and exponential growth.
- Have facility with algebra.
- Use calculators to explore and compute.
- Formulate and analyze situations using geometry and measurement.
- Analyze and describe data using statistics and probability.

College mathematics courses that cover the mathematical topics in the list above have been proposed and developed over the past century. For example, Allendoerfer [1947] described such a course in an article entitled "Mathematics for Liberal Arts Students." However, the bulk of the enrollments in college mathematics, even those for general education, remain in traditional courses in college algebra, trigonometry, and calculus.

References

[Allendoerfer 1947] C. B. Allendoerfer, "Mathematics for liberal arts students", *American Mathematical Monthly* **17** (1947), 573–578.

[Cohen 1982] P. C. Cohen, *A calculating people: The spread of numeracy in America*, Chicago: University of Chicago Press, 1982.

[Crosby 1997] A. W. Crosby, *The measure of reality: Quantification and Western society, 1250–1600*, Cambridge: Cambridge University Press, 1997.

[Steen 2004] L. A. Steen, *Achieving quantitative literacy*, Washington, DC: Mathematical Association of America, 2004.

[Wiggins 2003] G. Wiggins, " 'Get real!' Assessing for quantitative literacy", pp. 121–144 in *Quantitative literacy: Why numeracy matters for schools and colleges*, edited by B. L. Madison and L. A. Steen, Princeton: NJ: National Council on Education and the Disciplines, 2003.

Assessing Mathematical Proficiency
MSRI Publications
Volume **53**, 2007

Chapter 9
Learning from Assessment

RICHARD ASKEY

There are many types of assessment and books can and have been written about them. My aim is modest: to point out a few instances of what can be learned by looking at assessments given in the past.

General Considerations

One important use of assessment is to find out where education has been weak. As an example, consider the results in geometry in the 1995 Third International Mathematics and Science Study (TIMSS) assessment of knowledge of advanced students in the last year of high school [Mullis et al. 1997].[1] The average score for the 17 countries was 500. The scores ranged from 424 to 548. The four countries with lowest scores on advanced mathematics and percentage of students in advanced mathematics courses are as follows:

1995 TIMSS countries with lowest advanced mathematics scores

Country	Score	Students in advanced courses
USA	424	14%
Austria	462	33%
Slovenia	476	75%
Italy	480	14%

As can be seen, the percentage of students taking advanced mathematics in the last year of secondary school could vary widely between countries. In general, the better students take advanced mathematics in their final year of school, so a higher percent of students taking advanced math would tend to lower the average score. Thus the gap between 424 and 462, large as it is, underestimates

---

[1] For a sampling of TIMSS questions at various levels, see http://timss.bc.edu/timss1995i/items.html.

how much U.S. students are below the students in other countries which did not do very well. This gap is unacceptably large. Neal Lane, then director of the National Science Foundation, pointed this out in his comments in *Pursuing Excellence* [NCES 1998]. Unfortunately, this gap, which I consider strong evidence of a very poor geometry program in the United States, was not mentioned in *Principles and Standards for School Mathematics* [NCTM 2000].

To illustrate some of the problems in the knowledge of geometry, consider a classic simple problem from synthetic geometry [TIMSS 1995, problem K18].

> In the $\triangle ABC$ the altitudes BN and CM intersect at point S. The measure of $\angle MSB$ is 40° and the measure of $\angle SBC$ is 20°. Write a PROOF of the following statement:
>
> "$\triangle ABC$ is isosceles."
>
> Give geometric reasons for statements in your proof.

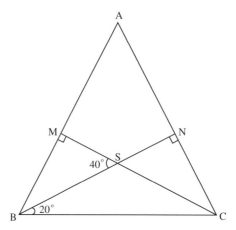

All that is needed to solve this problem is to know that the sum of the angles of a triangle is 180 degrees; and write that angle BSC is 140 degrees, so angle SCB is 20 degrees. Angle CSN is 40 degrees, so angles MBS and NCS are each 50 degrees. Thus the base angles are equal, so the triangle is isosceles. This is an easy example of a type of problem which is given in grade 8 in Hong Kong, Japan and Singapore. See, for example, [Chan et al. 2005, problem 38, p. 9.51; Kodaira 1992, problem 3, p. 121; Teh and Looi 2001, problems 1 and 3, p. 323] The first two are eighth grade books, the third is seventh grade. None of these East Asian countries participated in the 1995 TIMSS assessment of advanced students.

The international average of correct responses on this problem was not high, only 35%. The range was from 10% to 65% with the U.S. score the lowest at 10% correct. When partially correct answers were also considered, the U.S. percentage went up to 19% and the international average to 50%.

In the 2003 Trends in International Mathematics and Science Study [Mullis et al. 2004], it is interesting to compare the U.S. and Singapore scores:

2003 TIMSS grade 4 results for the U.S. and Singapore

Subject	U.S.	Singapore
Number	516	612
Patterns and relationships	524	579
Measurement	500	566
Geometry	518	570
Data	549	575
Overall average	518	584

The absolute difference has been mentioned by many people, but I want to point out a relative difference. In grade 4 there are five categories. The results for both of these countries were similar in four of these categories, but in both cases there was an outlier; for the U.S. in Data and for Singapore in Number. The Singapore program has a strong focus on arithmetic and multi-step word problems.

In *Beyond Arithmetic* the following claim is made [Mokros et al. 1995, p.74]:

> In the *Investigations* curriculum, standard algorithms are not taught because they interfere with a child's growing number sense and fluency with the number system.

The standard algorithms are taught in Singapore, and this is supplemented by other methods which work in specific instances. The standard algorithms do not get in the way of students learning mathematics. When taught well, they help develop knowledge of place value, the use of the distributive law when doing multiplication, and estimation when doing division. A focus on work with numbers and good word problems provides an excellent foundation for algebra. In grade 8 2003 TIMSS [Mullis et al. 2004], Singapore's average score went up to 605 while the U.S. average score went down to 504. The data in the preceding table and grade 8 TIMSS data, showing a large drop in the U.S. geometry score, suggest to me that we should rethink our priorities.

Examinations for Teachers

In his 1986 American Education Research Association Presidential Address, Lee Shulman [1986] started with some comments on content which he called "the forgotten part of education." To illustrate the importance of content long ago, he mentioned a test given to prospective grammar school teachers in California in the 1870s [California 1875]. (Grammar school was grades 1 to 8.) The exam covered a wide range of skills and knowledge teachers need. It had 20 sections worth a total of 1,000 points. Three parts were on mathematics, written arithmetic (100 points), mental arithmetic (50 points), and algebra (50 points). Here are some questions from each of the mathematics parts. These questions are among the more complicated ones, but are representative of a substantial fraction of the questions on these exams.

Written arithmetic

Prove that in dividing by a fraction we invert the divisor and then proceed as in multiplication.

If 18 men consume 34 barrels of potatoes in 135 days, how long will it take 45 men to consume 102 barrels? Work by both analysis and proportion. [The implied assumption is made that each man will eat the same amount each day.]

Mental arithmetic

A man bought a hat, a coat, and a vest for $40. The hat cost $6; the hat and coat cost 9 times as much as the vest. What was the cost of each?

Divide 88 into two such parts that shall be to each other as $2/3$ is to $4/5$.

Algebra

Three stone masons, A, B, C, are to build a wall. A and B, jointly, can build the wall in 12 days; B and C can build it in 20 days, and A and C in 15 days. How many days would each require to build the wall, and in what time will they finish it if all three work together? [There is an implied assumption that A will work at the same rate whether working alone or with others, and the same for B and C.]

Divide the number 27 into two such parts, that their product shall be to the sum of their squares as 20 to 41.

A May-pole is 56 feet high. At what distance above the ground must it be broken in order that the upper part, clinging to the stump, may touch the ground 12 feet from the foot?

There were 10 questions in each of these parts. The time allowed was not stated, but at least two hours should be allotted to written arithmetic and to written grammar, two of three topics which were worth 100 points. "Applicants who fail to reach fifty credits [points] in written arithmetic, or in grammar, or in spelling" fail. "Applicants reaching less than sixty credits in any one of these must receive only a third grade certificate, no matter what average percentage they may make."

One comment on the written arithmetic part is worth quoting completely.

Heretofore applicants have very seldom paid any attention to the special requirements of a paper. For instance, in written arithmetic, applicants are instructed that "no credits are to be allowed unless the answer is correct and the work is given in full, and, when possible, such explanations as would be required of a teacher in instructing a class; a rule is not an explanation." This instruction has generally been disregarded. In future this will be a ground for the rejection of the paper. [California 1875, p. 214]

Contrast these questions and remarks with some exams given now for teachers.

The Praxis tests seem to be the tests most commonly used by states as part of certification requirements. Here are two Praxis sample questions [ETS 2005] for candidates who want to teach middle school mathematics. (Middle school students are generally between ages 11 and 15.) These questions are representative of the posted questions.

x	y
-4	-2
-3	$-\frac{3}{2}$
-2	-1
-1	$-\frac{1}{2}$
0	0

1. Which of the following is true about the data in the table above?

(A) As x decreases, y increases.

(B) As x increases, y does not change.

(C) As x increases, y decreases.

(D) As x increases, y increases.

3. If there are exactly 5 times as many children as adults at a show, which of the following CANNOT be the number of people at the show?

(A) 102

(B) 80

(C) 36

(D) 30

This question is a much easier one than the analogous question in 1875. (A man bought a hat, a coat, and a vest for $40. The hat cost $6; the hat and coat cost 9 times much as the vest. What was the cost of each?). The Praxis tests are designed to help states ensure an adequate level of knowledge of content and professional practice for beginning teachers. It seems that what is considered adequate now would not have been adequate in 1875.

For experienced teachers, the National Board for Professional Teaching Standards (NBPTS) has developed a certification program, which is used to encourage teacher growth and reward those who show adequate knowledge on an examination and an adequate portfolio presentation of their teaching. Here are two questions on algebra for teachers teaching early adolescents, taken from the samples available at the NBPTS web site, www.nbpts.org, in 2005. (They and the NBPTS sample high school question on the next page have since been replaced by other sample questions.) NBPTS defines "early adolescents" as 11- to 15-year-olds, so these are students at the upper end of the ages of those taught by candidates taking California exams in the 1870s.

Use the table of coordinates below to respond to the following prompts.

x	−3	−1	0	1	4	6
y	3	1	0	1	4	6

a) Write an equation to model the given data set.

b) State the name of this function and describe the relationship between x and y.

c) Name and describe the specific type of transformation when y is increased by 5.

Use the graph below to respond to the following prompts.

COST OF TAXI RIDE

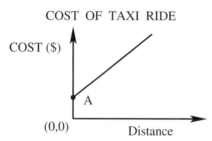

a) Explain the significance of point A within the context of the price of a taxi ride.

b) Given that the slope of the line is 0.5, explain the meaning of this value within the context of the situation modeled above.

Not only are these much easier questions, the second one does not model the cost of a taxi ride, since the cost goes up in discrete increments for certain fractions of a mile, so the graph should be a step function.

Here is a sample of the NBPTS test for high school mathematics teachers, again from www.nbpts.org.

Use the following problem to respond to the prompt below.

$$\text{Solve } |x - 3| + |x - 2| = 15.$$

A student submits the following response:

$$
\begin{array}{llll}
x - 3 + x - 2 = 15 & \text{or} & x + 3 + x + 2 = 15 \\
2x - 5 = 15 & & 2x + 5 = 15 \\
2x = 10 & & 2x = 10 \\
x = 5 & & x = 5
\end{array}
$$

a) Give a detailed explanation of the mathematical understandings and misunderstandings reflected in the student's response.

b) Identify the mathematical knowledge and skills the student needs in order to understand why the response is incorrect. Be specific about what mathematics the student needs to learn to enable her to avoid the same error when solving problems requiring similar conceptual and procedural knowledge.

In two of the nineteenth-century exams [California 1875], the following questions were in the section "Theory and Practice of Teaching":

How would you teach a class the reason for inverting the divisor in division of fractions?

(Dec. 1873, p. 224)

How would you require your scholars to explain the process of subtracting 7,694 from 24,302?

(Sept. 1874, p. 238)

Both of these are more important questions in elementary mathematics than the one involving absolute values is for high school mathematics. The California exams from the 1870s do not have high school questions, although some of the algebra questions deal with mathematics traditionally taught in high school. Some exams from Michigan [1899] have geometry questions that should be part of a good high school mathematics program. Here are two, one given in August, 1898 and the other in December, 1898. In both cases they are one of ten questions.

Demonstrate: The volume of a triangular pyramid is equal to one-third the product of its base and altitude.

(1899, p. 8)

Demonstrate: Parallel transverse sections of a pyramidal space are similar polygons whose areas are proportional to the square of the distances from the vertex to the cutting planes.

(1899, p. 17)

Nothing like this degree of knowledge seems to be asked on either the Praxis tests or the National Board of Professional Teaching Standards tests. Although only 50 points out of 1,000 in the California exams were in the section on the theory and practice of teaching, some of the questions in this section dealt with very important parts of teaching. Here are two, the second of which was mentioned in [Shulman 1986]. I had to learn how to improve my oral reading when starting to read to our children. Either the skill mentioned in the question below had disappeared by 1940 or I was a poor student in this area.

What is your method to teach children to discontinue the sing-song, or monotonous tones which many acquire in reading? Is the method original with yourself?

How do you interest lazy and careless pupils? [Answer in full]

Multiple-choice questions were not used 100 years ago. The National Board for Professional Teaching Standards does not use multiple choice items, but the Praxis mathematics content tests used by most states do. A method of asking questions so they can be machine graded yet cannot be solved by eliminating all but one of the answers will be described in the next section.

Shulman [1986] mentioned examining tests from Colorado, Massachusetts, and Nebraska in addition to tests from California and Michigan. Comparing such tests with current tests would be an important service that educators could do. State tests for teachers from the last part of the nineteenth century and the early part of the twentieth century can be found in reports by State Superintendents of Education. State historical societies or major university libraries are places to look for these reports. I thank William Reese for mentioning this to me. (See also the Acknowledgements section on page 134.)

Examinations for Students

A recent report [Ginsburg et al. 2005] compares and contrasts some aspects of mathematics education in Singapore with education in the United States. The comments on the level of questions asked of grade 6 students in Singapore and what is asked of grade 8 U.S. students in some state tests and of prospective teachers on the Praxis tests should be read by anyone who cares about mathematics education. The conclusions of this report, like the evidence from the test items cited above, make clear that U.S. expectations are set far too low.

Here is a problem for eighth-grade students mentioned in a March 20, 1998 guest column in the *Wisconsin State Journal* by M. Wilsman, then a member of the Board of the Wisconsin Mathematics Council. Twenty-six other members of the WMC helped in writing the column, which was meant as a response to an article that claimed mathematics education was not effective. Wilsman asserted that this problem was on the state test:

Eric hopes to sell $1/3$ of a dozen paintings that he has finished for the art fair. Which equation should you solve to find the number of paintings that he wants to sell?

(A) $1/3 = 12n$

(B) $3n = 12$

(C) $4n = 12$

(D) $1/3\,n = 12n$

Few people would solve an equation to find how many paintings Eric hoped to sell. Just divide 12 by 3 to find the answer. I tried to find out what was on this test. The test developers told me this question was not on the test. It had been submitted, but was not thought to be good enough for their tests. Instead, it was used as a sample problem in a test preparation guide [CTB/McGraw Hill 1997, p. 62]. I do not understand why the problem was not discarded rather than used as an illustrative example.

I wrote a response and mentioned a grade 6 Japanese problem that was available on the Web [Pacific Software 1995]. Contrast this problem for grade 6 students in Japan with the one which was suggested as a good problem for grade 8 students in Wisconsin. (Again, an assumption is made about a uniform working rate of people even in different size groups.)

A job takes 30 days to complete by eight people. How long will the job take when it is done by 20 people?

(A) 1

(B) 23

(C) 21

(D) 13

(E) 12

Another source of Japanese problems is a collection of translations of Japanese university entrance exam questions [Wu 1993]. All of the problems are interesting, but I want to just describe how the response format used allows machine grading without being multiple choice. Consider a problem from an 1875 California exam.

> Bought 4 lemons and 7 peaches for 13 cents; 5 oranges and 8 lemons for 44 cents; and 10 peaches and 3 oranges for 20 cents; what was the price of one of each kind?

As a multiple-choice problem, if the price of any one of the three is given the problem becomes trivial. Yes, one could ask what would be the cost of 13 lemons, 10 oranges and 16 peaches as a multiple-choice problem, but here is a much better way to ask for answers. The cost of a peach is A/B where next to A on the answer sheet is $+ - 0\ 1\ 2\ 3\ 4\ 5\ 6\ 7\ 8\ 9$ and test-takers are told that fractions should be recorded in lowest form. In this case, $A = 1$ and $B = 2$.

The Japanese use this format on their analogue of the SAT, and ask multi-step questions with individual parts having answers which are recorded. This way they can learn at what step in a multi-step problem many students have trouble. We in the United States need multi-step problems on our exams. We have turned too much of our assessing over to measurement people, and the level of mathematics being assessed shows this. Even the National Research Council, when it did a study of tests for prospective teachers, did not have content people looking at the tests [NRC 2001, p. 416].

The picture one gets of education from old exams is consistent with the picture I had formed on the basis of looking at many old textbooks. The common view is that long ago all that students were asked to do is learn a procedure. This is not borne out in the exams given to prospective teachers. Recall that candidates were expected to give reasons for what was being done, both in these tests and in their teaching. In addition, the level of problems which teacher candidates were expected to solve was higher than we now ask of prospective teachers. This higher level of questions also shows up in many old textbooks. We have a lot to learn from the past, including past assessments.

Further information on tests for teachers is given in [Burlbaw 2002].

Acknowledgements

Suzanne Wilson found the California exams which Lee Shulman used in his 1986 article. I asked an educational historian, David Labaree, where one might find such exams. He suggested the Cubberley Education Library at Stanford. Barbara Celone and Yune Lee of this library found the California exams and some exams from Michigan and sent me copies. I am grateful to all of these people for helping to bring to light some very interesting documents on education in the United States in the 19th century.

References

[Burlbaw 2002] L. Burlbaw, "Teacher test — could you pass the test?", presented at Mid-America Conference on History, Fayetteville, Sept. 2002. Available at http://www.coe.tamu.edu/-lburlbaw/edci659/Teacher%20Testing%20final.pdf. Accessed 17 Jan 2007.

[California 1875] *Sixth biennial report of the Superintendent of Public Instruction of the State of California, for the school years 1874 and 1875*, Sacramento, CA: G. H. Springer, 1875.

[Chan et al. 2005] M. H. Chan, S. W. Leung, and P. M. Kwok, *New trend mathematics* (S2B), Hong Kong: Chung Tai Educational Press, 2005.

[CTB/McGraw Hill 1997] *Teacher's guide to Terra Nova*, Monterey: CTB/McGraw Hill, 1997.

[ETS 2005] E. T. Service, *The PRAXIS series: Middle school mathematics* (0069), Princeton, NJ: ETS, 2005. Available at http://ftp.ets.org/pub/tandl/0069.pdf. Retrieved 26 Mar 2006.

[Ginsburg et al. 2005] A. Ginsburg, S. Leinwand, T. Anstrom, and E. Pollock, *What the United States can learn from Singapore's world-class mathematics system (and what Singapore can learn from the United States): an exploratory study*, Washington, DC: American Institutes for Research, 2005. Available at http://www.air.org/news/documents/Singapore%20Report%20(Bookmark%20version)% .pdf. Retrieved 19 Jan 2007.

[Kodaira 1992] Kodaira, K. (editor), *Japanese grade 8 mathematics*, Chicago: UCSMP, 1992. Translated from the Japanese.

[Michigan 1899] *Annual report of the Superintendent of Public Instruction of the State of Michigan, with accompanying documents, for the year 1898*, Lansing, MI: Robert Smith Printing, 1899.

[Mokros et al. 1995] J. Mokros, S. J. Russell, and K. Economopoulos, *Beyond arithmetic*, White Plains, NY: Dale Seymour, 1995.

[Mullis et al. 1997] I. V. S. Mullis, M. O. Martin, A. E. Beaton, E. J. Gonzalez, D. L. Kelly, and T. A. Smith, *Mathematics achievement in the primary school years: IEA's Third International Mathematics and Science Study* (*TIMSS*), Chestnut Hill, MA: TIMSS International Study Center, Boston College, 1997.

[Mullis et al. 2004] I. V. S. Mullis, M. O. Martin, E. J. Gonzalez, and S. J. Chrostowski, *Findings from IEA's Trends in International Mathematics and Science Study at the fourth and eighth grades*, Chestnut Hill, MA: TIMSS and PIRLS International Study Center, 2004. Available at http://timss.bc.edu/timss2003i/mathD.html. Retrieved January 17, 2007.

[NCES 1998] U.S. Department of Education: National Center for Education Statistics, *Pursuing excellence: A study of U.S. twelfth-grade mathematics and science achievement in international context* (NCES 98-049), Washington, DC: U.S. Government Printing Office, 1998. Available at http://nces.ed.gov/pubs98/twelfth/. Retrieved 26 Mar 2006.

[NCTM 2000] National Council of Teachers of Mathematics, *Principles and standards for school mathematics*, Reston, VA: Author, 2000.

[NRC 2001] National Research Council: Board on Testing and Assessment, Center for Education, Division of Behavioral and Social Sciences and Education, Committee on Assessment and Teacher Quality, *Testing teacher candidates: The role of licensure tests in improving teacher quality*, edited by K. Mitchell et al., Washington, DC: National Academy Press, 2001. Available at http://darwin.nap.edu/books/0309074207/html. Retrieved 26 Mar 2006.

[Pacific Software 1995] Pacific Software Publishing, *World math challenge*, vol. 1, 1995. Available at http://www.japanese-online.com; register and click on "lessons". Retrieved 26 Mar 2006.

[Shulman 1986] L. Shulman, "Those who understand: Knowledge growth in teaching", *Educational Researcher* **15** (1986), 4–14.

[Teh and Looi 2001] K. S. Teh and C. K. Looi, *New syllabus mathematics 1*, Singapore: Shing Lee Publishers, 2001. With the collaboration of P. Y. Lee and L. Fan, consultants.

[TIMSS 1995] *Third International Mathematics and Science Study — 1995: Released advanced mathematics items, population 3*, Boston College: International Study Center, 1995. Available at http://timss.bc.edu/timss1995i/TIMSSPDF/C_items.pdf. Retrieved 17 Jan 2007.

[Wu 1993] L. Wu (editor), *Japanese university entrance examination problems in mathematics*, edited by L. Wu, Washington, DC: Mathematical Association of America, 1993. Available at http://www.maa.org/juee. Retrieved 26 Mar 2006.

Assessing Mathematical Proficiency
MSRI Publications
Volume **53**, 2007

Chapter 10
When Assessment Guides Instruction
Silicon Valley's
Mathematics Assessment Collaborative

DAVID FOSTER, PENDRED NOYCE, AND SARA SPIEGEL

Standardized testing for the purpose of accountability continues to dominate our nation's schools. Since we first reported on the Mathematics Assessment Collaborative [Foster and Noyce 2004], states have responded to the stringent testing requirements of the No Child Left Behind legislation by expanding annual testing in reading and mathematics to all students in grades 3 through 8. Currently states are adding science tests and more tests in high school. Working under cost constraints, most states elect to use multiple-choice tests, while some commentators such as Peterson [2006] detect a "race to the bottom" — a tendency to lower standards and simplify tests as a way of ensuring that more and more students can be deemed proficient. (See also [Fuller et al. 2006].)

How has assessment been used to inform instruction? A number of districts, challenged urban districts in particular, have responded to the need to boost student scores by increasing the frequency of benchmark assessments. Some districts developed assessments aligned with local curricula to help ensure that coverage and learning across schools. Other districts invested in technology-based programs that offer quarterly updates on student progress along a linear scale, based on easily scored (but often skills-oriented) computer multiple-choice assessments. These programs, while they may reassure a school's staff about student progress or alert them to trouble ahead, do little to inform teachers about how students are thinking, what they understand, where they are falling down, and how, specifically, teachers might change their own instructional practices to address students' difficulties.[1]

---

[1] For an account of a different use of technology to inform teachers about student performance, see Artigue's description of diagnostic tests in France and the LINGOT project in this volume.

For the past nine years a group of school districts in California's Silicon Valley have taken a different approach to mathematics assessment. These districts have supplemented the state testing system with a coordinated program of support and learning for teachers based on a common set of assessments given to students. In this chapter, we briefly review the history of the Mathematics Assessment Collaborative (MAC). We describe how the results of the annual performance assessment are used to guide professional development. We offer additional examples of what the MAC is learning about student understanding across the grades. We review trends in student performance and discuss the relationship between student performance on the MAC assessment and on the state tests.

A Brief History of the Mathematics Assessment Collaborative

In 1996, the Noyce Foundation formed a partnership with the Santa Clara Valley Mathematics Project at San José State University to support local districts with mathematics professional development. The new partnership was dubbed the Silicon Valley Mathematics Initiative. Its early work focused on providing professional development, establishing content-focused coaching in schools, and collaboratively examining student work to inform teachers of pupils' understandings.

At that time, the state of California was beginning a long and turbulent battle over the establishment of new state curriculum standards [Jacob and Akers 2000; 2001; Jackson 1997; Schoenfeld 2002; Wilson 2003]. Following the state board's adoption of standards in mathematics, the governor pressed to establish a high-stakes accountability system. For the first time, California would require a test that produced an individual score for every student. Because developing a test to assess the state standards was expected to take several years, the state decided in the interim to administer an off-the-shelf, norm-referenced, multiple-choice test — Harcourt's Stanford Achievement Test, Ninth Edition (known as the SAT-9) — as the foundation for the California Standardized Testing and Reporting (STAR) program. In the spring of 1998, students in grades 2 through 11 statewide took the STAR test for the first time.

In an effort to provide a richer assessment measure for school districts, the Silicon Valley Mathematics Initiative formed the Mathematics Assessment Collaborative (MAC). Twenty-four school districts joined the collaborative, paying an annual membership fee.

Selecting an Assessment

MAC's first task was to create a framework characterizing what was to be assessed. Keeping in mind William Schmidt's repeated refrain that the U.S.

curriculum is "a mile wide and an inch deep,"[2] MAC decided to create a document that outlined a small number of core topics at each grade level. The goal was to choose topics that were worthy of teachers' efforts, that were of sufficient scope to allow for deep student thinking, and that could be assessed on an exam that lasted just a single class period. Using as references, standards developed by the National Council of Teachers of Mathematics, by the state of California, and by the local districts, teacher representatives from MAC districts met in grade-level groups to choose five core ideas at each grade level.

Once the core ideas document was created, the next task was to develop a set of exams that would test students' knowledge of these ideas. MAC contracted with the Mathematics Assessment Resource Service (MARS), creators of Balanced Assessment,[3] to design the exams. (For a description of Balanced Assessment's design principles and the work of MARS, see Chapter 6 in this volume.) Each grade-level exam is made up of five tasks. The tasks assess mathematical concepts and skills that involve the five core ideas taught at that grade. The exam also assesses the mathematical processes of problem solving, reasoning, and communication. The tasks require students to evaluate, optimize, design, plan, model, transform, generalize, justify, interpret, represent, estimate, and calculate their solutions.

The MARS exams are scored using a point-scoring rubric. Each task is assigned a point total that corresponds to the complexity of the task and the proportional amount of time that the average student would spend on the task in relation to the entire exam. The points allocated to the task are then allocated among its parts. Some points are assigned to how the students approach the problem, the majority to the core of the performance, and a few points to evidence that, beyond finding a correct solution, students demonstrate the ability to justify or generalize their solutions. In practice, this approach usually means that points are assigned to different sections of a multi-part question. (For an example of such a rubric, see Burkhardt, this volume.)

The combination of constructed-response tasks and weighted rubrics provides a detailed picture of student performance. Where the state's norm-referenced, multiple-choice exam asks a student merely to select from answers provided, the MARS exam requires the student to initiate a problem-solving approach to each task. Students may use a variety of strategies to find solutions, and most of the prompts require students to explain their thinking or justify their findings.

---

[2]William Schmidt, U.S. research director for the Third International Mathematics and Science Study, has made this statement in numerous places. See, for example, the press releases available on the Internet at http://ustimss/msu.edu.

[3]Balanced Assessment Packages [BAMC 1999–2000] of assessment items and sample student work were published by Dale Seymour Publications. Balanced Assessment tasks can be found at http://www.educ.msu.edu/mars and http://balancedassessment.gse.harvard.edu.

This aspect of the assessment seems impossible to duplicate by an exam that is entirely multiple choice. Details of the administration of the exams also differ from the state's approach, in that teachers are encouraged to provide sufficient time for students to complete the exam *without rushing*. In addition, students are allowed to select and use whatever tools they might need, such as rulers, protractors, calculators, link cubes, or compasses.

The Assessment in Practice

In the spring of 1999, MAC administered the exam for the first time in four grades — third, fifth, seventh, and in algebra courses — in 24 school districts. Currently the collaborative gives the exam in grades two through grade 8, followed by high school courses one and two. Districts administer the exam during March, and teachers receive the scored papers by the end of April, usually a couple of weeks prior to the state high-stakes exam.

Scoring the MARS exams is an important professional development experience for teachers. On a scoring day, the scoring trainers spend the first 90 minutes training and calibrating the scorers on one task and rubric each. After that initial training, the scorers begin their work on the student exams. After each problem is scored, the student paper is carried to the next room, where another task is scored. At the end of the day, teachers spend time reflecting on students' successes and challenges and any implications for instruction. Scoring trainers check random papers and rescore them as needed. Finally, as a scoring audit, 5% of the student papers are randomly selected and rescored at San José State University. Reliability measures prove to be high: a final analysis across all grades shows that the mean difference between the original score and the audit score is 0.01 point.

Along with checking for reliability, the 5% sample is used to develop performance standards for overall score reporting. The collaborative has established four performance levels in mathematics: Level 1, minimal success; Level 2, below standards; Level 3, meeting standards; and Level 4, consistently meeting standards at a high level. A national committee of education experts, MARS staff members and MAC leaders conducts a process of setting standards by analyzing each task to determine the core of the mathematical performance it requires. The committee examines actual student papers to determine the degree to which students meet the mathematical expectations of the task, and it reviews the distribution of scores for each task and for the exam as a whole. Finally, the committee establishes a cut score for each performance level for each test. These performance levels are reported to the member districts, teachers, and students.

Once the papers are scored, they are returned to the schools, along with a copy of the master scoring sheets, for teachers to review and use as a guide for further instruction. Each school district creates a database with students' scored results on the MARS exam, demographic information, and scores on the state-required exam. Using these, an independent data analysis company produces a set of reports that provide valuable information for professional development, district policy, and instruction.

How Assessment Informs Instruction

Over time, it has become clear that the tests, the scoring sessions, and the performance reports all contribute to MAC's desired outcome: informing and improving instruction. The scoring sessions are powerful professional development activities for teachers. To be able to score a MARS exam task accurately, teachers must fully explore the mathematics of the task. Analyzing different approaches that students might take to the content within each task helps the scorers assess and improve their own conceptual knowledge. The scoring process sheds light on students' thinking, as well as on common student errors and misconceptions. As one teacher said, "I have learned how to look at student work in a whole different way, to really say, 'What do these marks on this page tell me about [the student's] understanding?'" Recognizing misconceptions is crucial if a teacher is to target instruction so that students can clarify their thinking and gain understanding. The emphasis on understanding core ideas helps teachers build a sound sequence of lessons, no matter what curriculum they are using. All of these effects on instruction grow out of the scoring process.

The scored tests themselves become valuable curriculum materials for teachers to use in their classes. MAC teachers are encouraged to review the tasks with their students. They share the scoring information with their students, and build on the errors and approaches that students have demonstrated on the exams.

Tools for Teachers

Being data-driven is a common goal of school districts. In this day of high-stakes accountability, districts are awash with data, yet not much of it is in a form readily useful to teachers. To meet that need, MAC publishes *Tools for Teachers*, an annual set of reports derived from the results of each year's exam.

Along with broad performance comparisons across the collaborative's membership and analysis of the performance of different student groups, the reports provide a wealth of other information. A detailed portrait is compiled of how students approached the different tasks, with a description of common misconceptions and evidence of what students understand. The reports include student work samples at each grade level showing the range of students' approaches, successes, and challenges. (For examples, see Foster, this volume.) In addition, the reports educe implications for instruction, giving specific suggestions and ideas for teachers as a result of examining students' strengths and the areas where more learning experiences are required.

This set of reports is distributed to teachers throughout the initiative. Each October, MAC presents large-scale professional development workshops to introduce the new *Tools for Teachers*. Many teachers use these documents to plan lessons, determine areas of focus for the year, and fuel future formative assessment experiences for their classes.

Using Student Responses to Inform Professional Development

The Mathematics Assessment Collaborative provides a broad range of professional development experiences for teachers and leaders. A significant design difference between the professional development provided through MAC and other professional development is that the MAC experiences are significantly informed by the results from the annual exams. This translates into workshops that target important ideas where students show areas of weakness. Here are three examples.

Proportional reasoning is a central idea in middle school. In 2001, seventh-grade students were given a task called The Poster (see next page). The task assesses students' ability to apply their understanding of proportion to a visual scaling situation.

Only 37% of seventh graders were able to meet standard on the task, and only 20% could completely solve both questions in the task. Many (63%) of the students did not think of the problem as a proportional relationship; most used addition to find the missing measurement in the proportional situation. A typical misuse of addition in this problem is reproduced on the next page.

This student misunderstanding of proportional reasoning became a major focus of professional development planning for MAC middle school teachers. MAC institutes and workshops provided both content knowledge and pedagogical strategies for teaching proportional reasoning in middle school. These sessions for teachers made explicit the underlying concepts of ratios, rates, and proportions, including understanding proportions from a functional approach. At the professional development sessions, teachers practiced solving non-routine proportional reasoning problems. They made connections between representations and representatives[4] of these functions that used bar models, tables,

---

[4]Representatives of a function are descriptions that do not completely determine the function, for example, a finite table of values does not determine all values of a function that has infinitely many possible values.

The Poster

This problem gives you the chance to:
• calculate sizes in an enlargement

Photograph   Poster

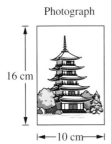

16 cm

←—10 cm—→

?

←———— 25 cm ————→

1. A photograph is enlarged to make a poster.
The photograph is 10 cm wide and 16 cm high.
The poster is 25 cm wide. How high is the poster?   _____
Explain your reasoning.

2. On the poster, the building is 30 cm tall.
How tall is it on the photograph?   _____
Explain your work.

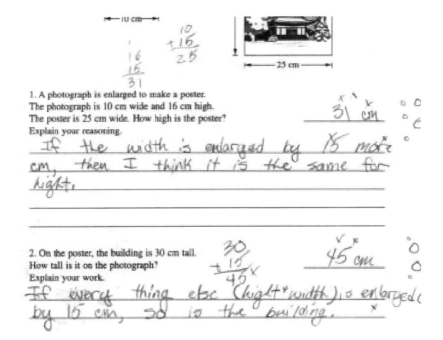

1. A photograph is enlarged to make a poster.
The photograph is 10 cm wide and 16 cm high.
The poster is 25 cm wide. How high is the poster?   31 cm
Explain your reasoning.

   If the width is enlarged by 15 more
cm, then I think it is the same for
hight.

2. On the poster, the building is 30 cm tall.
How tall is it on the photograph?   45 cm
Explain your work.

   If everything else (hight + width), is enlarged
by 15 cm, so is the building.

graphs, and equations. Teachers were encouraged to use non-routine problems in their classroom and to promote the use of different representations and multiple strategies to find solutions.

Four years later, seventh-grade students were given Lawn Mowing to assess proportional reasoning. The task involved making sense of rates, a special type of proportional reasoning that students traditionally struggle with.

Lawn Mowing

This problem gives you the chance to:
• solve a practical problem involving ratios
• use proportional reasoning

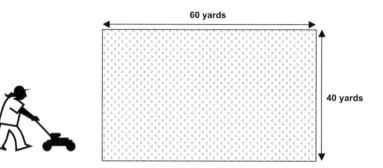

60 yards

40 yards

Dan and Alan take turns cutting the grass.
Their lawn is 60 yards long and 40 yards wide.

1. What is the area of the yard?   _____ square yards

Dan takes an hour to cut the lawn using an old mower.

2. How many square yards does Dan cut in a minute?   _____
   Show your work.

Alan only takes 40 minutes using a new mower.

3. How many square yards does Alan cut in a minute?   _____
   Show your calculation.

4. One day they both cut the grass together.
   How long do they take?   _____
   Show how you figured it out.

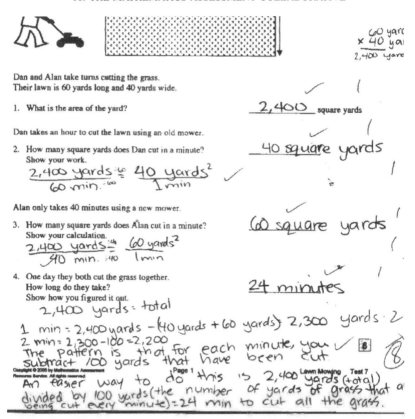

Dan and Alan take turns cutting the grass.
Their lawn is 60 yards long and 40 yards wide.

1. What is the area of the yard?

$\qquad$ 2,400 square yards

Dan takes an hour to cut the lawn using an old mower.

2. How many square yards does Dan cut in a minute?
   Show your work.

   $\dfrac{2,400 \text{ yards}}{60 \text{ min.}} = \dfrac{40 \text{ yards}^2}{1 \text{ min}}$

   40 square yards

Alan only takes 40 minutes using a new mower.

3. How many square yards does Alan cut in a minute?
   Show your calculation.

   $\dfrac{2,400 \text{ yards}}{40 \text{ min.}} = \dfrac{60 \text{ yards}^2}{1 \text{ min}}$

   60 square yards

4. One day they both cut the grass together.
   How long do they take?
   Show how you figured it out.

   24 minutes

   2,400 yards = total

   1 min = 2,400 yards − (40 yards + 60 yards) 2,300 yards. 2
   2 min = 2,300 − 100 = 2,200
   The pattern is that for each minute, you
   subtract 100 yards that have been cut.
   An easier way to do this is 2,400 yards (total)
   divided by 100 yards (the number of yards of grass that a
   being cut every minute) = 24 min to cut all the grass.

Fifty-nine percent of the students met standard on the task, a big improvement over students' performance on the proportional reasoning task in 2001. We see above work in which the student set up a ratio between the size of the lawn and the minutes it took to cut the lawn in order to find the unit rates. Then the student used two different reasoning strategies to determine and confirm the amount of time it takes for each person to cut the lawn. This paper is typical of the way students' approaches to proportional reasoning problems have improved over the years.

We believe, based on survey feedback and student achievement data, that MAC's explicit focus on proportional reasoning with middle school teachers contributed to this improvement in student achievement. Using student results from the MARS test to tailor and inform professional development for the following year has become a cornerstone of MAC strategy.

The MARS assessment provides a valuable perspective on students' understanding from grade level to grade level. This vertical view over the grades allows us to investigate how student performance related to a particular mathematical idea, such as patterns and functions, changes over time — or doesn't change. One worrisome trend is that students often are able to learn foundational skills in an area of mathematics in the early grades but then are unable to move beyond their basic understanding to apply, generalize, or justify solutions. This trend is illustrated in a set of similar tasks concerning linear functions in fourth grade, seventh grade, and Algebra 1 (Hexagon Desks, Hexagons, and Patchwork

---

Hexagon Desks

This problem gives you the chance to:
• find and extend a number pattern
• plot and use a graph

Sarah finds how many students can sit around a row of desks. The top surface of each desk is a hexagon, and the hexagons are arranged in rows of different shapes.

1 desk   6 students

2 desks   10 students

3 desks   14 students

4 desks

1. Complete Sarah's table.

Number of desks in a row	Number of students
1	6
2	10
3	
4	
5	
6	

2. On the grid, plot the results from the table you completed in question 1. The first two points have already been plotted for you.

3. Sarah says that 47 students can sit around a row of 11 desks. Without drawing the desks, explain how you know that Sarah is wrong.

_____

_____

_____

How many students can sit around a row of 11 desks?   _____

10

Hexagons

This problem gives you the chance to:
• recognize and extend a number pattern in a geometric situation
• find a rule for the pattern

Maria has some hexagonal tiles.
Each side of a tile measures 1 inch.
She arranges the tiles in rows; then she finds the perimeter of each row of tiles.

1 tile
perimeter = 6 in.

2 tiles
perimeter = 10 in.

3 tiles

4 tiles

Maria begins to make a table to show her results.

Number of tiles in a row	Perimeter in inches
1	6
2	10
3	
4	

1. Fill in the empty spaces in Maria's table of results.
What will be the perimeter of 5 tiles?   _____ inches

2. Find the perimeter of a row of 10 tiles.   _____ inches
Explain how you figured it out.

_____

_____

3. Write a rule or formula for finding the perimeter of a row of hexagonal tiles when you know the number of tiles in the row.
Let n = the number of tiles, and p = the perimeter.

_____

_____

_____

4. Find the perimeter of a row of 25 hexagonal tiles.
Show your work.   _____ inches

5. The perimeter of a row of hexagonal tiles is 66 inches.
How many tiles are in the row?   _____

10

Patchwork Quilt

This problem gives you the chance to:
• recognize and extend a number pattern
• express a rule using algebra

Sam is making a border for a patchwork quilt.

She is sewing black and white regular hexagons together.

Sam makes a table to show the number of black and white hexagons she needs.

Number of black hexagons	Number of white hexagons
1	6
2	11
3	16
4	21

1. How many white hexagons does Sam need for 6 black hexagons?
   Explain how you figured it out.

2. How many black hexagons does Sam need for 66 white hexagons?
   Explain how you figured it out.

3. Write a formula that will help you to find how many white hexagons (W) Sam needs for *n* black hexagons.

4. Use your formula to find how many white hexagons Sam needs for 77 black hexagons.
   _____ white hexagons
   Show your work.

8

Quilt). Each grade's task concerns a finite pattern to be examined, extended, and explored. Students are also asked questions about values in the domain or range of the linear function associated with the pattern. At the secondary level, students are asked to write a formula for this relationship.

MAC analyzed student papers at these three grades to determine where students were successful with functions and where they struggled. Here is the percentage of students successful on each element of the task:

	% successful		
Task element	Grade 4	Grade 7	Algebra
Extend the pattern	84%	82%	87%
Given a value in the domain, find the value	57%	58%	53%
Given a value in the range, find a value in the domain	40%	35%	68%
Write a formula	n/a	27%	27%

The exam results show that student performance on most elements of these tasks did not improve as the grade level increased. Students in all three grades were similarly successful at extending the pattern. The percentage of students successful at using a functional relationship to find a value was similar at all grade levels, although the algebra task asked students to use a larger value. The algebra students were more able to determine the inverse relationship but showed no more success in writing an algebraic equation than the seventh graders. These results — flat performance on function tasks across grade levels — point to a need to go beyond asking students to extend patterns by teaching them to reason and generalize at more complex levels. Without instruction asking students to go more deeply than their first solution, their mathematical development may stall at the basic level that they attained years earlier. These statistics help us understand the challenges of teaching upper-grade students. Not only do students need to learn more mathematical ideas and language, but they also need explicit learning experiences in reasoning, generalizing, and justifying. These higher-level thinking skills should be routinely addressed in mathematics classes.

Using the MARS exam has also helped MAC make inroads on student misunderstandings related to mathematical conventions, language, or notation. When we focus professional development on one of these misunderstandings, we often see dramatic changes in student responses. One common early problem, observed in student solutions involving multiple operations was the use of mathematical run-on sentences. Consider the problem of how to calculate the number of feet in a picture of a girl walking three dogs. A typical (incorrect) student response reads: $4 \times 3 = 12 + 2 = 14$. This is a "mathematical run-on sentence": 4×3 does not equal $12 + 2$. The solution steps should have been written out:

$$4 \times 3 = 12 \qquad 12 + 2 = 14$$

At first glance, this correction may seem like nit-picking. But the problem with the notation is more than just sloppiness; a run-on sentence betrays a common misconception. Instead of understanding that the equal sign indicates that expressions on the two sides of the sign have the same value, students using such run-on sentences take the equal sign to signal that an operation must be performed: "The answer is ..." [Siegler 2003]. This view contributes to further confusion as students learn to generalize and work with expressions containing variables in later grades.

We found that this error in notation occurred regularly throughout the tested population. On further investigation, we learned that teachers commonly allowed this notation to be used in classrooms, or even used it themselves when demonstrating solutions to multi-step problems. The assessment report for that year pointed out the problem and announced that solutions using run-on sentences would no longer receive full credit. Subsequent professional development showed teachers how such notation led to student misconceptions. Within a year, the collaborative noted a dramatic change in the way students in 27 districts communicated mathematical statements.

This matter of notation was just one example of how analyzing patterns of student error led to improvements in instructional practice. Other areas of improvement include differentiating between continuous and discrete graphs, noting and communicating the units in measurement problems, distinguishing between bar graphs and histograms, understanding correlation trends in scatterplots, and developing understanding of mathematical justifications. Examining MARS results has also led teachers to confront significant chunks of unfamiliar mathematical content. Discussing the tasks and student responses often uncovers the fact that, for many topics and concepts in algebra, geometry, probability, measurement, and statistics, teachers' understanding is weak. Uncovering these gaps in teachers' content knowledge is central to improving instruction.

MAC Assessment and State Testing

The quality of information that the Mathematics Assessment Collaborative has provided to its member districts has helped the districts maintain their commitment to professional development that concentrates on improving teacher understanding. California offers significant incentives and sanctions for student achievement on the state STAR exam, and many districts across the state are thus tempted to embrace narrow quick-fix methods of test prep (drill on practice tests and focus on strategies for answering multiple-choice tests) and "teaching to the test."

To counter this temptation, MAC has been able to show that, even when a significant number of students are improving on the state test, their success may not translate into greater mathematical understanding as demonstrated by success on the more demanding performance assessments. The statistics also indicate that, as students move up the grades, the disparity increases: more and more students who appear to be doing well on the state exam fail to meet standards on the performance exam. Conversely, success on the MARS exam becomes an ever *better* predictor of success on the state's STAR exam. By

	MARS	STAR	
		Basic or below	Proficient or above
grade 3	Below standard	23%	7%
	Meets or exceeds standards	12%	58%
grade 7	Below standard	46%	11%
	Meets or exceeds standards	6%	37%

grade 7, students whose teachers have prepared them to perform well on the MARS exam are extremely likely to perform above the fiftieth percentile on the STAR exam. The table on the preceding page compares success rates on the 2004 MARS and STAR exams for grades 3 and 7.

The Mathematics Assessment Collaborative has been able to demonstrate to the satisfaction of superintendents and school committees that high-quality professional development significantly enhances student achievement. District case studies show that students whose teachers participate in intensive MAC professional development achieve higher averages on both the state mathematics test and the MARS exam than students whose teachers who are less involved. As a result, districts have continued to invest in mathematics professional development and formative assessment. The number of students assessed and teachers and grade levels involved has grown every year, even as MAC has kept the number of member districts relatively constant. In 2006, more than seventy thousand students of 1300 teachers in thirty-five districts participated in the MARS exam.

The performance of MAC district students on the STAR exam has continued to rise. For example, while 53% of MAC district third graders performed above the fiftieth percentile on the state mathematics test in 1998, 68% met standard (Proficient or Advanced) on the more challenging California Standards Test for mathematics in 2005. Similar growth has occurred in the other grades, with a minimum of 52% of MAC district students meeting standard at each grade level.

There are a wide number of variables to consider when comparing student achievement statistics. In an effort to eliminate as many variables as possible and still compare performance on the STAR exam between students of teachers involved in MAC programs and students of other teachers, we analyzed statistics from nineteen districts in San Mateo County. Of the nineteen districts, ten are member districts of MAC.

The analysis compared student achievement on the 2005 STAR exam for students in second grade through seventh grade. The MAC students as a group are generally poorer than the comparison group, with 37% of the MAC students qualifying for the National School Lunch Program (NSLP) compared to 30% of the non-MAC students. Both groups have 26% English Language Learners. The data set consists of 21,188 students whose teachers are not members of MAC and 14,615 students whose teachers are involved in MAC programs. The figure at the top of the next page indicates that a larger percentage of students from MAC teachers met standards on the 2005 STAR exam than students from non-MAC teachers at every grade level except seventh grade, where the percentage was the same.

These statistics are encouraging because the population served by the MAC schools is slightly poorer, a demographic factor that has been shown to limit

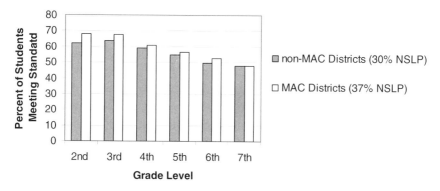

2005 student performance on the STAR exam

achievement. These findings are promising in that they reinforce that students of MAC teachers, who typically engage their students in high level problem solving and more open, constructed-response tasks, outperform other students on the more procedurally oriented STAR exam despite challenging economic factors.

Success on the state test is politically important, but knowing the percentage of students that exceeds a single score level tells us little about how scores are distributed. More telling as a measure of progress in student learning over the whole range of performance is the growth that students demonstrate on the MARS performance exam. Over time, grades 3 through 6 have shown considerable increases in the percentage of students meeting standard on the MARS exam. In addition to seeing more students in MAC achieving at the highest performance level, we also find fewer students who are performing at the lowest level. The figure below shows the percentage of students at each

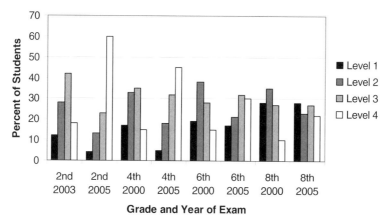

Performance on MARS Exam in year of implementation and in 2005

performance level in the first year of implementation versus the current year. In second grade, 60% reached level 4 in 2005, up from 18% in 2003. During that same period, the percentage of second graders in the lowest performance level decreased, with only 4% at the lowest level in 2005 compared to 12% in 2003. In fourth grade, there is a net change of 30% more students achieving level 4 in 2005 compared to 2000. More fourth graders have moved up from the lowest performance level, with numbers decreasing from 17% in 2000 to 5% in 2005. What is happening in practice at these elementary grades is an upward shift in performance for students at every achievement level. Even the lowest-achieving students are addressing the complex MARS tasks and demonstrating some level of understanding. In grades 6 and 8, on the other hand, there are gains in achievement at the highest level with an increase of 15% and 12% respectively, but there is little change in the lowest performance level.

These findings have convinced district leaders to embrace the theory central to our work. This theory states that when teachers teach to the big ideas (e.g., the core topics in the MAC framework), participate in ongoing content-based professional development, and use specific assessment information to inform instruction, their students will learn and achieve more.

Teachers benefit from this approach as much as students do. Just as formative assessment sends a different message to students than do final exams and grades, a collaborative system of formative performance assessment sends a different message to teachers than does high-stakes summative assessment. Teachers laboring to improve student performance on a high-stakes exam can come to feel isolated, beaten down, and mystified about how to improve. Because they are rewarded for getting higher percentages of students over one score bar in one year, they may be tempted to focus primarily on the group of students nearest to that bar or to grasp at a set of narrow skills and procedures that will allow students to answer a few more questions correctly. The exigencies of test security mean that teachers often receive little specific information about where their students' performance excelled or fell short. When high-stakes test results come back, often months after the exam, teachers can do little with the results but regard them as a final grade that marks them as a success or failure.

The Mathematics Assessment Collaborative fights teachers' sense of isolation and helplessness by sharing everything it learns about students. It identifies common issues and potential solutions. It helps teachers understand how learning at their particular grade level is situated within a continuum of students' growing mathematical understanding. It promotes communication across classrooms, schools, and grade levels. It encourages teachers to take a longer, deeper view of what they are working to achieve with students.

Assessment that requires students to display their work is not a panacea that suddenly transforms student learning. Rather, it is a tool for building the capacity of the teaching community to improve its work over time. The discipline of exploring together both the core mathematics we want students to know and the evidence of what they have learned is simultaneously humbling and energizing. Knowing that they are always learning and improving creates among educators a healthy, rich environment for change. To improve instruction requires that teachers become wiser about the subject they teach and the ways that students learn it. Performance assessment of students, with detailed formative feedback to teachers accompanied by targeted professional development, helps to build the teacher wisdom we need.

References

[BAMC 1999–2000] *Balanced Assessment elementary/ middle grades/ high school/ advanced high school assessment packages 1 & 2*, Balanced Assessment Project, White Plains, NY: Dale Seymour Publications, 1999–2000.

[Foster and Noyce 2004] D. Foster and P. Noyce, "The Mathematics Assessment Collaborative: Performance testing to improve instruction", *Phi Delta Kappan* **85**:5 (2004), 367–374.

[Fuller et al. 2006] B. Fuller, K. Gesicki, E. Kang, and J. Wright, *Is the No Child Left Behind Act working? The reliability of how states track achievement*, Working paper **06-1**: Policy Analysis for California Education, 2006. Available at http://pace.berkeley.edu/NCLB/WP06-01_Web.pdf. Retrieved 2 Jul 2006.

[Jackson 1997] A. Jackson, "The math wars: California battles it out over mathematics education", *Notices of the American Mathematical Society* **44** (1997), 695–702, 817–823.

[Jacob 2001] B. Jacob, "Implementing standards: The California mathematics textbook debacle", *Phi Delta Kappan* **83**:3 (2001), 264–272. Retrieved July 23, 2006 from http://www.pdkintl.org/kappan/k0111jac.htm.

[Jacob and Akers 2000] B. Jacob and J. Akers, "Research based mathematics education policy: The case of California 1995–1998", *International Journal for Mathematics Teaching and Learning* (2000). Available at http://www.cimt.plymouth.ac.uk/journal/bjcalpol.pdf. Retrieved 23 Jul 2006.

[Peterson and Hess 2006] P. E. Peterson and F. H. Hess, "Keeping an eye on state standards: A race to the bottom?", *Education Next* **6**:3 (2006), 28–29. Available at http://www.educationnext.org/20063/pdf/ednext20063_28.pdf. Retrieved 23 July 2006.

[Schoenfeld 2002] A. H. Schoenfeld, "Making mathematics work for all children: Issues of standards, testing and equity", *Educational Researcher* **31**:1 (2002), 13–25.

[Siegler 2003] R. S. Siegler, "Implications of cognitive science research for mathematics education", pp. 219–233 in *A research companion to Principles and Standards for School Mathematics*, edited by J. Kilpatrick et al., Reston, VA: National Council of Teachers of Mathematics, 2003.

[Wilson 2003] S. Wilson, *California dreaming: Reforming mathematics education*, New Haven: Yale University Press, 2003.

Section 4
The Case of Algebra

In this section and the next, we "crank up the microscope" to take a closer look at the examination of (a) specific subject matter, and (b) what different kinds of assessments can reveal. The subject of this section is algebra — mathematical subject matter that has always been important, but that has taken on increased importance in the U.S. in recent years. This is due to the confluence of two important social movements. The first is the "algebra for all" movement, an attempt to equalize opportunity-to-learn in American schools. In the past, many high school students were placed in courses such as "business mathematics" instead of algebra, or stopped taking mathematics altogether. Lack of access to algebra narrowed their options, both in curriculum and in employment. Hence policymakers and teachers in the U.S. declared that every student should have access to high-quality mathematics instruction, including an introduction to algebra in the early high school years. The second movement is a trend toward "high-stakes" testing. In various states around the nation, students must now pass state-wide assessments in mathematics in order to graduate from high school. Given that algebraic skills are now considered to be a part of quantitative literacy (as in "algebra for all"), state exams often focus on algebra as the gateway to high school graduation. In California, for example, the state's High School Exit Examination (the CAHSEE) in mathematics focuses largely on algebra. Students who do not pass the exam will not be granted a high school diploma.[1] Thus, understanding and improving student performance in algebra is a critically important matter.

In Chapter 11, William McCallum lays out a framework for examining algebraic proficiency. As he notes, "in order to assess something, you have to have some idea of what that something is." Thus he begins by clarifying what he believes is important about thinking algebraically. His approach is grounded in the perspective outlined in the National Research Council's volume *Adding It Up: Helping Children Learn Mathematics*. That is, proficiency in any mathematical arena includes conceptual understanding, procedural fluency, strategic competence, adaptive reasoning, and having a productive disposition toward mathematics. He goes on to exemplify these.

---

[1] As of June 2006, the question of whether or not passing the CAHSEE can be required for graduation is being adjudicated in the courts. The requirement is in place as this introduction is being written.

In Chapter 12, David Foster provides a framework for thinking about big ideas in algebra and a set of assessment items to match. (This is one aspect of the Silicon Valley Mathematics Initiative (SVMI) agenda, which was described in Chapter 10.) Foster's first task was to settle on high priority themes for assessment in algebra. What really counts? What do you want to make sure gets teachers' and students' attention? This is an important in general, but it is especially important for the SVMI, in the context of California's state Standards. The Standards, captured in the 2006 *California Mathematics Framework for California Public Schools, Kindergarten Through Grade Twelve*, provide a long list of skills that students are to master. How is one to prioritize? Which items on the list reflect isolated skills; which ones reflect core mathematical ideas? How do you build assessments that get at the core ideas? Foster elaborates on five main algebraic themes: doing-undoing, building rules to represent functions, abstracting from computation, the concepts of function and variable, and the concepts of equality and equation. For each, he provides specific assessment items. As one reads the items, it becomes clear how much the devil is in the details. For example, if you want to see whether students can explain their thinking, you have to provide them opportunities on the assessment to do so. (And if you want them to get good at explaining their thinking, you need to provide them ample opportunities, in class, to practice this skill. Thus tests can be diagnostic and suggest modifications of classroom practices.)

In Chapter 13, Ann Shannon gives some examples that serve to "problematize" the issue of context. As we saw in Section 3, context makes a difference: "knowing" the mathematics and being able to use it in specific situations can be two different things. Shannon takes a close look at a number of tasks, showing that "context" is anything but a zero-or-one variable: small changes in problem statements can produce large differences in what people do or do not understand about the contexts that are represented in the problems. In her fine-grained analyses, Shannon points to the complexities of drawing inferences about what people understand from their responses on tests. This presages the discussion in the next section, on what assessments can reveal or obscure.

Assessing Mathematical Proficiency
MSRI Publications
Volume **53**, 2007

Chapter 11
Assessing the Strands of Student Proficiency
in Elementary Algebra

WILLIAM G. MCCALLUM

Algebra and Functions

In order to assess something, you have to have some idea of what that something is. Algebra means many different things in today's schools (for a discussion, see [RAND 2003, Chapter 4]). In particular, the study of algebra is often blended with the study of functions. Although it is true that the notion of a function is lurking behind much of beginning algebra, and that there is an algebraic aspect to many of the tasks we want our students to carry out with functions, the conglomeration of algebra and functions has considerably muddied the waters in the teaching of algebra. Therefore, I'd like to spend some time clarifying my own stance before talking about how to assess proficiency. In the process, while acknowledging other possible uses, I will use the word "algebra" to mean the study of algebraic expressions and equations in which the letters stand for numbers. I include in this study consideration of the relationship between the algebraic form of expressions and equations and properties of their values and solutions. I make a distinction, however, between this study and the study of functions.

In the progression of ideas from arithmetic to algebra to functions, there is an increase in abstraction at each step, and the increase at the second step is at least as large as that at the first. In the step from arithmetic to algebra, we learn to represent numbers by letters, and calculations with numbers by algebraic expressions. In the step from algebra to functions we learn of a new sort of object "function" and represent functions themselves by letters. There are two complementary dangers in the teaching of algebra, each of which can cause students to miss the magnitude of this step.

The first danger is that functions and function notation appear to the students to be mostly a matter of using a sort of auxiliary notation, so that if a function f is defined by $f(x) = x^2 - 2x - 3$, say, then $f(x)$ is nothing more than a short-hand notation for the expression $x^2 - 2x - 3$. This results in a confusion between functions and the expressions representing them, which in turn leads to a broad area of confusion around the ideas of equivalent expressions and transforming expressions. For example, we want students to understand that $(x + 1)(x - 3)$ and $(x - 1)^2 - 4$ are equivalent expressions, each of which reveals different aspects of the same function. But without a strong notion of function as an object distinct from the expression defining it, the significance of equivalence and transformation is lost in a welter of equals signs following the $f(x)$. Students can't see the forest for the trees.

A complementary danger is that algebra appears as a sort of auxiliary to the study of functions. It is common to use multiple representations of functions in order to make concrete the notion of a function as an object in its own right. Functions are represented by graphs, tables, or verbal descriptions, in addition to algebraic expressions. By looking at the same object from different points of view, one aims to foster the notion of its independent existence. In this approach, algebra provides one way of looking at functions, but functions are not regarded as purely algebraic objects. This solves the problem of students not seeing the forest for the trees, but risks them not being able to see the trees for the forest. Overemphasizing functions when teaching algebra can obscure algebraic structure, and too much attention to graphical and numerical approaches can subvert the central goal of thinking about symbolic representation.

In considering assessment, I limit myself to the narrower meaning of algebra that I have been using, and interpret proficiency in algebra as largely a matter of proficiency with symbolic representations. I understand proficiency to include the five strands of mathematical competence described in [NRC 2001]:

- *conceptual understanding*: comprehension of mathematical concepts, operations, and relations
- *procedural fluency*: skill in carrying out procedures flexibly, accurately, efficiently, and appropriately
- *strategic competence*: ability to formulate, represent, and solve mathematical problems
- *adaptive reasoning*: capacity for logical thought, reflection, explanation, and justification
- *productive disposition*: habitual inclination to see mathematics as sensible, useful, and worthwhile, coupled with a belief in diligence and one's own efficacy.

Proficiency in Algebra

Consider the following common algebraic error, the like of which everybody who teaches calculus has seen at one time or another.

$$\int \frac{1}{2x^2 + 4x + 4}\,dx = \int \frac{1}{x^2 + 2x + 2}\,dx$$

$$= \int \frac{1}{(x+1)^2 + 1}\,dx$$

$$= \arctan(x + 1) + C$$

The student has factored a 2 from the denominator, and then lost it. Students often regard such errors as minor slips, like a wobble while riding a bicycle, and instructors often indulge this point of view with partial credit. The suggested remediation for students who make this sort of error often is lots of drill; lots of practice riding the bicycle. The error may be regarded as a matter of *procedural fluency*; either the student does not know the rules of algebra, or is insufficiently practiced in their execution.

Of course, without further information, it is impossible to diagnose the error precisely. However, it is worth drawing on experience to consider possible diagnoses that involve other strands of proficiency. For example, is it possible that it could be an error of *conceptual understanding*? Experience talking to students about their mistakes suggests the possibility that this was not an unintentional procedural slip, but rather a confusion about what procedures are permissible under what circumstances. Indeed, there is a situation in which it is quite permissible to lose the factor of 2, namely in solving the equation

$$2x^2 + 4x + 4 = 0.$$

For a student with a weak grasp of the difference between equations and expressions, the superficial procedural similarity between solving equations and transforming expressions is a snare. Both procedures involve doing something to expressions connected by equals signs. We should consider the possibility that the failure to appreciate the fundamental difference between the two situations is a failure of understanding, not a failure in manipulative skills.

As for *strategic competence*, there is an important aspect of this error which the partial credit mentality fails to weigh sufficiently: in addition to making the original slip, the student failed to correct it. Stepping back from the solution and contemplating it as a whole suggests an obvious strategy for checking the solution, namely differentiating the answer. Even without actually carrying out the differentiation, a student with a strategic frame of mind might well wonder where all the necessary 2s and 4s in the denominator will come from.

What of *adaptive reasoning*? Attention to the meaning of the equals sign and a tendency to reflect on the assertion being made each time you put one down on paper might have led the student to notice that the first line in the solution effectively declares a number to be half of itself.

Finally, there could be an overall failure of *productive disposition*. Experience suggests that erroneous solutions like the one described here are often the product of a lack of any purpose other than moving the symbols in the correct way. The bicyclist has a fundamental sense of purpose that corrects the wobble; this is often lacking in algebra students.

Assessing the Strands of Proficiency

Many algebra assessments concentrate on assessing procedural fluency. Indeed, the common view of algebra among students is that it is nothing but procedural fluency. The idea that there are ideas in algebra comes as a surprise to many. How would we go about assessing some of the other strands? One way is to ask more multi-step questions and word problems that involve all strands at once. But such questions rarely make their way onto standardized assessments. There is also a need for simple questions that assess the non-procedural aspects of algebraic proficiency. Here are a couple of suggested questions.

Conceptual understanding

Question: In the following problems, the solution to the equation depends on the constant a. Assuming a is positive, what is the effect of increasing a on the value of the solution? Does the solution increase, decrease, or remain unchanged? Give a reason for your answer that can be understood without solving the equation.

A. $x - a = 0$

B. $ax = 1$

C. $ax = a$

D. $x/a = 1$

Answer:

A. Increases. The larger a is, the larger x must be to give 0 when a is subtracted from it.

B. Decreases. The larger a is, the smaller x must be to give the product 1.

C. Remains unchanged. As a changes, the two sides of the equation change together and remain equal.

D. Increases. The larger a is, the larger x must be to give a ratio of 1.

The equations in this problem are all easy to solve. Nonetheless, students find this question difficult because they are not being asked to demonstrate a procedure, but rather to frame an explanation in terms of what it means for a number to be a solution to an equation. It is possible for students to learn the mechanics of solving equations without ever picking up the rather difficult concept of an equation as a statement of equality between two expressions whose truth is contingent upon the value of the variables, and the process of solving an equation as a series of logical inferences involving such statements. Asking students to reason about equations without solving them could assess this area of proficiency.

Strategic competence

Question: A street vendor of t-shirts finds that if the price of a t-shirt is set at $\$p$, the profit from a week's sales is

$$(p-6)(900-15p).$$

Which form of this expression shows most clearly the maximum profit and the price that gives that maximum?

A. $(p-6)(900-15p)$

B. $-15(p-33)^2 + 10935$

C. $-15(p-6)(p-60)$

D. $-15p^2 + 990p - 5400$

Answer: B. Because $(p-33)^2$ is a square, it is always positive or zero, and it is only zero when $p = 33$. In the expression for the profit, a negative multiple of this square is added to 10,935. Thus the maximum profit is $10,935, and the price which gives that profit is $33.

In this problem the student is not asked to perform the manipulations that produce the different forms, but rather to show an understanding of why you might want to perform those manipulations, and which manipulations would be the best to choose for a given purpose. A similar question could be asked about form C, which shows prices that produce zero profit, and form A, which exhibits the profit as a product of two terms each of which has meaning in terms of the context of the original problem, allowing a student to infer that the production cost per item is 6$, for example, or that the demand function is given by the expression $900-15p$.

Such questions invite students to contemplate the form of an algebraic expression and to formulate ideas about the possible purposes to which that form is adapted. Too often, students feel they must instantly do something to an expression, without having formulated a purpose.

Conclusion

Algebra is at core about proficiency with symbolic representations. Assessment of this proficiency should include fluency with symbolic manipulations, but should also include the other four strands of mathematical competence. Such a richer assessment is often unfortunately delayed until functions have been introduced, since the different ways of representing functions, and their numerous applications in real-world contexts, provide good ground for the formulation of more conceptual and strategic questions. Simple questions of this nature at the level of elementary algebra are hard to come by, with the result that it is often taught as a purely procedural skill, with its more conceptual and strategic aspects either ignored or veiled behind the more abstract concept of function.

References

[NRC 2001] National Research Council (Mathematics Learning Study: Center for Education, Division of Behavioral and Social Sciences and Education), *Adding it up: Helping children learn mathematics*, edited by J. Kilpatrick et al., Washington, DC: National Academy Press, 2001.

[RAND 2003] RAND Mathematics Study Panel, *Mathematical proficiency for all students: Toward a strategic research and development program in mathematics education*, Santa Monica, CA: RAND, 2003.

Assessing Mathematical Proficiency
MSRI Publications
Volume **53**, 2007

Chapter 12
Making Meaning in Algebra
Examining Students' Understandings and
Misconceptions

DAVID FOSTER

Students often get confused and lost when they take Algebra 1.[1] The underpinning of the course is the generalization of the arithmetic they have previously studied and communication about mathematical ideas in a language that is rich in symbolic notation. Therefore it is not surprising for students to find algebra as abstract and unconnected to the real world.

Over the past twenty years, there has been a movement in the United States to make algebra more concrete. Specialized manipulatives have been invented in order to provide "hands-on" materials for students to use. Algebra tiles, positive and negative counters, and balance apparatuses, to name a few, are commercially available materials for "concrete learning" of algebra.

Although concrete materials may be helpful for students to learn algebra, it is not the materials themselves that provide algebraic meaning or understanding. Some educators assume that from the use of hands-on materials, students will automatically jump to abstract understanding of algebra. Research informs us that this is not the case. Physical knowledge is knowledge of objects observable in external reality, while mathematical knowledge is the mental relationships students construct in their heads. The source of mathematical knowledge is thus in the student [Kamii and DeClark 1985].

It is equally true that if students are taught abstract ideas without meaning, there will be no understanding. Students need experiences with a concept to develop meaning for themselves. If we want students to know what mathematics is as a subject, they must understand it. Knowing mathematics — really knowing it — means understanding it. When we memorize rules for moving

---

[1] In the United States, algebra is often taught in two year-long courses called Algebra 1 and Algebra 2.

symbols around on paper we may be learning something, but we are not learning mathematics [Hiebert et al. 1997].

In order to provide learning experiences that students can use to develop understanding of important algebra concepts, teachers must learn what students understand, and they must be sensitive to possible misconceptions held by the students. The learning experiences should allow students to confront their misconceptions and build upon knowledge they already understand. If students are unfamiliar with symbols, vocabulary, representations, or materials, the meaning students gain might differ significantly from what the instructor intends. Mathematical tools should be seen as support for learning. But this learning does not happen instantly: it requires more than watching demonstrations, but working with tools over extended periods of time, trying them out, and watching what happens. Meaning does not reside in representations and concrete materials; it is constructed by students as they use them [Hiebert et al. 1997].

What is Central to Algebra?

Habits of mind. Identifying what students must learn is paramount. One formulation of important ideas of algebra is in terms of "habits of mind" that students need for algebra. Examples of habits of mind are *doing and undoing, building rules to represent functions*, and *abstracting from computation*.

> *Doing-undoing.* Effective algebraic thinking sometimes involves reversibility: being able to undo mathematical processes as well as do them. In effect, it is the capacity not only to use a process to get to a goal, but also to understand the process well enough to work backward from the answer to the starting point....
>
> *Building rules to represent functions.* Critical to algebraic thinking is the capacity to recognize patterns and organize data to represent situations in which input is related to output by well-defined rules....
>
> *Abstracting from computation.* This is the capacity to think about computations independently of particular numbers that are used. One of the most evident characteristics of algebra has always been its abstractness. But, just what is being abstracted? To answer this, a good case can be made that thinking algebraically involves being able to think about computations freed from the particular numbers they are tied to in arithmetic — that is, abstracting system regularities from computation.
>
> [Driscoll 1999]

For example, if algebra tiles are used to teach factoring of quadratic polynomials, students first need to understand how an area model represents a multiplication, then abstract from arithmetic computation (a habit of mind). This supports students in learning to understand how area is a model for making sense of the relation between factors and product, and how even when actual values are unknown the generalization holds. In the case of factoring, students also need to understand the habits of mind of Doing and Undoing, since factoring is the inverse of multiplying two numbers.

Variable and function.   An important concept in algebra is the concept of variable. Often students are first introduced to the term "variable" when they are asked to solve for an unknown in an equation. For example, in the equation $3x + 4 = 19$ there is only one value of x that makes the equation true, namely 5. Although in situations like this, x is often called the variable in the equation, its value does not vary.

It is important that students understand the notion of variable as something that can vary. In prototypical functional relationships, the unknowns vary because a change in the input affects the output. Students can start to understand functions and variables by drawing on their experiences. For example, students can consider a sliding door as an example of a relationship of two variables in which one is a function of the other. The width of the opening varies as the door moves on its track. The distance moved by the door can be considered as the input and the width of the opening as the output. Students should examine other situations in order to elaborate their understanding of variables and functions. When students are asked to make lists of situations that involve variables and functional relationships, they may come up with:

The length of a candle varies with the amount of the time it burns.

The height of a child varies as its age increases.

The temperature of a liquid varies with the amount of time it cools in a freezer.

The distance an object moves varies with the force of a push.

The weight of a piece of a given rope varies with its length.

Students benefit from developing a connection between concrete experiences and the abstract concepts important to algebra.

Equality and equation.   Another important idea in algebra is equality. In the United States, although students use the equals sign early in their school careers, they often use it to mean "the answer follows." For example, in $45 - 23 = ?$, the equals sign can be and often is interpreted as a signal to execute an arithmetic operation [Siegler 2003]. When used in an equation, the equals sign indicates that the expressions on the left and right sides have the same value. This can be a stumbling block for students who have learned that the equals sign means "the answer follows."

The idea of a balance scale may be used to help students understand how the equals sign is used in equations and what kinds of operations on equations are permissible. Students can connect representations of a balanced scale with operations that preserve equalities in an equation. The equals sign is synonymous with the center of the scale. If a scale is in balance and a weight is added to the pan on the left, an equal weight must be added to the right pan in order for the scale to remain balanced. Experiences of adding or removing "weights" from both sides of a scale representation may help students develop the concept of equality and strategies for solving for unknowns.

For example, suppose a student is asked to solve $3x + 5 = 11$ for x. The student uses a representation of a balanced scale. On the left side are three boxes, each representing the unknown x, and five marbles. On the right side are eleven marbles.

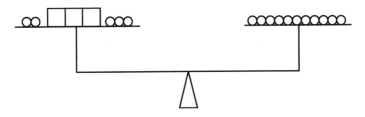

Five marbles may be removed from each side of the scale. This corresponds to the process of subtracting equal quantities from both sides of an equation. When only the unknowns are left on one side of the equation, then the boxes and marbles may be partitioned to determine the unknown.

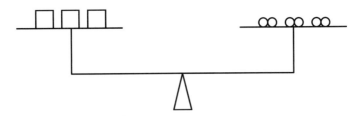

The importance of using a representation is to emphasize the meaning of the mathematical ideas being investigated. The scale representation is only valuable if students understand that an equation involves the same principles as a balanced scale — that both sides are equal. Students must understand the correspondence between the arithmetic operations and their scale counterparts of adding objects, removing objects, or partitioning objects. With these understandings, students can solidify meanings of solving equations.

An example of a student utilizing the meaning of the equals sign and the power of habits of mind is illustrated by a third-grade student who is determining what number could be substituted for d in:

$$345 + 576 = 342 + 574 + d$$

SAM: There's 300 on each side, so I took them off, and you can do the same thing with the 500s. That gave me 45 plus 76 equals 42 plus 74 plus d. Now you can do the same thing again with the 40s and the 70s, and that leaves 5 plus 6 equals 2 plus 4 plus d. And that's 11 and 6, so d has to be 5 to make that side 11.

MS. V: Sam, how do you know that you can do that?

SAM: If something is the same on both sides of the equal sign, you don't even have to think about it, you can just get rid of it. When you get rid of what's the same, the numbers get smaller and then it gets real easy to tell what d is equal to.

[Carpenter et al. 2003]

Although examples have been presented in this chapter, there is no one good way to present and teach a topic. Often, one student in a class may immediately construct mathematical meaning from an experience while another student actually confuses the meaning. Students need to share ideas and have varied experiences involving important concepts. Multiple representations arise both from mathematics and from students. Because they think differently from one another, students see problems in a range of ways and represent and solve them in multiple ways [Ball 1999]. Experiences with ideas, technology, or materials are important tools teachers need to use to foster students' understanding of algebra.

Assessment Tasks Focused on Algebraic Ideas

Given the importance of these themes, the Silicon Valley Mathematics Initiative has included in its assessments a number of tasks designed to help teachers understand how well their students are making sense of them. (See Chapter 10 in this volume for more details about this project.) The value of these assessments is found in the responses of students. Through examining student work on these tasks, teachers can focus on important ideas in algebra and identify common misconceptions. This feedback process is a powerful strategy for improving algebra instruction. We give here examples of tasks from the Mathematics Assessment Resource Service aimed at the five major themes discussed in this chapter. Each task is accompanied by a sample of student work and commentary about approaches, understanding, and misconceptions adapted from the Mathematics Assessment Collaborative of the Silicon Valley Mathematics Initiative's *Tools for Teachers*.

Assessment Task Focused on Doing-Undoing

Number Machines

This problem gives you the chance to:
• work with number chains
• explain your reasoning

Here are two number machines:

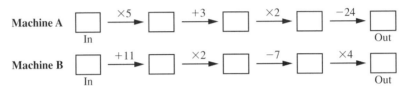

When you put an integer into the "In" box of a machine and do the operations shown in order, an answer appears in its "Out" box.

For example, if you put 3 into the "In" box of Machine A, 12 appears in its "Out" box.

Machine A   [3] —×5→ [15] —+3→ [18] —×2→ [36] —−24→ [12]
                  In   Out

1. Ray puts 13 into the "In" box of both machines.

(a) Which machine gives the largest number in its "Out" box? _____

(b) What is the largest answer he gets when he puts in 13? _____

2. Ray puts an integer into the "In" box of one of the machines. The integer 196 appears in its "Out" box.

(a) Which machine did he use? _____ Which integer did he put in? _____

(b) Explain how you can tell that he did not use the other machine.

3. Leila puts an integer into the "In" box of each machine.
She finds that the answer that appears in the "Out" box of Machine A is the same as the answer that appears in the "Out" box of Machine B.

What was the answer produced by both machines? _____
Show how you figured it out.

The student whose work is shown on the next page successfully answers question 1 by working his way forward: he fills in the boxes on the top two rows with the appropriate numbers, and concludes that Machine B, with a result of 164, wins out. He can also solve question 2 by using a working backward strategy — specifically, he writes *above* the boxes the result of applying each step backwards, from right to left:

Machine A: 13 $\xrightarrow{\times 5}$ 65 (107) $\xrightarrow{+3}$ 68 (110) $\xrightarrow{\times 2}$ 136 (220) $\xrightarrow{-24}$ 112 (190)

Machine B: 13 (17) $\xrightarrow{+11}$ 14 (28) $\xrightarrow{\times 2}$ 48 (56) $\xrightarrow{-7}$ 41 (49) $\xrightarrow{\times 4}$ 164 (190)

In — Out

As an answer to part (b) of this question, he writes: "I can tell he didn't use the other machine because, while I was working backwards, when I got to the last part I noticed 5 did not go evenly into 107. So it couldn't be that one."

The student does not attempt to answer question 3.

Assessment Task Focused on Building Rules to Represent Functions

Square Patterns

This problem gives you the chance to:
• work with a sequence of tile patterns
• write and use a formula

Mary has some white and gray square tiles.
She uses them to make a series of patterns like these:

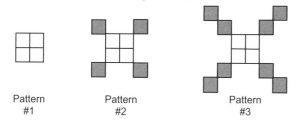

Pattern #1   Pattern #2   Pattern #3

1. How many gray tiles does Mary need to make the next pattern?

2. What is the total number of tiles she needs to make pattern number 6?

   Explain how you figured it out.

3. Mary uses 48 tiles in all to make one of the patterns.

   What is the number of the pattern she makes?

   Show your work.

4. Write a formula for finding the total number of tiles Mary needs to make pattern # n.

Mary now uses gray and white square tiles to make a different pattern.

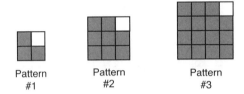

Pattern   Pattern   Pattern
#1   #2   #3

5. How many gray tiles will there be in pattern # 10?   _____

    Explain how you figured it out.

6. Write an algebraic formula linking the pattern number, P, with the number of gray tiles, T.

The student is willing to experiment with patterns in a variety of ways to help make sense of the relationships. In the beginning he uses drawings to solve for parts 2 and 3:

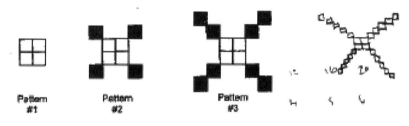

Pattern   Pattern   Pattern
#1   #2   #3

1. How many gray tiles does Mary need to make the next pattern?

                                                    ___12___ ✓

2. What is the total number of tiles she needs to make pattern number 6?

    Explain how you figured it out.   ___24___ ✓

    _____I___drew___the__pattern_for_#6.✓_____

3. Mary uses 48 tiles in all to make one of the patterns.

    What is the number of the pattern she makes?   ___12___ ✓

    Show your work.

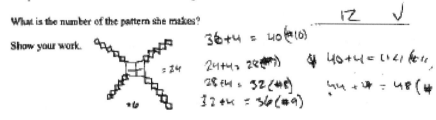

36+4 = 40 (#10)

24+4 = 28 (#7)   & 40+4 = 44 (#11)

28+4 = 32 (#8)   44+4 = 48 (#12)

32+4 = 36 (#9)

Then he notices a pattern of adding four every time and uses this relationship to verify the drawing. By doing all this thinking about the relationships, the student is then able to come up with an algebraic expression in part 4 that works for any pattern number:

4. Write a formula for finding the total number of tiles Mary needs to make pattern # n.

<u> $4n$ = total number of tiles.   ✓</u>

In part 5, the student first notices the differences increases by consecutive odd numbers. With further thinking about the pattern number, he sees that you square one more than the pattern number and subtract 1. It is interesting to see all the stages of the student's thinking.

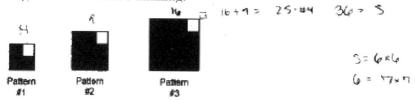

5. How many gray tiles will there be in pattern # 10?   120 ✓   1

   Explain how you figured it out.

<u>   I used the formula $(p+1)^2 -1$.   ✗   </u>

6. Write an algebraic formula linking the pattern number, P, with the number of gray tiles, T.

<u> $(p+1)^2 - 1$ ✓   1</u>

Assessment Task Focused on Abstracting from Computation

Multiples of Three

This problem gives you the chance to:
• test statements to see if they are true
• find examples to match a description
• explain and justify your conclusions

If a number is a multiple of three, its digits add up to a multiple of three.

For example, 15 is a multiple of three ($15 = 3 \times 5$)
and $1 + 5 = 6$, which is a multiple of three.

Also, the number 255 is a multiple of three ($255 = 3 \times 85$)
and $2 + 5 + 5 = 12$, which is a multiple of three.

1. Use the above rule to test whether 4721 is a multiple of three or not **and** explain how you figured it out.

4721 is not a multiple of 3 because its numbers (4, 7, 2, and 1) do not add up to a multiple of 3 (14), and it cannot be divisible by 3, meaning it cannot be multiplied to get 4721.

(margin annotation: a number)

2. Use the above rule to find a 5-digit multiple of three **and** explain how you know you are correct.

12345 is a multiple of 3 because its #'s (1, 2, 3, 4, & 5) when added up, is 15, which is divisible by 3. also, 4115 × 3 = 12345, so it is a multiple of 3.

(margin work: 1+2+3+4+5, 6, 9, 15, 4115, 12345, -12, 34, -3, 15)

3. Zara says, "If you add two multiples of three you always get another multiple of three."

Is Zara correct? __yes__

Explain how you decided.

If one number is a multiple of 3, and the other one is, then you can say x·3 = multiple of 3 + y·3 = multiple of 3 so (x+y) 3 = multiple of 3. Basically you can make them one number and it will still be a multiple of 3.

(margin annotation: 2 2)

4. Phil says, "If you add two multiples of three you always get a multiple of six."

Is Phil correct? __NO__

Explain how you decided.
(multiples of)
6 are every other multiple of 3 (starting from 6). You could have a number that was just a multiple of 3 and add it to a number that was a multiple of 3 and 6, and it would only be a multiple of 3, because you would be 3 short / 3 over a multiple of 3.

Assessment Task Focused on Variable and Function

ROPE

This problem gives you the chance to:
• Interpret information from a graph

The six points on the graph represent pieces of climbing rope.

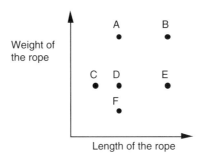

(a) Which pieces are the same length as rope D?   _____

(b) Which have the same weight as rope D?   _____

The ropes are made from the same material, but there are 3 different thicknesses:

(c) Which points represent 'thin' rope?   _____

(d) Which points represent 'thick' rope ?   _____

Explain how you know.

This student illustrates his knowledge of the functional relationship between the length of the rope and the weight of the rope. He is able to identify sets of points corresponding to same weight, same length, and — using lines drawn from the origin through sets of points — same weight/length ratio.

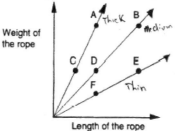

Thus he correctly completes parts (a)–(d) of the task. To explain how he determined the pairs of points that represent thick, medium and thin pieces of rope, he writes:

I drew lines in the graph and by the positions of the dots, connected them to each other and to the origin in a straight line. Then, I labeled the lines with thin, medium, & thick. Thin being the least steep & thick being the most steep line. And with the lines & them labeled, everything else was easy.

Assessment Task Focused on Equality

Party Flags

This problem gives you the chance to:
• find sizes by interpreting a diagram
• express a function by a formula

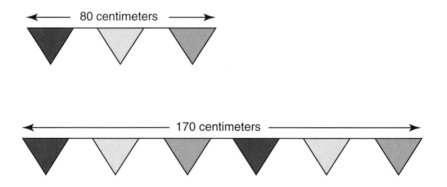

Erica is putting up lines of colored flags for a party.
The flags are all the same size and are spaced equally along the line.

1. Calculate the length of the sides of each flag, and the space between flags.
Show all your work clearly.

2. How long will a line of n flags be?
Write down a formula to show how long a line of n flags would be.

The student uses a visual strategy to make sense of the given information, formulating an algebraic expression and solving for the missing values for the space between the flags and their sides. The focus on the physical model also allows the student to generalize a formula for any size string of flags.

Erica is putting up lines of colored flags for a party.
The flags are all the same size and are spaced equally along the line.

1. Calculate the length of the sides of each flag, and the space between flags.
Show all your work clearly.

- If 3 triangles & space between flag's total equals 20 cm.
- Then if there's 6 triangles & space between flags total equals 170v
  you will use this formula to find the answer $20 \times 6 + 10 \times 5 = 120 + 5$

The sides of each flag measure ___20___ cm.

6 triangle   5 space

The space between flags measures ___10___ cm.

2. How long will a line of *n* flags be?
Write down a formula to show how long a line of *n* flags would be.

$20cm \times n + 10cm \times n = $ Answer (Total)

length of flag   # of flags   space between flags   # of space between flags

References

[Ball 1999] D. L. Ball, "Crossing boundaries to examine the mathematics entailed in elementary teaching", pp. 15–36 in *Algebra, K-theory, groups, and education: on the occasion of Hyman Bass's 65th birthday*, edited by T. Lam, Contemporary Mathematics **243**, Providence, RI: American Mathematical Society, 1999.

[Carpenter et al. 2003] T. Carpenter, M. Franke, and L. Levi, *Thinking mathematically: Integrating arithmetic and algebra in elementary school*, Portsmouth, NH: Heinemann, 2003.

[Driscoll 1999] M. Driscoll, *Fostering algebraic thinking: A guide for teachers grades 6–10*, Portsmouth, NH: Heinemann, 1999.

[Hiebert et al. 1997] J. Hiebert, T. P. Carpenter, E. Fennema, K. Fuson, D. Wearne, H. Murray, A. Olivier, and P. Human, *Making sense: Teaching and learning mathematics with understanding*, Portsmouth, NH: Heinemann, 1997.

[Kamii and DeClark 1985] C. Kamii and G. DeClark, *Young children reinvent arithmetic: Implications of Piaget's theory*, New York: Teachers College Press, 1985.

[Siegler 2003] R. S. Siegler, "Implications of cognitive science research for mathematics education", pp. 219–233 in *A research companion to Principles and Standards for School Mathematics*, edited by J. Kilpatrick et al., Reston, VA: National Council of Teachers of Mathematics, 2003.

Assessing Mathematical Proficiency
MSRI Publications
Volume **53**, 2007

Chapter 13
Task Context and Assessment

ANN SHANNON

Introduction

In this chapter, I will explore the impact of task context on assessment in mathematics. It is nontrivial to determine the understandings measured by a given assessment, so a close examination of some tasks and what they reveal is the main focus of this paper. Before considering these contemporary explorations, and in order to establish for the reader that the context of a mathematics task is indeed a salient feature, I will review findings from research in mathematics education and psychology. I will show that the role of context in mathematics assessment is a complex issue that involves much more than capturing the interest and harnessing the motivation of the student.

Background

Keeping it real.   Over the past several decades, many different researchers and educators have pointed out the benefits of setting mathematical tasks in rich, attractive, and realistic contexts (e.g., [de Lange 1987; Freudenthal 1983]). Realistic contexts are generally regarded as referring to aspects of the "real" social or physical world as well to fictional, imaginary, or fairy-tale worlds. Specifically, there are no restrictions on the contexts that can be called realistic as long as they are meaningful, familiar, appealing, and morally appropriate for students. In the literal sense, it is not the degree of realism that is crucial for considering a context as realistic, but rather the extent to which it succeeds in getting students involved in the problem and engages them in meaningful thinking and interaction.

Realistic contexts are recommended for two main reasons. On the one hand, it is thought that a realistic context will facilitate student success by intrinsically motivating students and thus increasing the likelihood that they will make a serious effort to complete the problem. On the other hand, a realistic context may facilitate performance by helping students to make a correct representation of the problem and to formulate and implement a feasible solution strategy by activating the use of prior knowledge specific to that context that is helpful for understanding and solving the problem.

There is growing evidence, however, that realistic contexts can have a negative impact on the performance of students on mathematics tasks. For example, [Boaler 1994] analyzed the performance of 50 students from a school with a traditional approach to mathematics education on two sets of three questions intended to assess the same content, but set in different contexts. Interestingly, the girls from the school that Boaler studied tended to attain lower grades on an item that had been cast in the context of "fashion" than they did on isomorphic items that were cast either in an abstract context or even in a realistic context that was thought to be inherently less appealing to the girls (such as football — "soccer" in U.S. terms). According to Boaler, the girls' relatively poor performance on the fashion item was caused by the problem context distracting them from its deeper mathematical structure.

More recently, De Bock, Verschaffel, Janssens, Dooren, and Claes [2003] also found that context could have a negative impact on students' performance. These authors analyzed student responses on tasks set in a context that included Lilliputians (of Swift's *Gulliver*). The authors suggest that, just like the girls in Boaler's study, their students' emotional involvement with the Lilliputians may have had a negative rather than positive influence on student performance. Thus, at the same time as we observe textbook writers and test developers infuse their new curriculum and assessment materials with sought-after realistic contexts, current educational research evidence is making it increasingly clear that the underlying case for realistic contexts has neither been well made nor well understood.

Assessing reasoning. In the early 1970s, a flurry of investigations into the role of context in reasoning was motivated by Piaget's theory of formal operations. Briefly, Piaget's earliest rendering of the stage of formal operations described this level of thought as unshackled by either content or context of a problem. Instead, it was believed that with the onset of the *formal operational* stage of thinking, problem solvers were guided by propositional logic and the problem's structure, rather than its content or context.

The individual and combined work of Wason, Johnson-Laird, and others within the British and U.S. cognitive psychology community developed what is affectionately known today as the Four-Card Problem [Wason 1969]. This work showed very clearly that when people solve a problem they usually rely upon its contextual features rather than solving the problem by abstracting form from content as Piaget's initial theory of formal operations had suggested.

The problem was often presented as follows:

Four-Card Problem

Here are four cards. You know that each has a number on one side and a letter on the other. The uppermost face of each card is like this:

The cards are supposed to be printed according to the following rule:
If a card has a vowel on one side, it has an even number on the other side.

Which among the cards do you *have* to turn over to be sure that all four cards satisfy the rule?

Before you read any further, try to answer the question. You probably decided that you needed to turn over the card with the A, because if the number on the other side turned out not to be even, you would have disproved the rule. In the 1970s, most people almost always got this right. Similarly, most people knew that they did not need to turn over the card with the P, because the rule says nothing about consonants. In the 1970s, however, the research participants were divided as to whether or not it was necessary to turn over the card with the 6. The correct answer is that you do not need to turn over the card with the even number. This is because it might have either a vowel or a consonant on the other side and neither card will violate the rule. With regard to the card with the number 3 on it, most of the research participants decided that they did not need to turn it over in order to check the rule. The correct answer, however, is that the card with the number 3 must be turned over to verify that the other side does not contain a vowel. If the other side contains a vowel then the rule will not be satisfied.

Due to the low incidence of correct responses on the Four-Card Problem, researchers created different versions of the task that employed thematic contexts. For example, during the 1970s in England and parts of Ireland, there were two main rates for mailing envelopes — first and second class, first class being more expensive. At that time, if you sealed your envelope you had to put a first-class stamp on it, but if your envelope was left unsealed then a second-class stamp would do. For some people living in England and parts of Ireland during that time, the context was real and relevant and thus an envelope version of the Four-Card Problem was created in order to assess reasoning in a real-world context [Johnson-Laird et al. 1972].

Four-Envelope Problem

Here are four envelopes. Each has a stamp on one side and each is either sealed or left unsealed.

The uppermost face of each envelope is like this:

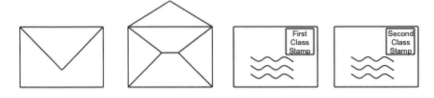

The envelopes are mailed according to the following Post Office rule:
If an envelope is sealed then it must have a first-class stamp on it.

Which among the envelopes do you have to turn over to be sure that all four envelopes satisfy the Post Office's rule?

The Four-Envelope Problem is identical to the Four-Card Problem in structure. In the early research in Britain, participants found the envelope version of the task easy to solve and made few errors. Compared to the Four-Card Problem, the Four-Envelope Problem was almost a trivial exercise, because the solvers' understanding of the familiar context carried them to a successful conclusion. Thus, because of the context, participants were able to say that the sealed envelope and the one with the second-class stamp were the envelopes that needed to be checked. Participants also responded correctly that neither the unsealed envelope nor the one with the first-class stamp needed to be checked because if people wanted to waste a first-class stamp on an unsealed envelope then that was their business. The Post Office rule did not stipulate the kind of stamp that needed to go on an unsealed envelope nor did it say anything about sealing an envelope with a first-class stamp. Johnson-Laird et al. [1972] reported what they called a "thematic-materials effect," given the way in which the postal-rule context facilitated improved performance compared to the vowels and numbers context.

In the 1980s, members of both the U.S. and British cognitive psychology community had difficulty replicating the results of that particular experiment. Researchers in the U.S. could not replicate it because the postal regulations of the Four-Envelope Problem have never existed in that country [Griggs and Cox 1982]. British researchers could not replicate this effect when their participants were too young to have experienced the obsolete postal regulations (Golding, 1981, as cited in [Griggs and Cox 1982]). In response to the U.S. undergraduates' poor performance on the Four-Envelope Problem, Griggs and

Cox ingeniously invented their own thematic version. The context for this task is one that is undoubtedly as real and as relevant today as it was in the 1980s.

Four Drinkers Problem

You are in charge of a party that is attended by people ranging in age. The party is being held in a state where the following law is enforced:

If you are under 21 you cannot drink alcohol.

Your job is to make sure that this law is not violated.

Understandably, you want to check only those people who absolutely need to be checked. At one table there are four people drinking. You can see the IDs of two of these people: One is under 21 and one is older than 21. You do not know what these two are drinking. You do however know what the other two people at the table are drinking: One is drinking soda and the other is drinking beer. You cannot see the ID of either of these two people. So to summarize your problem:

Under 21	Over 21	Drinking soda	Drinking beer

Of these four, who do you need to check in order to make sure that the law is not broken?

Within this context, you can probably see at a glance that you need to check both the drink of the person who is under 21 and the ID of the person who is drinking beer. But you probably would not think of checking the drink of a person who is older than 21, because the law does not say anything about the drinking habits of a person who is over 21. Similarly, you would not think of checking the ID of a person drinking soda because the law does not have a legal age for drinking soda.

I have chosen to take the reader through just a few of the highlights of this now decades-old thread of psychological research because the original Four-Card Problem and its subsequent recastings cogently "problematize" the issue of task context and assessment. These highlights clearly show that context can aid or impede the solver, and also show that context can sometimes change a task so substantially so as to lead one to ask, "Is this still a math task?"

From the highlights, you can see that we have a well-tested example of three tasks with the same mathematical structure, which appear to be solved quite differently. For most people, the Four-Card Problem is essentially a problem in pure logic. On the other hand, for some people, the envelope and drinking age versions of the problem may not be problems in pure logic. In the latter variations, the success of the problem solver has been shown to depend heavily on whether the context of the problem is familiar and meaningful to the solver. Given sufficient familiarity with context, it seems that the solver can be carried along by the meaning of the context, and so is prevented from making the logical errors that trip up participants on the Four-Card Problem.

In the envelopes and drinking age scenarios, the familiarity of the situation rather than the mathematical structure facilitated success. But with regard to assessment, what does participant performance on these tasks tell us about the participants' understanding of mathematical logic? More specifically, what do these performances tell us about what the participants might have learned about propositional logic or about analysis of propositions of the form "If P then Q"? Interestingly, the cognitive researchers found that prior success and experience with familiar contexts such as envelopes and drinking age did not readily transfer to the abstract Four-Card Problem involving vowels and consonants [Cox and Griggs 1982; Johnson-Laird et al. 1972; Griggs and Cox 1982; 1993]. Thus, it seems that if our aim were to assess learning about propositional logic, it would not make sense to deploy a thematic context. When it comes to propositional logic, thematic context can lead to false-positive or false-negative results; see, for instance, [Cox and Griggs 1982].

This discussion is not meant to suggest that task context has nothing to with the assessment of mathematics. The issue of assessment is just far too complex for that [Boaler 1994]. To the contrary, I will argue that assessment of important mathematics can be facilitated by tasks with appropriate real-world context, provided that the task context and the mathematics to be assessed are sufficiently integrated. This can be accomplished if problem solvers are invited to use the context to demonstrate some aspect or aspects of their mathematical prowess.

Thus, the relevant question that this cognitive research raises in relation to the more contemporary problem of teaching and learning school mathematics is: How can familiar, real, and relevant contexts be used effectively to assess mathematics? Some insight into this question is afforded by another series of studies that I carried out for the Balanced Assessment and New Standards projects [Shannon and Zawojewski 1995; Shannon 1999; 2003].

Beyond Interest and Motivation

As part of my work for the Balanced Assessment Project (a task development project funded by the National Science Foundation) and later for New Standards, I conducted a series of mini-studies focused on a group of three very similar tasks involving linear functions in real-world contexts [Shannon 1999]. The three tasks were called Shopping Carts, Shopping Baskets, and Paper Cups. In each task, students were presented with diagrams of common objects that can be nested when stacked, and were asked to measure the diagrams and develop linear functions describing how the length or height of a stack would vary with the number of objects in the stack.

Shopping Carts

The diagram shows a drawing of a single shopping cart. It also shows a drawing of 12 shopping carts that have been nested together. The drawings are 1/24th real size.

Create a formula that gives the length of a row of nested shopping carts in terms of the number of carts in that row.

Define your variables and show *how* you created your formula.

Shopping Baskets

The diagram [next page] shows a drawing of a single shopping basket. It also shows a drawing of 7 shopping baskets that have been nested together. The drawings are 1/10th real size.

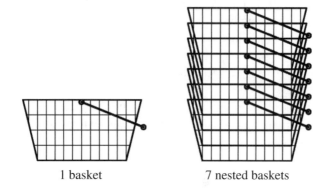

1 basket   7 nested baskets

Create a formula that gives the height of a stack of shopping baskets in terms of the number of baskets in a stack.

Define your variables and show *how* you created your formula.

Paper Cups

The diagram shows drawings of one paper cup and of six paper cups that have been stacked together. The cups are shown half size.

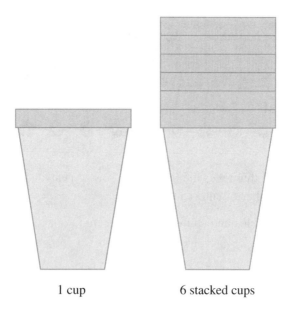

1 cup   6 stacked cups

Create a formula that gives the height of a stack of cups in terms of the number of cups in the stack.

Define your variables and show *how* you created your formula.

The three tasks described here are clearly all variants of a common theme, and they tap into the same area of mathematics — linear functions and arithmetic sequences (linear functions of positive integers). Despite their common mathematical structure, the three tasks proved to present three separate levels of challenge for most students. Shopping Carts was the most difficult, followed by Shopping Baskets, and then Paper Cups. To appreciate the nature of the challenges that these tasks offered to students, I analyzed the responses from comparable groups of students on Shopping Carts and Shopping Baskets, then analyzed responses from comparable groups of students on Shopping Baskets and Paper Cups.

Shopping carts versus shopping baskets. In using the pictures of the stacks to make the necessary measurements, I have found that students are considerably more successful in using the diagram of the baskets than they are in using the more detailed diagram of the carts. For example, it has seemed that the wheels and handles of the carts present students with a much more complicated diagram to work with, when compared with the relatively more straightforward picture of the baskets. Thus, the authenticity of the diagram of the stack of carts presented students with many extraneous details, and students had to identify those aspects of the structure of the stack of carts that were relevant to the problem and those that were not relevant.

There are also scale differences between these two versions of the task. The carts were drawn as $1/24$ of the actual size, while the baskets were drawn to $1/10$ scale. Students have proved to make fewer computational errors using the scale factor of $1/10$ for Shopping Baskets than when using the scale factor of $1/24$ needed for Shopping Carts, and they have also seemed better positioned to know how to use this information in the case of the baskets than in the case of the carts.

Finally, there are differences of orientation. The length of the stack of carts increases in a horizontal direction while the height of the stack of baskets increases in a vertical direction, and it has seemed as a result that students are better able to visualize the increasing stack of baskets than they do the increasing stack of carts. While working on Shopping Baskets, for example, students were observed gesturing with their hands as if to show how the height of a stack of baskets might increase with increasing number, but no similar student actions were observed when students were working on Shopping Carts.

Shopping baskets versus paper cups. The baskets were drawn as $1/10$ of the actual size and the cups were drawn as $1/2$ of the actual size. There was no evidence that either of these scale factors provided a greater challenge to students. Similarly, students were no more nor less successful in using the picture of the stack of baskets to make measurements than in using the picture of the stack of cups. Both the stack of baskets and the stack of cups increased in a vertical direction. Thus, it seemed that the stack of baskets presented no more visual or measurement-related complications than did the stack of cups.

Despite these similarities, however, students were more successful in expressing the height of a stack of cups in terms of the number of cups in the stack than in expressing the height of a stack of baskets in terms of the number of baskets in the stack. This has appeared to be because the cups were depicted as having a discernible lip and base, and students have then found it fairly straightforward to decompose the height of the stack of cups into that of "one base and n lips." The structure of the baskets does not invite a similar decomposition, probably because the part of each basket that protrudes above the previous basket cannot readily be depicted as a separate entity, and cannot easily be named or conceptualized in the same way as the lip and base of the cup can be identified and named. Thus, students have to think of the stack of baskets as one basket and $n - 1$ "stick outs." This small difference in the structure of the two stacks has seemed to make for a much larger impact on the complexity of the algebraic prowess that is needed to create a formula for each stack.

With the cups, students could finesse having to deal with $n-1$ lips and could instead write the formula as $h = nl + b$ (where h represents the height of the stack of cups, n represents the number of cup-lips, and l and b represent the actual measurements of the cup-lip and cup-base, respectively).

With the baskets, students cannot finesse having to deal with $n-1$ "stick outs" and instead have to find their way to expressing the formula as $h = p(n-1) + B$ (where h represents the height of the stack of baskets, n the number of baskets, B the actual height of a basket, and p the amount that each basket protrudes above the one below). There is no doubt that this difference in challenge is not trivial to students whose grasp of algebra is still fragile and not yet flexible.

When I present these findings at professional development workshops or at conferences, participants invariably attribute the relative success of students on Paper Cups to the familiarity of a paper cup in the everyday life of students. Workshop participants expressed the view that it made sense that a paper cup would be more familiar to students than a shopping basket. On the other hand, the widely held view that a familiar context will facilitate success on a mathematics task may lead many participants to interpret the finding that students are more successful on Paper Cups as evidence in and of itself that the paper cup must be more familiar to students than is the shopping basket.

My analysis of the student work suggests that the issue of the relative familiarity of a paper cup and a shopping basket is something of a red herring. It does however allow us to glimpse at the firmly held but perhaps erroneous beliefs that are held about the role of task context in the learning of mathematics. Based on my analysis of student work, I want to suggest that it is the specific geometry of the stack of cups that facilitates success with this version of the problem, rather than students' assumed greater prior experience with this everyday object. From the student work, it has seemed that it is simply easier in the case of the cups for students to translate from a visual representation (the diagram of the stack) to an algebraic representation (a formula). In later versions of these tasks, I have asked students to engage in a variety of related activities: express the height of the stack in terms of the number of cups or baskets in a stack; make a graph of that relationship; find the slope of the associated line; and finally, interpret the slope in terms of the initial situation. Students have proved much more successful in interpreting the slope when it represents the increase in height per cup (length of a cup-lip), as compared to the situation when it represents the increase in height per basket (length of the part of the basket that protrudes above the previous basket). Again, it seems reasonable to suggest that this is because (unlike "the part of a shopping basket that protrudes above each previous basket"), the cup-lip is a tangible object that can be named and understood with relative ease. There is clearly also a visualization aspect. The height of a cup lip is a vertical line segment on the diagram, but there is no line segment on the diagram to show the height of a basket's protrusion.

Discussion

The responses of students to tasks such as Shopping Carts, Shopping Baskets and Paper Cups suggests that the importance of these tasks for teaching and learning mathematics lies not in their authenticity or their familiarity for students, but in the opportunities that each of these structures provides to students in translating among different representations and in affording students the opportunity to engage in mathematical abstraction. Thus, I am suggesting that it is the geometry of the various stacks that make these important mathematics tasks, rather than the fact that the structures comprise mundane, and more or less familiar, or everyday objects.

It is interesting to note that tasks of this caliber proliferate in the curriculum of "Railside," the astoundingly successful urban high school mathematics department described by [Boaler 2004]. Boaler notes that the teachers in this highly successful mathematics department do not select the context of mathe-

matics tasks so as to promote equity, or to relate to the students' own culture —
rather, they select mathematics tasks so as to promote abstract mathematical
discussions.

The capacity of a task to promote abstract mathematical discussion and thus
precipitate learning depends on how the task is used. Boaler's research makes
it clear that the effectiveness of the mathematics tasks assigned in Railside is
due largely to the high expectations that teachers place on students as they work
to complete the tasks. Specifically, Boaler's research, reports from the Railside
teachers, and classroom visits to Railside all suggest that the Railside teachers'
demand that all students engage in mathematical *justification* is instrumental in
precipitating learning. For example, [Boaler 2004] discusses a video clip where
students are asked to find the perimeter of a simple structure created using Lab
Gear.[1] The structure is by no means intrinsically interesting, nor would it be
familiar to the students in Railside's sheltered algebra class. However, working in groups and with varying degrees of difficulty, the students find that the
perimeter is given by the expression $10x + 10$ or its equivalent. Boaler's video
data show that the teacher is not satisfied by this correct answer and insists that
each student explain in terms of the structure "where the 10 is." The teacher's
tenacity is noteworthy, she persists until each student has justified the algebraic
expression in terms of the structure created using Lab Gear materials. Thus,
while the initial task simply asks students to represent the Lab Gear structure
in terms of an algebraic expression, the teacher asks the student to go further,
and to interpret the algebraic expression that they have created in terms of the
underlying geometry. This approach provides a means for the teacher to ensure
that the task provides the opportunity for students to translate among multiple
representations, and challenges the students to do so in ways that are far from
perfunctory. This short video clip communicates a clear sense that learning has
taken place in the group, because students who struggle with the teacher's demands can be seen going on to tackle and accomplish subsequent, more difficult
tasks with enthusiasm and confidence.

The examples discussed by Boaler make it clear that a task's capacity to
precipitate student learning of mathematics will be highly dependent on how
it is used in the classroom and on the particular efforts that are made to keep
the cognitive demands of the task high [Schoenfeld 1988; Stein et al. 2001;
Stein et al. 1996] and promote mathematical abstraction [Boaler 2004]. It is
not an overstatement to say that if its cognitive demands are not maintained, the
mathematics task — contextualized or not — will function no better than a series
of "drill sheets." In particular, tasks with the potential to be worthwhile can be

---

[1] Lab Gear is a manipulative for algebra designed by Henri Picciotto.

rendered worthless if students are not given the opportunity to grapple with the intrinsic complexities of the underlying mathematics [Ball and Bass 2003].

This particular idea is one that is often lost in what seems to be a rush to wrap assessment tasks in a context in order to motivate students. Unfortunately, the seemingly firmly held belief that task context is a good thing often leads directly to poor quality assessment tasks. Consider this contemporary grade 6 example from a state testing program:

> There are 30 pencils left at a store after Shilo buys a certain number of pencils, p.
>
> Delia buys 4 times as many pencils as Shilo. The expression below shows the number of pencils remaining at the store after Delia buys her pencils.
>
> $$30 - 4 \times p.$$
>
> How many pencils remain at the store if Shilo bought 3 pencils?
>
> A. 14   B. 18   C. 78   D. 104

Among the difficulties with this task are the following:

(a) The mathematics itself is low-level: the problem simply asks, "What is the value of $(30 - 4 \times p)$ when $p = 3$?"
(b) The most important aspect of mathematizing, generating the formula, is not part of the task; in this connection see Chapter 7 in this volume.
(c) The linguistic complexity of the task far overshadows the mathematical complexity; see also Chapters 19 and 20 in this volume.

As point (a) indicates, this task can be completed by ignoring the context entirely and simply "plugging 3" into the given expression. This runs the risk of teaching students that context is not important — an unfortunate message to send to students since task context is extremely important when used correctly [Boaler 2004; Shannon 1999]. From an assessment standpoint tasks of this type, even if written as "free response" rather than multiple-choice questions, are problematic: if a student were to give the wrong answer it would be difficult to diagnose what caused the student to have problems. From an equity standpoint the task is problematic because the context clearly places extraneous reading demands on students. Thus it places unnecessary burdens on the shoulders of English learners and others who might find reading difficult. In sum, using context in a problem statement without examining its impact on students' problem-solving processes can be problematic.

Concluding Remarks

In looking for mathematics assessment tasks it is not necessary to look for a real-world context *per se*, but to look instead for the opportunities that the task provides for the student to formulate his or her own approach to the task, represent the solution in some appropriate and mathematically abstract form, and then interpret the salient components of the solution in terms of the initial task. The role of context is a complex and a subtle one, but there is no doubt that it plays a critical role in creating student access to worthwhile and important mathematics.

Acknowledgment

My sincere thanks to Meghan Shaughnessy for her help with reading drafts, locating references, and many other things.

References

[Ball and Bass 2003] D. L. Ball and H. Bass, "Making mathematics reasonable in school", pp. 27–44 in *A research companion to Principles and Standards for School Mathematics*, edited by J. Kilpatrick et al., Reston, VA, 2003.

[Boaler 1994] J. Boaler, "When do girls prefer football to fashion? An analysis of female underachievement in relation to realistic mathematic contexts", *British Educational Research Journal* **20**:5 (1994), 551–564.

[Boaler 2004] J. Boaler, "Promoting equity in mathematics classrooms: important teaching practices and their impact on student learning", paper presented at the International Congress of Mathematics Educators, July 2004.

[Cox and Griggs 1982] J. R. Cox and R. A. Griggs, "The effects of experience on performance in Wason's selection task", *Memory and Cognition* **10**:5 (1982), 496–502.

[De Bock et al. 2003] D. De Bock, L. Verschaffel, D. Janssens, W. Dooren, and K. Claes, "Do realistic contexts and graphical representations always have a beneficial impact on students' performance? Negative evidence from a study on modeling nonlinear geometry problems", *Learning and Instruction* **13** (2003), 441–465.

[Freudenthal 1983] H. Freudenthal, *Didactical phenomenology of mathematical structures*, Dordrecht: Reidel, 1983.

[Griggs and Cox 1982] R. A. Griggs and J. R. Cox, "The elusive thematic-materials effect in Wason's selection task", *British Journal of Psychology* **73**:2 (1982), 407–420.

[Griggs and Cox 1993] R. A. Griggs and J. R. Cox, "Permission schemas and the selection task", *The Quarterly Journal of Experimental Psychology* **46A**:4 (1993), 637–651.

[Johnson-Laird et al. 1972] P. D. Johnson-Laird, P. Legrenzi, and M. S. Legrenzi, "Reasoning and a sense of reality", *British Journal of Psychology* **63**:3 (1972), 395–400.

[de Lange 1987] J. de Lange, *Mathematics, insight and meaning. Teaching, learning and testing of mathematics for the life and social sciences*, Utrecht: OW&OC, 1987.

[Schoenfeld 1988] A. H. Schoenfeld, "When good teaching leads to bad results: The disasters of 'well taught' mathematics classes", *Educational Psychologist* **23** (1988), 145–166.

[Shannon 1999] A. Shannon, *Keeping score*, National Research Council: Mathematical Sciences Education Board, Washington, DC: National Academy Press, 1999.

[Shannon 2003] A. Shannon, "Using classroom assessment tasks and student work as a vehicle for teacher professional development", in *Next steps in mathematics teacher professional development, grades 9–12: Proceedings of a workshop*, Washington, DC: National Academy Press, 2003.

[Shannon and Zawojewski 1995] A. Shannon and J. Zawojewski, "Mathematics performance assessment: A new game for students", *Mathematics Teacher* **88**:9 (1995), 752–757.

[Stein et al. 1996] M. K. Stein, B. Grover, and M. Henningsen, "Building student capacity for mathematical thinking and reasoning: An analysis of mathematical tasks used in reform classrooms", *American Educational Research Journal* **33** (1996), 455–488.

[Stein et al. 2001] M. K. Stein, M. Smith, M. Henningsen, and E. Silver, *Implementing standards-based mathematics instruction: A case book for professional development*, New York: Teachers College Press, 2001.

[Wason 1969] P. C. Wason, "Regression in reasoning?", *British Journal of Psychology* **60**:4 (1969), 471–480.

Section 5
What Do Different Assessments Assess?
The Case of Fractions

"How do I understand thee? Let me count the ways." Well, that's not what Elizabeth Barrett Browning said, but if she were a mathematician, teacher, or mathematics educator referring to any topic in mathematics, she might have. And if she were an assessment specialist she might have noted that there are countless ways to explore and document those understandings.

This section provides two detailed explorations of one mathematics topic, fractions. Broadly speaking, it addresses two main issues: what does it mean to understand fractions (at, say, the sixth-grade level), and what is the potential of various kinds of assessments to reveal those understandings?

Let us start with procedural fluency. Certainly, one expects sixth graders to be fluent in adding, subtracting, multiplying, and dividing fractions. They should be able to convert fractions to decimals, and place both fractions and decimals on the number line; consequently, they should be able to compare the magnitudes of various fractions. They should have a sense of magnitude, and be able to answer questions like: "Which of the numbers 0, 1, or 2 is the sum $\frac{7}{8} + \frac{12}{13}$ closest to?"

A next level of performance consists of being able use one's knowledge of fractions and be able to explain why what one has done makes sense. Here are some items from the *Mathematics Framework for California Public Schools* [California 2006, pp. 77–78]:

> Your after-school program is on a hiking trip. You hike $\frac{3}{4}$ of a mile and stop to rest. Your friend hikes $\frac{4}{5}$ of a mile, then turns around and hikes back $\frac{1}{8}$ of a mile. Who is farther ahead on the trail? How much farther? Explain how you solved the problem.

> Jim was on a hiking trail and after walking $\frac{3}{4}$ of a mile, he found that he was only $\frac{5}{8}$ of the way to the end of the trail. How long is the trail? Explain.

These are applications. One level deeper involves understanding what fractions are and various representations of them. The following item from the *California Framework* [California 2006, p. 78] begins to explore this territory:

Draw a picture that illustrates each of the following problems and its solution. Explain how your drawings illustrate the problems and the solutions.

1. $\dfrac{3}{4} \times \dfrac{1}{2}$ 2. $\dfrac{3}{4} + \dfrac{1}{2}$ 3. $2 \times \dfrac{3}{4}$

But this just scratches the surface. Does the student understand that all of the n-ths in the fraction m/n must be the same size? (Recall the sample assessment item from Chapter 1.) That if the numerator of a fraction is kept constant and as the denominator increases, the magnitude of the fraction decreases? Can the student work with representations of fractions on the number line? As parts of a "pie chart"? When the "whole" is, say, a pie and a half? Can the student explain why $\frac{2}{3}$ is equivalent to $\frac{4}{6}$, using any of these representations? Even this is just a first step into the domain of fraction understanding: the literature on what it means to understand fractions is immense. (A Google search on the phrase "understanding fractions" gives 22,000 hits.)

Once one has a sense of the terrain, there is the question of how one finds out what any particular student knows. This is enormously complex — and fascinating, as the three contributions in this section demonstrate. In Chapter 14, Linda Fisher shows the kinds of insights that well-crafted assessment tasks can provide into student thinking. She presents a collection of tasks that have been used by the Silicon Valley Mathematics Assessment Collaborative, and discusses the ways in which student responses reveal what they understand and do not. Such information is of use for helping teachers develop deeper understandings of student learning, and (when one sees what students are or are not making sense of) for refining curricula. In Chapter 15, Deborah Ball cranks up the microscope one step further. No matter how good a paper-and-pencil assessment may be, it is static: the questions are pre-determined, and what you see in the responses is what you get. At the MSRI conference, Ball conducted an interview with a student, asking him about his understandings of fractions. Like a written assessment, the interview started off with a script — but, when the student's response indicated something interesting about his understanding, Ball was in a position to pursue it. As a result, she could delve more deeply into his understanding than one could with a test whose items were fixed in advance, and also get a broad and focused picture of how his knowledge fit together. In Chapter 16, Alan Schoenfeld reflects on what such interviews can reveal regarding the nature of student understanding, and about the ways in which skilled interviewers can bring such information to light.

[California 2006] *Mathematics framework for California public schools, kindergarten through grade twelve*, Sacramento: California Department of Education, 2006.

Assessing Mathematical Proficiency
MSRI Publications
Volume **53**, 2007

Chapter 14
Learning About Fractions from Assessment

LINDA FISHER

Assessment can be a powerful tool for examining what students understand about mathematics and how they think mathematically. It can reveal students' misconceptions and gaping holes in their learning. It can also reveal the strategies used by successful students. Student responses can raise questions for teachers: Of all the strategies that students are exposed to, which ones do they choose to help make sense of a new situation? Which can they apply accurately? Which aspects of these strategies might be helpful for other students? Assessments can also, with guidance and reflection, provide teachers with a deeper understanding of the mathematical ideas and concepts that underlie the rules and procedures frequently given in textbooks. A good assessment raises questions about changing or improving instruction. The diagnostic and curricular information afforded by assessments can provide powerful tools for guiding and informing instruction.

Over the past seven years, the Mathematics Assessment Collaborative has given formative and summative assessments to students and used their responses in professional development for teachers and to inform instruction. This chapter will share some examples of how assessment can reveal student thinking and raise issues for instructional planning and improvement.

Learning from Formative Assessment

A group of teachers engaged in lesson study[1] was interested in the idea of focusing on students' use of mathematical representations. The teachers had looked at lessons and representations from a variety of sources, including an intriguing lesson from *Singapore Primary Mathematics 3B* which uses bar models (diagrams which use rectangular "bars" to represent quantities in a problem),

---

[1] Lesson study is a process in which teachers jointly plan, observe, analyze, and refine classroom lessons which are known as "research lessons." See http://www.lessonresearch.net/

triangles, pentagons, hexagons, squares, and circles to represent part-whole relationships. The middle part of that lesson tries to get students to see the following
generalization: when the numerators of a collection of fractions are the same,
the size of the portion represented by each fraction is inversely related to the
numerical value of the denominator. The lesson study group at first considered
designing its research lesson as an introduction to bar models. After lively debate, it opted to investigate what representations students would use to solve
a problem for themselves without further instruction. The lesson study group
then went into several classrooms to collect data about the representations that
students already used.

The teachers asked students the question, "Which is greater, one fifth or one
third? We are curious about how you think this out. So please show us in words
and pictures." As students finished their thinking, the teachers selected student
work and asked students to come to the board and explain their representations.

The lesson study teachers were surprised that all but one student used circles to represent fractions. Here are four typical representations used by third
graders:

Alice   Adam   Ariana   Alfred

Their work suggests that Alice and Adam do not have the understanding
that when one third (or one fifth) are represented as one of three (or five) parts
of a whole, that each part has equal size. Ariana, and many other students,
had difficulty drawing fifths. Her solution was to ignore the extra "one fifth."
That piece *is* one fifth, but not one-fifth of the "one" represented by the circle.
Instead, the piece is one-fifth of the unit represented by the unshaded pieces
in her drawing. Does Ariana know that? Finally, some students, like Alfred,
used a matching strategy. Alfred explained, "Two fifths is about the same size
as one third. The next two fifths is the same size as one third, so there is one
piece left in each circle. So one third and one fifth are the same size." Alfred
apparently doesn't see any problem in saying that one-third and one-fifth of the
same object are equal. Notice that while all the drawings attempt to represent

denominators, none of these third-grade students attempted to shade the part of the representation that corresponded to the "one" in one fifth and one third.

This type of assessment provides many insights for the classroom teacher about the complexity of understanding fractions and what types of experiences students need to make sense of the concepts that make up this understanding. It raises issues about the difference between identifying fractional parts in a diagram in a textbook and being able to make a representation of fractional parts accurate enough to help in solving a particular problem. The student work raises many pedagogical issues. How can teachers help students to develop the idea of equal parts? How can teachers get students to understand the idea of a unit — that you can't remove a piece of the object that represents the unit and have the remaining object represent the same unit? Is it appropriate to use odd denominators at this grade level and have students struggle to draw equal pieces or should students only be given amounts that are convenient to draw? Do students learn enough from examples and explanations or is there a quantifiable benefit from confronting and discussing misconceptions? What is gained and what is lost by these approaches? These are the types of questions raised by good assessments. Such assessments provide evidence of gaps in student understanding that are not revealed through computational exercises. They provide windows into what is missing when instruction focuses solely on procedures.

Looking at Summative Assessment: Sharing Pizza

A common way to introduce fractions to students is in the context of sharing. Consider the grade 5 task Sharing Pizza. This task gives students the opportunity to identify fractional parts, combine common unlike fractions, draw representations of fractions, and use fractions in a sharing context.

Sharing Pizza

Aretha, Beth, Carlos, and Dino go into a pizza shop and order three different pizzas. They divide the pizzas so that they each end up with the same amount to eat. Aretha can't eat seafood. The other friends like all the pizzas.

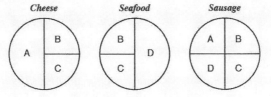

Aretha gets all the pieces labeled A. Beth gets those labeled B.
Carlos gets those labeled C. Dino gets those labeled D.

1. What fraction of the *Cheese* pizza does Aretha eat?   _____

What fraction of the *Sausage* pizza does Aretha eat?   _____

How much pizza does Aretha eat?   _____

2. Complete the diagram below to show how **five** friends—Aretha,
Beth, Carlos, Dino, and Erica—would divide the **three** pizzas.
Remember that each person gets the same amount to eat.
Remember that Aretha can't eat seafood, but the other friends like all three pizzas.

Cheese   *Seafood*   *Sausage*

◯   ◯   ◯

How much pizza does Aretha eat this time? Explain._____

This task was one of five given in a summative test to approximately 11,000 fifth graders in March, 2001. On this task 68% of the students were able to calculate $\frac{1}{2} + \frac{1}{4}$ in Part 1. However, only 33% were able to demonstrate any proficiency with any of the mathematics in Part 2.

Some students could give responses to Part 2 of the task — share the three pizzas among five friends — such as this:

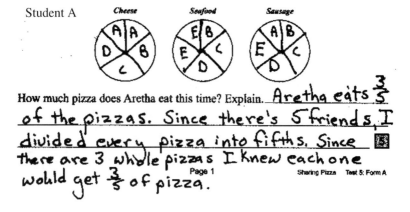

Student A can clearly articulate the part-whole relationship and describe each student's portion. The student is easily able to deal with the constraint that Aretha doesn't like seafood, without it affecting her overall serving size.

Student B has also solved the task, by partioning each pizza into ten pieces:

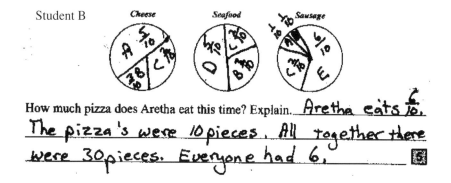

How much pizza does Aretha eat this time? Explain. _Aretha eats $\frac{6}{10}$._ _The pizza's were 10 pieces. All together there were 30 pieces. Everyone had 6._

Student C's response shows the importance of digging beneath the surface of the answers that students provide. Giving constructed-response questions provides an opportunity for students to reveal what they know and what they do not. Student C writes "$\frac{1}{2} + \frac{1}{5} = \frac{7}{10}$," a correct calculation, although not one that expresses Aretha's share of the pizzas. However, look at the rest of the student's work:

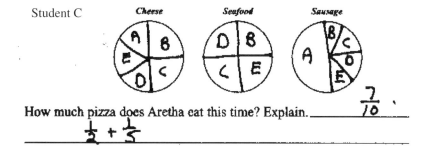

How much pizza does Aretha eat this time? Explain. ___$\frac{7}{10}$___
___$\frac{1}{2} + \frac{1}{5}$___

Here the student shows no evidence of knowing that unit fractions with the same denominator must each be represented by parts of equal size. The slice labeled A in the cheese pizza is not a representation of one fifth. Student C cannot adequately deal with the constraint of Aretha not liking seafood. The student does not seem to have noticed that Aretha's, Carlos's, and Dino's shares are not represented as having the same size. A and C both get a half-pizza and another piece. C's half is composed of a fourth of the cheese pizza and a fourth of the seafood pizza, and A gets half of the sausage pizza. But the portions being added to those halves are not the same size: A gets a share that looks like one sixth of the cheese pizza, and C gets approximately one eighth of the sausage pizza. D gets an even smaller "piece of the pie." Student C appears to be using ideas similar to those of the third graders, Alice and Adam: fractions of the form 1/*n* are represented by any one of *n* pieces of an object, but each piece is not necessarily of the same size. In this case, a correct computation may not include an increased understanding of fractional parts and their underlying relationships.

Teachers who only look at the written answer may think that Student D has a firm grasp of the material:

Student D

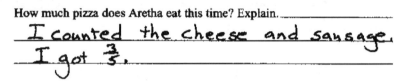

A response like this can be misleading, however. Good assessments allow teachers to probe further into issues of student understanding. Looking at Student D's drawing (below), one might wonder: Does the student have a grasp of equal-size pieces, or is the student just counting parts? How has the student made sense of "Aretha can't eat seafood"? Is this approach correct? What should be the "whole" for this task? This response also raises questions about experience with fractions: How often has Student D had the opportunity to grapple with the idea of the whole being several objects rather than one object?

Assessments allow us to carefully examine work across grade levels for trends. If we see many misconceptions from third grade still appearing in fifth grade, this raises important questions about instruction and instructional materials. Are the representations presented in textbooks helping students to develop the intended ideas? Would different representations help? Should we look at some of the materials from Japan and China, which quickly move from fractions of one object like a pizza, or of a collection of objects, like marbles, to fractions of quantities like distances or cups of sugar?

Middle School Work with Fractions

What happens in middle school when ideas of part-whole relationships are not fully developed? How do student conceptions progress as students move through the curriculum? Consider the grade 7 assessment task Mixing Paints.

Mixing Paints

Wayne is mixing paint.

He makes six quarts of brown paint by mixing equal quantities of yellow paint and violet paint. The violet paint is made from one-third red paint and two-thirds blue paint.

1. How much red paint does he use? _____ quart(s)

2. How much blue paint does he use? _____ quart(s)

3. What percentage of the brown paint is made from blue paint? _____

Explain how you figured it out.

On the surface, this task appears trivial. The first sentence says that there are three quarts each of yellow and violet paint; the second sentence says that the three quarts of violet paint were made from one quart of red paint and two quarts of blue paint. Two of the six quarts of paint used to make the brown paint were blue, so the answer to Question 3 is $33\frac{1}{3}\%$. Thus, from our perspective as adults the numbers are easy and the mathematics appears straightforward.

Yet, in a sample of almost 11,000 seventh graders, more than 50% of the students scored *zero points* on this task. That is, more than half the students were unable to identify the quantity of either red or blue paint. About 44% of the students could only find the quantity of red and blue paint. Only 18% of the students could do this and convert the amount of blue paint into a percentage of the total amount of brown paint — they had difficulty identifying the whole, which was the six quarts of brown paint. (Generally, students wrote that the answer to Question 3 was 66%. This may have been a conversion of the "two thirds" in the statement that the violet paint is two-thirds blue paint.)

Assessment items such as Mixing Paints provide a way for teachers to gain a sense of how students are thinking, and to see how students are (or are not) making sense of part-whole relationships, and what kinds of representations might help students better grasp those relationships. How do various representations facilitate student learning and deepen student thinking over time? Mathematics Assessment Collaborative teachers have come to appreciate the bar models used in Singapore and Russia, which help students visualize fractional relationships and make sense of rates. A benefit of bar models is that they can be modified to represent percents and other mathematical topics.

A question related to the part-whole issue explored in Mixing Paints is, "How do students' difficulties in understanding part-whole relationships and identifying the whole come into play when students are trying to make sense of percents, particularly percent increase or percent decrease?" Consider this grade 8 task:

Traffic

The daily average of motor vehicles using a certain freeway during 1997, 1998, and 1999 is shown in the table below.

	Year		
	1997	1998	1999
Average number of motor vehicles each day	54,800	61,700	73,400

1. In 1997, 15% of the motor vehicles traveling along the freeway were taxis. Calculate the average number of taxis traveling along the freeway each day in 1997.

Show your work.

2. Find the percentage increase in the average number of motor vehicles using the freeway each day from 1997 to 1999. Round your answer to the nearest whole number.

Explain how you figured it out.   _____

3. From 1999 to 2000, there was an average increase of 20% in the number of motor vehicles using the freeway.
Find the daily average of motor vehicles using the freeway in 2000.
Round your answer to the nearest hundred.   _____

Explain your reasoning.

In a sample of more than 6,000 eighth graders, 41% scored no points on this task. Slightly less than 50% of the students could successfully find 15% of the traffic in Part 1. Overall, less than 10% of the students could calculate accurately using percents. Many students who scored zero points tried to divide instead of multiply in Part 1:

Student E

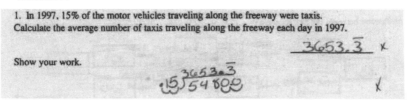

Other students learn rules for dealing with percent problems, such as "divide" or "move the decimal point." But what understanding of the underlying concepts do they show when they use those rules? Consider the work of Students F and G:

Student F

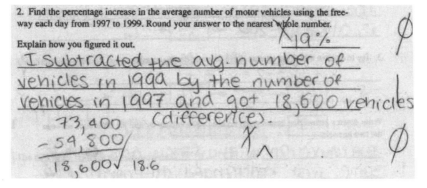

Student G

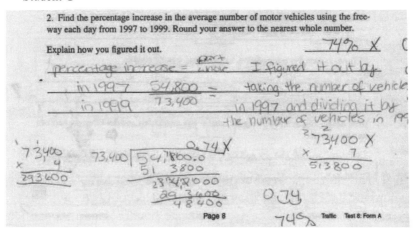

Often interesting assessment tasks can provide insight into the strategies used by successful students. For example, consider the task Fractions.

Fractions

Tom has four number cards.

1. He arranges his four cards to make fractions less than 1. Using each number card only once, make two fractions that have the same value.

and

2. Find a way to use two number cards to make a fraction less than $\frac{1}{2}$.

3. Find two ways to use two number cards to make fractions between $\frac{1}{2}$ and 1.

and

Although students were not asked to explain their answers, many students provided clear drawings. These drawings help to provide insights into what they know and understand, and how they make sense of fractions.

Some students used number lines to locate fractions and to judge relative sizes; others used pie diagrams. Note that the student whose work is shown at the bottom of this page also uses words to indicate that all the constraints in the problem are being met by the solution.

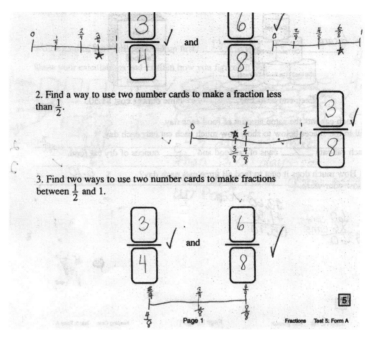

Fraction comparison via number lines, reduction to common denominator

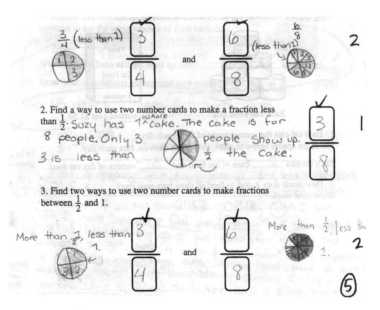

Fraction comparison via pie diagrams

A few students reduced fractions to lowest terms to check size, but still relied on drawings. Another strategy was to convert fractions to decimals.

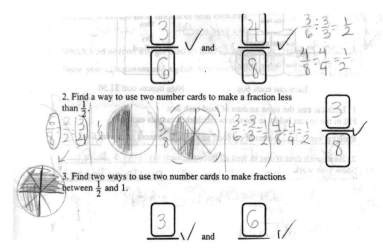

Fraction comparison via pie diagrams, reduction to lowest terms

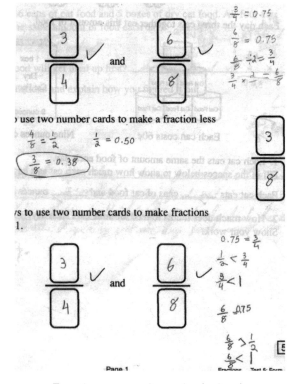

Fraction comparison via decimals

Some students have a clear understanding that "half of" a quantity means dividing that quantity by 2, and they use that understanding to compare fractions, writing, for example: "Half of eight is four and three is less than four, so $\frac{3}{8}$ is less than $\frac{1}{2}$ $(= \frac{4}{8})$."

As noted above, students' responses to assessment tasks yields insights into their thinking and into the advantages and disadvantages of particular curricula. As some of the previous examples indicate, giving students experience with a wide variety of strategies and representations provides them with a range of tools they can use to make sense of problem situations.

In summary, high quality assessments can provide teachers with useful information about student misconceptions related to: the role of equal parts in the definition of fractions; using and interpreting representations of fractions that display the whole; and understanding equality. These assessments can provide a window into student thinking, and help teachers think about their own instructional practices and concrete ways in which they might be improved. Good assessments help teachers reflect upon issues like: "What does it mean to understand an mathematical idea and how is assessing it different from looking at computational fluency? What is the value of having students draw representations themselves versus interpret representations provided in textbooks? How is the learning different?" They also give teachers the chance to look across grade levels and see how student thinking is or is not getting deeper and richer over time. Mathematically rich performance assessments provide insight into the strategies used by successful students. Of all the tools available to a student, which one does the student use to solve a complicated problem? When a teacher sees this, how can he or she turn these strategies into tools for all students?

Multiple-Choice and Constructed-Response Tasks: What Does a Teacher Learn from Each?

Two of the released items from the California Standards Test are given below.

1. What fraction is best represented by the point P on this number line?

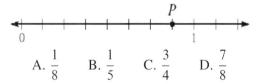

    A. $\frac{1}{8}$ B. $\frac{1}{5}$ C. $\frac{3}{4}$ D. $\frac{7}{8}$

2. Which fraction represents the largest part of a whole?

    A. $\frac{1}{6}$ B. $\frac{1}{4}$ C. $\frac{1}{3}$ D. $\frac{1}{2}$.

Let us view these tasks through the lens developed in the first part of this chapter. The question is: What does student work on these tasks tell the classroom teacher? For example, if you know that your student missed the first task, that tells you that the student may be struggling with number lines and the relative size of numbers represented as fractions. If the teacher knows that the student picked answer A, the teacher might understand that the student is making sense of partitioning into equal parts but is not understanding the role of the numerator. Yet the teacher can't be sure; perhaps the student merely guessed or picked that answer for a different reason. The reality of high-stakes testing is that the teacher does not see the task or receive information about which distractor the student selected. Feedback comes in the form of an overall rating: "your student is a basic (one step below proficient)" or "your student is not doing well in 'decimals, fractions, and negative numbers.'" This provides the teacher with little information that could be used to think about changing instruction or helping the student in question.

Now consider a very similar task in a constructed-response format.

Here is a number line.

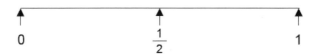

1. Mark the position of the two fractions $\frac{2}{3}$ and $\frac{2}{5}$ on the number line.

2. Explain how you decided where to place $\frac{2}{3}$ and $\frac{2}{5}$ on the number line.

3. Which of the two fractions, $\frac{2}{3}$ or $\frac{2}{5}$, is nearer to $\frac{1}{2}$? _____

   Explain how you figured it out.

A great deal can be learned when such tasks are used for purposes of formative (as well as final) assessment. First, as discussed above, the written prompts provide an opportunity for individual students to reveal their thinking, and for collective scores to reveal the strengths and weaknesses of the current curriculum. A scoring rubric can assign points for particular competencies. (For an example of such a rubric, see Hugh Burkhardt's chapter in this volume.) Individual scores on the problem can indicate individual strengths and weaknesses, and cumulative distributions of scores can reveal areas of the curriculum where students need help. School districts working with the Mathematics Assessment Collaborative generate graphs showing student score distributions. This allows the district to pinpoint places where curricular attention is needed, and to examine progress over the years. (See Elizabeth Stage's chapter in this volume for a similar discussion of what tests can reveal.)

Beyond the statistical data, actual student work can be used for purposes of professional development. Teachers can look at student work and try to infer the strategies used by students — thus developing a better understanding of what the students have learned. Consider, for example, what might be gained from a group of teachers looking at student work as follows:

Examine the work done by students H, I, and J in the next three figures.

- How would you characterize the strategies being used by these students?

- What did students need to understand about fractions to use these strategies? Could you give a name to these strategies?

Here is a number line.

0                    $\frac{1}{2}$ 1

1. Mark the position of the two fractions $\frac{2}{3}$ and $\frac{2}{5}$ on the number line.

2. Explain how you decided where to place $\frac{2}{3}$ and $\frac{2}{5}$ on the number line.

   I made all the fractions to have the same denominator and I put the fractions either before $\frac{1}{2}$ or beyond $\frac{1}{2}$. I got $\frac{12}{30}$ for $\frac{2}{5}$, $\frac{20}{30}$ for $\frac{2}{3}$, and $\frac{15}{30}$ for $\frac{1}{2}$. I compared those two numbers with $\frac{1}{2}$ or $\frac{15}{30}$.

3. Which of the two fractions, $\frac{2}{3}$ or $\frac{2}{5}$, is nearer to $\frac{1}{2}$? _____ $\frac{2}{5}$

   Explain how you figured it out.

   I made the fractions to have the same denominator and I compared them. $\frac{2}{3}$ was $\frac{5}{6}$ off of $\frac{1}{2}$. $\frac{2}{5}$ was $\frac{1}{10}$ off of $\frac{1}{2}$. So $\frac{2}{5}$ is closer.

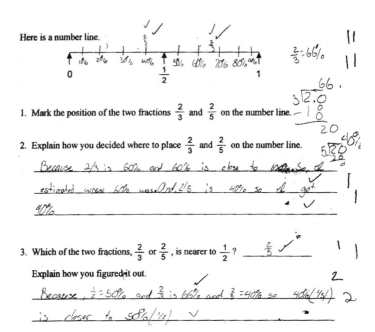

Here is a number line.

2/3 = 66%

1. Mark the position of the two fractions $\frac{2}{3}$ and $\frac{2}{5}$ on the number line.

2. Explain how you decided where to place $\frac{2}{3}$ and $\frac{2}{5}$ on the number line.

   Because 2/3 is 60% and 60% is close to ... So, I estimated where 60% was. And 2/5 is 40% so I got 40%

3. Which of the two fractions, $\frac{2}{3}$ or $\frac{2}{5}$, is nearer to $\frac{1}{2}$? ___ 2/5 ___

   Explain how you figured it out.

   Because, $\frac{1}{2}$ = 50% and $\frac{2}{3}$ is 66% and $\frac{2}{5}$ = 40% so 40% (2/5) is closer to 50% (1/2)

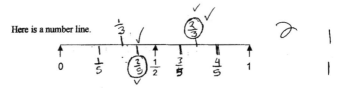

Here is a number line.

1. Mark the position of the two fractions $\frac{2}{3}$ and $\frac{2}{5}$ on the number line.

2. Explain how you decided where to place $\frac{2}{3}$ and $\frac{2}{5}$ on the number line.

   I divided the top part of the line into thirds and marked $\frac{2}{3}$ and divided the bottom part of the line into fifths and marked $\frac{2}{5}$.

3. Which of the two fractions, $\frac{2}{3}$ or $\frac{2}{5}$, is nearer to $\frac{1}{2}$? ___ 2/5 ___

   Explain how you figured it out.

   On the number line the dash that marks $\frac{2}{3}$ is farther from $\frac{1}{2}$ then the dash that marks $\frac{2}{5}$.

In discussions of such student work, teachers become increasingly attuned to diagnosing student thinking — to seeing not only what students got right or wrong, but how and why they got those things right or wrong. These understandings can lead to more effective teaching and improved student learning.

Chapter 15
Assessing a Student's Mathematical Knowledge by Way of Interview

DEBORAH LOEWENBERG BALL
WITH
BRANDON PEOPLES

Deborah Loewenberg Ball conducted the following interview with Brandon Peoples, a sixth grader, on March 8, 2004 at the first MSRI Workshop on Critical Issues in Mathematics Education.

This interview assessment of a student's mathematical understanding, conducted live in front of the assembled workshop participants, provides an immediate and vivid case of student thinking and exemplifies the interview assessment, an important mechanism for accessing student thinking.

———————————————

1 Ball: [To audience] Brandon and I are going to pretend you're not in the room, so I'm going to stop talking to you and we don't actually care that you're there, so, uh, goodbye. And uh, just be as quiet as you can because we actually want to do the work that we're setting out to do here. So, okay?

[To Brandon] So, you know some of the stuff we were doing, before people came, we might go back to some of that just because some of it was interesting and we didn't really finish talking about it. But I wanted to just start by, talking to you just a little bit about school. Do you remember when I called you the other day? We had a few minutes and I asked you a few things about your school and what you're working on this year? So can we just go back to that, because we were – we didn't have very much time that day. I was just wondering, like what you're doing in math right now in school.

2 Brandon: Really, we're learning about fractions.

3 Ball: Mm-hmm.

4 Brandon: And, umm, like – like we're doing – we're going on to like more – to more advanced of it instead of just – we're just moving on up the line of the fr – of math.

5 Ball: So can you give me an example of something you've worked on, I don't know, recently in fractions? Can you think of something you've been working on?

Time: 00:02:00

6 Brandon: Okay.

7 Ball: Do we have plenty of paper around? Do we have blank paper?

8 Brandon: Yeah.

9 Ball: Where is it? Oh, okay. No, you don't have to change, I just wanted to make sure we had enough.

[Brandon starts to give his example.]

10 Ball: Okay, so what did you write?

[Brandon has written $\frac{1}{2} \div \frac{1}{2}$ on the overhead.]

11 Brandon: One-half divided by one-half.

12 Ball: And you've been working on this?

13 Brandon: Mm-hmm.

14 Ball: So, what have you learned to do with that?

15 Brandon: See, it's the part – you have to – you have to multiply these two in, but instead you flip it upside down...

16 Ball: Uh-huh.

17 Brandon: So you, instead of one-half it'll be two over one.

18 Ball: Okay, can you show me?

19 Brandon: Mm-hmm.

[Brandon writes $\frac{1}{2} \times \frac{2}{1}$ on the overhead.]

20 Ball: Okay. And after that there is another step that you take? Or what do you do next?

21 Brandon: You multiply –

22 Ball: Uh-huh.

23 Brandon: – and then you get your answer.

24 Ball: Can you do that?

25 Brandon: Mm-hmm.

[Brandon writes $\frac{2}{2} = 1$ on the overhead.]

26 Brandon: Which equals this to this or one whole.

27 Ball: Okay, umm, is there – did you make any pictures of that or anything like that or did you just kind of learn how to do it the way you just showed me?

28 Brandon: Well it's – I learned different ways. I tried to come up with different strategies, but this was the better – this just adding and stuff is a better idea.

29 Ball: Well maybe we'll talk about that a little bit more later. Do you remember – have any recollection or memory of when you first started to work on fractions at school? Like what grade it was?

30 Brandon: I think we started to – I think we started on in fifth grade –

31 Ball: Mm-hmm.

32 Brandon: because we didn't – we didn't learn – our teacher in fourth grade didn't – we really didn't learn fractions.

33 Ball: Mm-hmm. And what's the new stuff this year? Like this I guess, right?

34 Brandon: Yeah, and, umm, this thing of my other math teacher, he calls it cross-canceling.

35 Ball: Mm-hmm... You mentioned that on the phone. What is that exactly?

36 Brandon: It's like when instead of, like, going through all this reducing, instead it's like two over – two over four times, umm, three sixths.

[Brandon writes $\frac{2}{4} \times \frac{3}{6}$.]

37 Ball: Mm-hmm.

38 Brandon: See, you'll want it – you'll want to reduce that to maker it smaller instead so – and then you get – you come out with a fraction.

39 Ball: Mm-hmm. How do you do that?

40 Brandon: So I think you... And then you multiply.

[Brandon has written $\frac{\cancel{2}^{1}}{4} \times \frac{3}{\cancel{6}_1} = \frac{3}{4}$.]

41 Ball: Okay, so can you explain what you just did?

42 Brandon: Mm-hmm. Umm, two-sixths can be reduced to? I mean oops.

[Brandon corrects what he has written: $\frac{\cancel{2}^{1}}{4} \times \frac{3}{\cancel{6}_3} = \frac{3}{4}$.]

43 Brandon: Three – two sixths can be reduced to one-third, so I put a cross through the two, I put a one and cross this out and put a third. Multiplied one time – by three, and multiplied four times one.

44 Ball: But you have a three there now.

45 Brandon: Oh. Hold on, I think I didn't do it right? three-twelfths, so it should be? Hmm.

[Brandon crosses out $\frac{3}{4}$ and writes $\frac{3}{12}$ directly to the right of it.]

46 Brandon: Let me see it's three-twelfths but it can be reduced to I think it's one-fourth or one-third. How did I get to ... ? Wait, hold on. Something isn't right.

[Brandon writes $\frac{1}{4}$ directly to the right of $\frac{3}{12}$.]

47 Ball: Is that chair too high?

48 Brandon: Yeah, I think so ...

49 Ball: So, do you want to leave this right now or do you want to work on this some more right now?

50 Brandon: We can work on it more.

51 Ball: You want to go back to the beginning then and tell me what you were trying to show me?

52 Brandon: I think I was trying to redu – reduce it, but then I got the wrong answer. I redu – I got – I add – I multiplied it, got an answer, but the answer could be reduced too.

53 Ball: What was the original problem that you wrote down?

54 Brandon: Was two-fourths times three-sixths.

55 Ball: Uh-huh. And do you – do you already have a sense of what you think the answer's supposed to be to that?

56 Brandon: Yeah.

57 Ball: What is it you think it's supposed to be?

58 Brandon: Well, I thought it was three-fourths, but it wasn't real, so I, it was ...

59 Ball: Why did you think it would be three-fourths?

60 Brandon: Because, umm, I put a one instead of a three.

61 Ball: Mm-hmm.

62 Brandon: I thought because, umm, two-sixths was supposed to be reduced to one-third, so –

63 Ball: Uh-huh.

64 Brandon: – but can re – it's redu – when it's reduced to one third.

65 Ball: But you wrote one fourth. Do you mean one-fourth or one-third?

66 Brandon: One fourth. Yeah.

67 Ball: One-fourth. So now you're no – which answer are you saying is correct? The three-fourths you originally wrote or the one-fourth you just wrote?

68 Brandon: One-fourth.

69 Ball: Okay. Maybe – let's save this one and maybe – maybe if there's some time when we've done some other work we'll come back to it. Is that okay? 'Cause right now we're just trying to get a sense of the different things you've been working on.

70 Brandon: Mm-hmm.

71 Ball: Do you ever use fractions any place besides in school?

72 Brandon: Mm-hmm, not really.

73 Ball: Does anybody? Do you know anybody who uses fractions anyplace other than in school?

74 Brandon: Mm-mm.

75 Ball: Uh-huh. So, why do you think you learn about fractions?

Time: 00:08:23

76 Brandon: I think so it's like – it's kinda – it kinda takes, umm – it's kinda in relation to percentages.

77 Ball: Mm-hmm.

78 Brandon: Because like – because like one-fourth would be to – in percentage, it would be 25 percent.

79 Ball: Mm-hmm.

80 Brandon: And in, umm, decimals it would be zero-point-two-five

81 Ball: Mm-hmm.

82 Brandon: So it's – I think it's one way we can tell, umm, how much is something.

83 Ball: Okay.

84 Brandon: Or – or like how much, or what is – what percentage of it is.

85 Ball: Okay.

86 Brandon: Like – it's like different things contain different things. So like one part of something would be a percent – a percentage or a fraction. Same thing.

87 Ball: Okay. Umm, I think we're going to shift over and do some other problems and questions and things like that and I just wanted to remind you of something I told you before we started today which is anything, you've kind of already doing it, but anything I ask you, you can draw, you can write, you can use words. Any – anything that you've ever used before you can do, or anything that you think you can use – the main thing I'm trying to understand is how you're actually thinking and what you're thinking about. So even if I don't, you know, suggest using these blocks, or I don't suggest using a

drawing, you're totally free to do that. You can use the board, we can use this thing. Okay, do you remember when I was talking about that?

88 Brandon: Mm-hmm.

89 Ball: And, um, it's really not a test, I'm actually just trying to do some mathematics problems with you and see the ways that you think about them, kind of like you've already started to do. So you're doing exactly what I was hoping you would do. And the more you can tell me about kind of how you're thinking and why you're thinking certain things, the more helpful that will be in what I'm trying to learn about watching you work. Okay?

90 Brandon: Mm-hmm.

91 Ball: So, umm, I think I'll start with, umm – can you hand me one of those green pieces of paper? And maybe we'll need clean paper to write on too, so can I have another one?

92 Brandon: Mm-hmm.

93 Ball: Thanks. Okay. Can you fold that piece of paper in half?

[Brandon folds paper in half.]

Okay. So can you just explain how you knew what – what to do when I asked you that? Like, how did you decide how to do it?

94 Brandon: Well, umm, you take it like this and then draw down the line so – I know that it would be half 'cause it's only two –

95 Ball: Mm-hmm.

96 Brandon: – it's only two uh, two major things. So like 'cause if it was like – if it was like this, then it would be in fourths [referencing his folded paper]. But since it's – I only see two sides then I know it's just half. They're both halves.

97 Ball: If I asked a little kid to do that, like a five year old, do you think they would be able to do it too? Or do you think they might make a mistake when they did it?

98 Brandon: I think they would get, like, how to fold it in half, but I don't think they could explain how would it be half.

99 Ball: Can you think of any mistakes a little kid might make? In folding it? 'Cause you did it right and I'm curious if you can think of anything a little kid might do that would be a mistake that you didn't make.

100 Brandon: They would probably – they would probably fold it a different way –

101 Ball: Mm-hmm.

102 Brandon: – 'cause they might not know what a half is, or what like, what would – what would this be.

103 Ball: Like, what if they did something like this? Can you picture a little kid doing that?

[Ball folds a sheet of paper incorrectly: One side is obviously larger than the other.]

104 Brandon: Yeah, 'cause –

105 Ball: Why would they do that?

106 Brandon: 'Umm, probably 'cause I think they probably might not know what half is. They might just fold it. They might just fold it in their way.

107 Ball: Mm-hmm.

108 Brandon: Instead of like, er, fold it not like – not like this but like one side would be bigger and one would be smaller.

109 Ball: So another thing you – you – you were trying to do what to make the parts be the same?

110 Brandon: Mm-hmm.

111 Ball: Okay [referencing the folded paper]. So can you write down how much of the piece of paper is on this side, or can you just tell me how much that is? Just this amount of the paper. How much of the paper is that?

112 Brandon: Wh – half or, umm, that's like fifty percent?

113 Ball: Can you write those down?

114 Brandon: Okay.

115 Ball: You can use this paper to write with.

[Brandon writes 50% on the left side of the paper.]

116 Ball: Okay. And what's the other thing?

117 Brandon: 50 percent.

[Brandon writes 50% on the right side of the paper.]

118 Ball: Okay. And you also said half. How would you write half?

119 Brandon: Half. Umm . . .

[Brandon writes on the right and left side of the paper.]

120 Ball: Okay. Is there any other way you could write it?

121 Brandon: Yeah.

122 Ball: Oh I see, you're writing it for each side, right? Is that what you're doing?

123 Brandon: Mm-hmm.

[Brandon writes 0.50 on the right and left side of the paper.]

124 Ball: Okay. You mentioned – A minute ago you mentioned that you could fold it into fourths. Can you show me that one? How you'd do that?

[Brandon folds the same paper into fourths.]

125 Ball: Okay, so what did you do to that?

126 Brandon: Umm, it was already like this from the half, so I folded that into – into right here –

127 Ball: Mm-hmm.

128 Brandon: – and turned it into fourths.

129 Ball: Why did it turn into fourths?

130 Brandon: Cause since it – it already had a line right here from the – from half, so if I fold it in this one it'll have another line right here and that separates them into fourths.

131 Ball: Okay. So what – how would you – how could you explain – how could you write down what amount that each of those portions of the paper are? Do you want that pen?

    Can you put it up here? [Referring to the projector.]

132 Brandon: Oh, sure.

[Brandon moves paper up to projector. He writes 0.25 and $\frac{1}{4}$ on the top left quarter of the paper.]

    Or umm?

[Brandon writes 25% on the top left quarter of the paper.]

133 Ball: Okay. Now when you think about, umm, the paper you just had – this one – are there any other fractions you could use to represent how much? Actually we should – let's fold this one. I think I should've just let you do it on the one you had. Okay, are there any other fractions you can use to write down how much this part of the paper is? What else could you write?

134 Brandon: Umm, you could write other – You could write, umm, one number and what would be half of it.

135 Ball: Like what?

136 Brandon: Like, umm, say for all of this could be – like, all of this could be thirty-eight, but, like only – only nineteen of it – this is nineteen and this is nineteen.

137 Ball: So how would you write that fraction?

138 Brandon: Ummm.
[Brandon writes $\frac{19}{38}$ on the right and left side of the paper.]

139 Ball: Okay, and – go ahead. Okay, and is there another one you can do?

140 Brandon: Hmm. I'm thinking of one.

[Brandon writes $\frac{32}{74}$ on the right and left side of the paper.]

141 Ball: So how did you do that one?

142 Brandon: 'Cause, umm, wait? No, actually no, it's thirty-seven. Thirty-seven.

[Brandon corrects both $\frac{32}{74}$ fractions to $\frac{37}{74}$.]

143 Ball: So how – how did you do this one?

144 Brandon: Umm. 'Cause this – all of this could be equal to – is seventy-four, but half of it would be thirty-seven.

Time: 00:16:44

145 Ball: So what's the general thing you're doing? You seem to have some way you can always write a fraction that's that amount. What's the general thing you're doing?

146 Brandon: It's like, umm, see the whole would be seventy-four, but umm... like that if you – Since it's one side is this and one side is that, when you add these two together it equals that, so – so right there you know that half of it is thirty-seven, the other half is...

147 Ball: Okay. Can you do similar things with one-f – with the fourths? What would be an example of something like that? Can you write another fraction for one-fourth?

148 Brandon: Okay.

149 Ball: How would you do it with fourths? Just a minute. Okay. [adjusts paper on the projector]. Okay, so what did you do this time?

[Using the paper folded into fourths, Brandon has written $\frac{2}{8}$ on the overhead.]

150 Brandon: Umm, two-eighths 'cause it takes four – it take four twos to equal eight, so two would be 25 percent –

151 Ball: Okay.

152 Brandon: – or one-fourth.

153 Ball: Okay. Let's go to a different... can you get another piece of paper? Let's do one more of these paper-folding things. Do you have a blank one over there somewhere?

154 Brandon: Yeah.

155 Ball: Could you fold the piece of paper in thirds?

156 Brandon: Thirds? I think so.

157 Ball: I'm sorry that chair is too tall. Do you remember how we adjusted it?

158 Brandon: Yeah...

159 Ball: Oh, I know. It's that lever over on your left – on your right. I think if you – If you pull on that . . . Are you uncomfortable? Okay.

[Brandon folds paper in thirds. He estimates the first fold, and sees that the next fold will not come out even with the paper's edge. So he then adjusts the first fold to correct before finishing.]

160 Ball: You can put it on the desk if you want. Okay, so how did you do this one?

161 Brandon: Umm . . . One – I put one right here and then folded it.

162 Ball: It seemed like this was a little bit more difficult than folding the half or the fourth. Why was that?

Time: 00:19:00

163 Brandon: 'Cause you – you can't exactly – See it – with thirds it's like you try to fold it right – try to fold it right – in right where you, umm, put the crease into it so you know that all the sides would be even, but with fourths all you have to do is just fold it like this –

164 Ball: Mm-hmm.

165 Brandon: – and – but half is just like this.

166 Ball: What do you get if you fold it like you did – remember when you had the half and you folded it this way? What do you get if you do that this time?

167 Brandon: You get sixths?

168 Ball: Okay. Can you write, umm, like [we just worked – we'll spend a minute on this] can you write what fraction would you put for that portion?

169 Brandon: I think this would be . . .

[Brandon writes $33\frac{1}{3}$ in each third of the paper.]

170 Ball: Thirty-three and a third is . . .

171 Brandon: Mm-hmm.

172 Ball: So there – this is . . .

173 Brandon: Because when – if you just – 'Cause all this is – I think it's one-hundred, but when you put thirty three into it, it's – it's – it – it'll go on and on with threes –

174 Ball: Mm-hmm.

175 Brandon: – so what would be better is to just put one-third.

176 Ball: Okay. Are there any other fractions you can write for this?

177 Brandon: I don't think . . . Mm-mm.

178 Ball: No? How much would you write for this much of it; for two of the sections?

179 Brandon: So that'll be six, so . . . Umm. I'm not sure.

180 Brandon: Umm.

[Brandon folds the paper into sixths as he is working on finding an answer.]

181 Brandon: I think sixteen.

182 Ball: You think what?

183 Brandon: Sixteen.

184 Ball: Sixteen?

185 Brandon: Mm-hmm.

186 Ball: How would you write it?

187 Brandon: 'Cause – since this is thirty-three, then I think it would be sixteen because I just thought that if fifteen – it's – it's almost, umm, half of thirty-three, so if this was sixteen ... So add it all together they would equal – I think they would equal ninety-si –

[Brandon writes 16 into each sixth of the paper.]

188 Ball: Mm-hmm.

189 Brandon: No. Ninety-six?

190 Ball: Mm-hmm.

191 Brandon: So in – if I put seventeen then they'll be one-hundred and something, so I'll be over.

192 Ball: Okay. So let's just keep these available for a minute. So you have this one – here's the one you made in halves,

193 Brandon: Mm-hmm.

194 Ball: okay, here's the one you made in fourths, and this one has what? Thirds and sixths?

195 Brandon: Mm-hmm.

196 Ball: Okay, so I'm going to write a fraction down and then I'd like to see if you can show me what part of one of these pieces of paper you could use for that fraction.

197 Brandon: Mm-hmm.

198 Ball: So give me a clean piece of paper please. I'll show you – we'll start with one that's easy.

199 Ball: So here's one you already did, just you see what I'm saying. If I write that fraction, can you – here, let's put these on your side. Can you pick up a piece of paper and show me which – how to use one of those pieces of paper to show one-half?

[Ball writes $\frac{1}{2}$ on the paper.]

200 Brandon: This one?

201 Ball: Okay. So now I'm going to write a different one, okay? Can you use one of your pieces of paper to show that?

[Ball writes $\frac{2}{4}$ on the paper.]

202 Brandon: I think this one solves both of those.

203 Ball: Okay, how?

204 Brandon: Because when one-half is two-fourths – when it's reduced – so it's the same thing, only just the numbers are bigger.

205 Ball: Okay. What about this?

[Ball writes $\frac{2}{3}$.]

206 Brandon: I think [when it] – this in th – in thirds . . .

207 Ball: Mm-hmm

208 Brandon: . . . this in thirds, so it would be these two – these two parts would be two-thirds

209 Ball: Okay. What about . . . ?

[Ball writes $\frac{4}{4}$.]

210 Brandon: That would equal a whole.

211 Ball: Okay. Okay so how do you know that was a whole?

212 Brandon: Because, umm, all fractions are pieces of a whole, so – just – I knew that half wouldn't be, so – but all of those just equal into one.

213 Ball: Okay. So what if I wrote down this fraction? Could you show me that with the paper?

[Ball writes $\frac{3}{2}$.]

214 Brandon: Wait, how would it go into it?

215 Ball: Could you show me that much paper with the . . . 'Cause you've been showing me one-half of a piece of paper, you showed me two-thirds of pa – of a piece of paper. Can you show me that much paper?

216 Brandon: Mm-hmm.

217 Ball: Can you put your paper up here?

218 Brandon: Mm-hmm.

219 Brandon: You would have to put two into three in order to find what – how mu – how mu – how much a piece of pach – paper would be. So it'd be one and a half.

[Brandon has written out the long division of 3 by 2:]

$$2\overline{)3} \atop \begin{array}{r} 1 \\ \underline{2} \\ 1 \end{array}$$

220 Ball: Okay. So how would you show that with the paper?

221 Brandon: It's the whole and this part would be a half.

222 Ball: Okay. How would you read that number that I wrote? Can you read that number?

223 Brandon: Three-twos, I think.

224 Ball: Okay. So, umm, let's, uh, work on the board for a few minutes. It's getting kind of boring sitting here. Umm, where's our chalk? I thought we had some chalk somewhere. Here. So, umm, here's where we have a bunch of fractions we were just talking about. Can you, umm, take, let's say one-half and two thirds. Could you write them up there?

[Brandon writes $\frac{1}{2}$ and $\frac{2}{3}$ on the chalkboard.]

225 Ball: Okay. So now which one of those two do you think is great – is larger?

226 Brandon: Larger? Umm. I think one-hal – I think two-thirds?

227 Ball: Why do you think two-thirds is larger?

228 Brandon: Because half would just be half of something, but one-and-a-half is – is half a third, so – but it's two – so, two would be bigger.

229 Ball: Okay. What about if we add, uh, three-fourths to that? Where would you put three-fourths?

230 Brandon: Let me see.

[Brandon writes $\frac{3}{4}$ to the right of $\frac{1}{2}$ and $\frac{2}{3}$.]

231 Ball: Okay. So what are you saying about three-fourths?

232 Brandon: I think this would be a little bit bigger.

233 Ball: Than?

234 Brandon: Than ... Three-fourths would be bigger than two-thirds 'cause, umm, it's just – this is just over one-and-a-half by a half, but this is over – this – 'cause this is two thirds. Two-thir – I mean two fourths would be half, but three is over two so it's over half by one ...

Time: 00:27:20

235 Ball: Okay. All right, let's sit back down again for a second. I'm going to show you some pictures of some different things. We'll get away from the paper

folding for a minute. And then maybe we'll do a little more comparing, like we were – remember we were doing it before this morning.

236 Brandon: Mm-hmm.

237 Ball: So, umm, like this for example. Can you tell me if this whole thing is the whole?

[Ball shows Brandon this figure:]

238 Brandon: No.

239 Ball: Okay. So can you tell me what that fraction would be?

240 Brandon: It would be . . .

241 Ball: The shaded part is what I'm talking about.

242 Brandon: Oh.

243 Ball: Sorry.

244 Brandon: Sixty percent?

245 Ball: Fifty percent?

246 Brandon: Sixty . . .

247 Ball: . . . Percent?

248 Brandon: Mm-hmm.

249 Ball: Why do you think it's sixty percent?

250 Brandon: 'Cause five – it's five whole things and all those equal to one, so each one is worth twenty percent.

251 Ball: Mm-hmm.

252 Brandon: So since it's three of those it's sixty.

253 Ball: Okay. Is there a fraction you could write for that?

254 Brandon: Mm-hmm.
    [Brandon writes $\frac{60}{100}$.]

255 Ball: Okay. What if somebody wanted to write a fraction that had a five in the denominator since, like you said, there's five parts? What fraction could you write if that was a five in the denominator?
    [Brandon writes $\frac{3}{5}$.]

256 Ball: Okay. What if there was a ten in the denominator? Could you make a fraction that had a ten in the denominator?

257 Brandon: Ten? Would you – would you call these ten?

258 Ball: No, these would – You have five parts. How could you make it so that, umm, you represent it as a number of tenths? You can change the drawing if you want to.

259 Brandon: I think you can dr – you draw a line through this.

260 Ball: Okay, go ahead. You can do that.

[Brandon draws a line through the figure:]

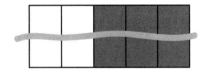

261 Ball: Okay. So now could you write a fraction that was over ten? To –

262 Brandon: Yeah.

263 Ball: – represent the shaded part?

264 Brandon: Mm-hmm.

[Brandon writes $\frac{6}{10}$.]

265 Ball: Okay. How did you decide how to do that?

266 Brandon: 'Cause if – when I put a line through it, it makes it ten so this, even though these – all this is – is six, it's six of them shaded with the line going through it. And without the line it's – it's just, umm, three over five.

267 Ball: Okay. Umm, of these three fractions you've written, which one's the largest?

268 Brandon: Three-fifths?

269 Ball: Why is it the largest?

270 Brandon: 'Cause three-fifths is like – like we said earlier, it – 'cause one-hun-dredths are really small . . .

271 Ball: Mm-hmm

272 Brandon: I mean it – In – I – my opinion it's not – it's not about the numera-tor, I think it's about the denominator.

273 Ball: Tell me – I know you were telling me that earlier, so keep going 'cause we didn't really get to talk about that.

274 Brandon: Okay. So this would –

275 Brandon: That would be fifths

[Brandon draws:]

276 Brandon: and – this is tenths,

[Brandon draws:]

277 Brandon: so – so it's six of that, even though the numer – the numerator's bigger than this numerator, it – my opinion is that the denominator – how big
the denominator determines how – how big the fraction – the whole fraction
is.

278 Ball: So, what were we talking about earlier? Do you – is that paper around
still? We were comparing, umm, let's pursue this a little bit. Is that one-half
and seven-eights, or something like that? Or one-quarter and seven-eighths?

279 Brandon: Oh. They were on the – they were on those little cards.

280 Ball: Oh yeah, that's right. Is it over by you?

281 Brandon: I'm not sure.

Ball and Brandon transition from working on the projector to working on the
chalkboard. In this next segment Ball continues to lead Brandon through a
series of activities where he draws pictorial representations of selected fractions
to explain his understanding of the numerator and the denominator.

---

282 Ball: Here they are. Yeah. Here, why don't – You want to put them up on the
board?...Okay. So I asked you, I think, to put them up and put the smaller
one on the, uh, right. I meant on the left, yeah. So you put them up, like...
oh okay, is that what you meant before?

283 Brandon: No.

284 Ball: Okay.

285 Brandon: 'Cause this would be...

[Brandon places the $\frac{7}{8}$ card on the left of the $\frac{1}{4}$ card on the board.]

286 Ball: Okay, you put them sm – okay, right, sorry. Okay. So then tell me again what you were explaining, because I didn't f – I didn't completely – we didn't get to really finish talking about it.

287 Brandon: Umm. What I understand is that fourths – fourths would be like that [Brandon draws:]

288 Brandon: and eights is like this.
[Brandon draws:]

289 Ball: Okay.

290 Brandon: So s – Like, I said, umm, I think that f – the d – the, umm, denominator determines how big the whole fraction is.

291 Ball: Okay. So where does that lead you with these two fractions?

292 Brandon: 'Cause fourths – 'cause, like, fourths are bigger than eights so –
[Brandon colors the circle:]

293 Brandon: so even though it's just...
[Brandon colors the circle:]

294 Brandon: So I think that –

295 Ball: So that's your seven-eighths...

296 Brandon: Uh-huh. And this is one-fourth.

297 Ball: Okay. But when I look at your drawings, it looks like you've shaded more for your picture of seven-eights than for one-fourth. That's the part I'm not completely understanding.

298 Brandon: 'Cause – because it's – you have fourths – I meant – I mean eighths is – eighths are a lot smaller, so seven of them would have to – you have to shade in 'cause you couldn't put seven into four.

299 Ball: Okay. So then you have this piece of paper from earlier you – remember where you divided the paper into fourths and you told me that one of these could also be – you could write it as two-eighths, right?

300 Brandon: Mm-hmm

301 Ball: So what if we say instead of one-fourth, I take – 'cause you said that would be the same, right?

302 Brandon: Mm-hmm

303 Ball: Do you s – do you still think that's the same?

304 Brandon: Right.

305 Ball: So what if I put two-eighths here instead?

[Ball moves the $\frac{1}{4}$ card down and writes $\frac{2}{8}$ to the right of the $\frac{7}{4}$ card.]

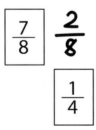

306 Ball: Now how would you compare seven-eighths and two-eighths? 'Cause then the denominators aren't different?

307 Brandon: I think that seven-eighths would be bigger.

308 Ball: Okay.

309 Brandon: Because they're – they both have eighth, but, umm, si – I mean seven would be bigger than two. They have the same denominator.

310 Ball: Okay. So in general, when they have the same denominator, how do you compare them?

311 Brandon: By the numerator.

312 Ball: But when the no – denominators are different, what is it you do?

313 Brandon: I compare them by the denominator.

314 Ball: Okay. So then this is the part I'm not – I don't think get completely, 'cause if I understood you correctly, you said two eights was the same as one-fourth. Or were you not saying that?

315 Brandon: Umm. It's –

316 Ball: Do you have a picture of two-eighths up on the board? Can you make a picture of this? [She points to the $\frac{2}{8}$ written on the board.]

317 Brandon: Okay.

[Brandon draws on chalkboard:]

318 Ball: Okay.

319 Brandon: So ...

[Brandon finishes his drawing.]

320 Ball: Okay. So now this is your picture of one-fourth. Can you label it?

321 Brandon: Mm-hmm.

322 Ball: Your picture – this is a picture of ... Okay. So now, what I'm trying to understand is when you compare these two [pointing to Brandon's drawings of $\frac{1}{4}$ and $\frac{2}{8}$], what do you conclude about which one's bigger?

323 Brandon: How do I determine it?

324 Ball: Yeah. How do you decide? Like, which one is bigger? 'Cause I thought you were saying they were the same 'cause you labeled this one one-quarter and then you labeled this one two-eighths, so I thought you were saying those are different ways to write the same amount? [Showing Brandon the paper he had written on previously showing that $\frac{1}{4}$ is the same as 0.25]

325 Brandon: No, it's 'cause this would be bigger, but it can be – what I'm saying is that it can go smaller.

326 Ball: Say more about – what do you mean, "it can go smaller"?

327 Brandon: Like ... I can't really decide on how big they are when they're reduced.

328 Ball: Mm-hmm. When you look at these two pictures, which one do you think is greater: two-eighths or one fourth?

329 Brandon: Umm ... One-fourth?

330 Ball: Why do you think one-fourth?

331 Brandon: Umm. 'Cause it has – it has bigger chunks into it to make fourths, so – but these are all, like li – sorta small, so just one out of four is bigger than two out of eight.

332 Ball: Okay. Let's, umm, let's go over and use this line. Remember when I talked to you on the phone and you said you don't usually use the line so much?

333 Brandon: Oh a number line?

334 Ball: Yeah. Or do you – did you say you did use it?

335 Brandon: I don't think we used it.

336 Ball: Okay. I'm just going to mark a few points and then we can go from there. So I'm going to call that point zero, and then I'm going to call this one one, okay? And then I'll call this one two, and that's all I'm going to put on there for right now, okay?

[Ball draws on the board:]

337 Brandon: Mm-hmm.

338 Ball: So. Like where would three be if we were to put it on?

[Brandon adds a 3 to the number line:]

339 Ball: Okay. Now do you think there's some numbers between these?

340 Brandon: Mm-hmm

341 Ball: Like what?

342 Brandon: Umm...

[Brandon begins to draw on chalkboard. He adds three lines between 0 and 1.]

343 Brandon: So I think that this part would be zero...

344 Ball: Mm-hmm

345 Brandon: And this would be – I think it would be – so what number would you declare it to be whole?

346 Ball: Well why don't we – what would you call this number? Do you a n – a name – what you would call this [pointing to the line Brandon has drawn in

the middle, between 0 and 1]? If this is one – this still has to be one and that's zero so what would this be? [Again, referring to the middle line.]

347 Brandon: I think this would be half.

[Brandon writes:]

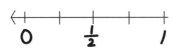

348 Ball: Okay. So then...

349 Brandon: Just half.

350 Ball: Just half. And why did you think that would be one-half?

351 Brandon: Because this would be, umm, this would be...

[Brandon writes:]

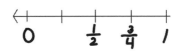

352 Brandon: So it's – so like this is half [pointing to the middle line]...

353 Ball: Mm-hmm

354 Brandon: This is like – this – this side is bigger – I mean the same – they have the same side, but this is closer to one, so...

355 Ball: Okay. Then what's that one there then?

[Brandon writes:]

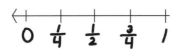

356 Brandon: One-fourth.

357 Ball: Okay. And how did you decide to label this one one-fourth and this one three fourths?

358 Brandon: 'Cause one fou – I mean... 'Cause four of these equal one, so this is just one section of it, so it'd be one, and this would be half point, and this is fourths, and then this would equal one.

359 Ball: What did you mean four of them equal one? What did you mean by four of them?

360 Brandon: Four is... So – I probably should change it. I need an eraser.

Time: 00:38:34

[Brandon changes his number line:]

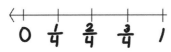

361 Ball: Why did you change that?

362 Brandon: 'Cause, umm, I wanted them all to be fourths...

363 Ball: Oh.

364 Brandon: ...so we wouldn't get confused, so...

365 Ball: How did you decide the denominator should be fourths? Why not fifths or sevenths or something like that?

366 Brandon: Well I think... Because in some numbers you can't – you can't put the numerator into the denominator...

367 Ball: Mm-hmm

368 Brandon: ...so – but this would be easier.

369 Ball: So, was one-half wrong where you had it? I mean is – is this point – could you call this one-half also?

370 Brandon: Mm-hmm

371 Ball: So why don't you put it back, like underneath it?

[Brandon had erased the $\frac{1}{2}$ he had written and replaced it with $\frac{2}{4}$. Ball was now asking that he add $\frac{1}{2}$ back to his number line, positioning it just below the $\frac{2}{4}$.]

372 Brandon: Okay.

[Brandon writes:]

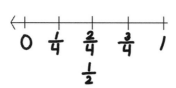

373 Ball: Is there any other fraction you could call that point besides two-fourths and one-half?

374 Brandon: Umm, yeah. I call it – you can call it fifty-out-of-a-hundred, or...

375 Ball: So why don't you write fifty-out-of-a-hundred too, then.

[Brandon writes:]

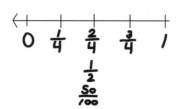

376 Ball: So this is – goes back to the thing you were telling me earlier in general. Like you can take some number and then half of it and then...

377 Brandon: Mm-hmm

378 Ball: Okay. So, uh, what number would be here? [She draws a mark on the number line, half way between the 1 and the 2.] Here, I'll use a different color from you. What number would you say would be here?

379 Brandon: This would be half point 'cause it's the same size or bigger, so, umm...

[Brandon writes:]

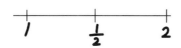

380 Brandon: So it's like – it's not exactly right here [pointing to a space between $\frac{1}{2}$ and 2] where it's like three out of four and then it's not exactly one out of four [pointing to a space between 1 and $\frac{1}{2}$], so it's half.

381 Ball: Are you saying it's the same number as that number? [She points to the $\frac{1}{2}$ mark between the 1 and the 2 and refers back to the $\frac{1}{2}$ mark between the 0 and 1.]

382 Brandon: Yeah.

383 Ball: 'Cause you have one-half there.

384 Brandon: Mm-hmm.

385 Ball: But this number is greater than one. See, it's past the one [indicating that the new mark between 1 and 2 lies to the right of the 1 mark].

386 Brandon: Yeah.

387 Ball: So, is it just one-half?

388 Brandon: It'll be one-and-a-half.

[Brandon writes:]

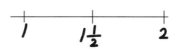

389 Ball: Oh, it'd be one-and-a-half. Okay, so if we repeat the same thing you did before, could you label that one and that one?

390 Brandon: Mm-hmm.

391 Ball: What would that be?

[Brandon writes:]

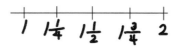

Time: 00:40:42

392 Ball: Okay. So, umm, is there anything else you could write – you know how you wrote different ways to write one-half? Are there for one-and-a-half?

393 Brandon: Yeah.

[Brandon writes:]

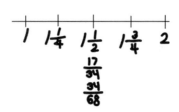

394 Ball: Do you need the one in front of it, or you don't need the one in front of it?

[Brandon shakes his head in the negative.]

395 Ball: So this is just seventeen thirty-fourths?

396 Brandon: Mm-hmm. It's – that's half.

397 Ball: Okay.

398 Brandon: Mm-hmm.

399 Ball: But it's greater than one, right? So...

400 Brandon: Mm-hmm.

401 Ball: So...

402 Brandon: It's one.

[Brandon writes:]

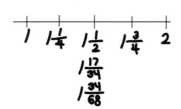

403 Ball: Okay. All right. So, umm, can you think of any other fractions that you could put up here? Like, is there any fraction that goes between one-fourth and two-fourths, for example?

404 Brandon: Umm, mm-mm. No.

405 Ball: No? Er, have we put all the fractions up here that we can?

406 Brandon: Yeah.

407 Ball: Well what if we wanted to put one-third up there? You think it couldn't – we couldn't put it on there?

408 Brandon: Sure, but we would probably have to put them into thirds instead of fourths.

409 Ball: Okay, so you can do that. Here, let's – what color were you using? Pink? Here, just use this.

410 Brandon: Okay.

411 Ball: Can you – can you mark it so that you'd be able to put thirds one there?

412 Brandon: Do I need to erase this?

413 Ball: You don't have to erase it.

414 Brandon: Okay.

415 Ball: You can kind of ignore what you've done; now concentrate on this. Okay, so what would you label those two points?

[Brandon writes:]

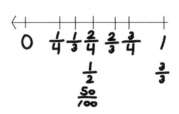

416 Brandon: This section would be one-third. This would be two-thirds. And then this would be three-thirds right here.

417 Ball: Okay.

418 Brandon: Whole.

419 Ball: If you wrote this with – in halves, what number would you write there? So you wrote three-thirds, could you write some number of halves there? Something over two?

420 Brandon: Mm-mm

421 Ball: Could you write something over four there?

422 Brandon: Over four? Yeah.

423 Ball: What would you write over four?

424 Brandon: For half?

425 Ball: ...yeah. For one. Like instead of three thirds, could you write a fraction that's...

426 Brandon: That's...

427 Ball: ...over four, like that?

[Ball adds $\frac{6}{6}$ to the number line:]

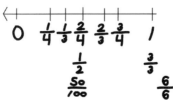

Time: 00:42:53

428 Brandon: No, 'cause this is all one, so three-thirds or four-fourths, or...

429 Ball: Okay. You can write four-fourths.

430 Brandon: Or...

[Brandon writes:]

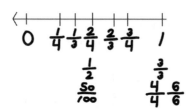

431 Ball: Okay. That's supposed to go, like here? It's the same?

432 Brandon: Mm-hmm.

433 Ball: Okay. So now I'm going to give you some cards with fractions on them, and you don't have to find exactly where they would go, but I'd like you to – here I'll show you. Where're those ones with the magnets that we had? These? Okay. So here I just want you to put it sort of in the right place where you think it would go. Like is it more than one, is it less than one? Is it more than a half, is it less than a half? It'll be hard to tell exactly where it should go, but about where do you think that fraction goes? What is it? Can you read it to me?

434 Brandon: Eight-ninths.

435 Ball: Okay. So wh – approximately where do you think you would put that? For example, is it – which side of one does it go on?

436 Brandon: It goes right here because if it's – it's not – it's not – it's – it's not nine-ninths...

[Brandon places the $\frac{8}{9}$ card directly to the left of the 1 on the number line.]

437 Ball: Okay.

438 Brandon: ...so it wouldn't be one.

439 Ball: So nine-ninths would be one? And how do know it's more than a half? 'Cause you put it – like here's a half. You put it quite far over from a half.

440 Brandon: Yeah. 'Cause four-and-a-half is half of nine, so – but this is eight so it's over –

441 Ball: Okay.

442 Brandon: it's over...

443 Ball: All right. Where would you put this one? Can you read that?

444 Brandon: Three-fifths.

445 Ball: Yeah. Where would you put that one?

446 Brandon: [right] two, three, four… Hold on.

447 Ball: I mean again roughly. Like is it close to one? Is it close to zero? Is close to a half?

448 Brandon: So these lines right here would be representi – representative –

449 Ball: So about where do you think it goes? Which side of one-half does it go on?

450 Brandon: I think it's a li – it's a little over half.

[Brandon places the $\frac{3}{5}$ card directly to the right of the $\frac{1}{2}$ on the number line.]

451 Ball: Okay.

452 Brandon: 'Cause two-and-a-ha – two-and-a-half is five – is half of five, so…

453 Ball: Okay.

454 Brandon: …but it's three so it – it would be over.

455 Ball: Where would you put that one?

456 Brandon: Right here.

[Brandon places the $\frac{19}{19}$ card directly under the 1 on the number line.]

457 Ball: What is that?

458 Brandon: Nin – nineteen-nineteenths.

459 Ball: Okay. Why does that go there?

460 Brandon: 'Cause it's whole.

461 Ball: Okay. And where would you put this one? What is that fraction?

462 Brandon: Three-twos?

463 Ball: Okay. Where would you put it?

464 Brandon: So, okay. So which one is the whole number? Which…

465 Ball: It's divided into parts that are halves. So…

466 Brandon: To find this answer – to find this answer you would have to divide two into three. So…

[Brandon writes on chalkboard:]

467 Brandon: So it would be one-and-a-half.

468 Ball: Okay. So where would you put it?

469   Brandon: [Gap in audio] – put this right here.

470   Ball: Okay. I have a – one or two more to show you.

[Ball goes for more cards.]

471   Ball: Oh. Where would you put that one. What is that?

472   Brandon: Zero-fourths.

473   Ball: Yeah. Where would you put that?

474   Brandon: I'd put this right here.

[Brandon places the $\frac{0}{4}$ card directly under the 0 on the number line.]

475   Ball: Why?

476   Brandon: Because, umm, it's not a percentage yet, so its –

477   Ball: It's not a what?

478   Brandon: I mean it's not a part of it.

479   Ball: Uh-huh.

480   Brandon: It's not like – the numerator doesn't have a number on top of it.

481   Ball: And what does that tell you?

482   Brandon: It tells me that it's not part of a fraction yet, so it would just be – it would just be zero.

483   Ball: Okay. Do you want to a really strange one, and then we can stop with this? What does that say?

484   Brandon: Two – eight – two-hundred-eighteenths over two-hundred-six?

485   Ball: Now I just want to know roughly, do you think it's less than one? More than one? Is it more than two?

Time: 00:47:05

486   Brandon: I think it's more than one 'cause it's – 'cause two-hundred-and-six can't go into two-hundred-eighteen two times, so . . .

487   Ball: Okay. So where would you put it, about?

488   Brandon: I would put it – just randomly I would put it right here.

[Brandon places the $\frac{218}{206}$ card directly to the right of the $1\frac{1}{2}$ on the number line.]

489   Ball: Why did you put it so close to two? Like how did you decide to put it all the way over there?

490   Brandon: 'Cause I'm not really sure 'cause it's not really – it's not – I'm not sure if it's over half or before half.

491   Ball: So what if I asked you to write a fraction, let's say, that was – well let me write it in a different spot. What if here where you've been writing these

other fractions I asked you to write a fraction that was the same as one that had two-hundred-six in the denominator? What would you put in the numerator if you wanted it to be the same amount as all these other numbers here?

[Brandon writes $\frac{206}{206}$.]

492 Ball: Why did you write that?

493 Brandon: If I wanted it to be whole, then it would have to be two-hundred-and-six.

494 Ball: Okay. So now if you look at this one, does that help you know about how big that one is, or not really?

495 Brandon: Not really.

496 Ball: Okay. Why don't – You want to switch and do something else for a while?

497 Brandon: S – sure.

Time: 00:48:18

498 Ball: Did you say you hadn't been working with number lines or you had worked with number lines?

499 Brandon: We haven't.

500 Ball: How – was that hard to do, what we were doing?

501 Brandon: No, 'cause – 'cause most of – I think – in my opinion it's sort of like common sense.

502 Ball: Uh-huh.

503 Brandon: It's like – if it's one then it would just be – if you put, like, four points – it depends on how much points you put.

504 Ball: Mm-hmm.

505 Brandon: If you put five then we'll know that it'll take a certain amount – five to equal one, or four – you put four points then it'll be –

506 Ball: Well, you know this thing we were having over here about the eighths?

507 Brandon: Mm-hmm.

508 Ball: I wonder how that looks on here. So you said, umm, you were trying to talk about two-eighths and one-fourth. So where would you put two-eighths on this number line?

509 Brandon: Well, let me see.

[Brandon divides the sections between 0 and 1 on the number line into eighths.]

510 Ball: Okay, so . . .

511 Brandon: The white points would be – these are eighths.

512 Ball: Okay. So can you label them?

513 Brandon: Mm-hmm

[Brandon adds more fractions to the number line, but the number line begins to appear cluttered.]

Time: 00:50:01

514 Ball: I think we're getting too many things on here, maybe.

515 Brandon: Probably.

516 Ball: It's hard to keep track of. Umm, maybe let's just make a clean one 'cause I think we have too many things on here and I think y – it was hard to see what you were doing. Let's come back – well let's go over here and we'll just work on a little piece of it and see if we can talk about this eighths. So... There's zero, and I'll just do it between zero and one this time, okay?

517 Brandon: Mm-hmm.

518 Ball: So I'm just going to put back a couple things we had. We had one half, and we had th – uh, what was this?

519 Brandon: Umm, three-fourths

520 Ball: Three-fourths. We also had two-fourths. And we had one-fourth. That's all I'm going to put on for right now, okay? Or maybe we'll put a couple of your whole ones.

521 Brandon: Mm-hmm.

522 Ball: What did you have? Four-fourths?

523 Brandon: Yeah.

524 Ball: Okay. So is that enough now just to get oriented?

525 Brandon: Mm-hmm.

[Ball has drawn:]

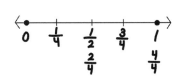

526 Ball: So now try to make the eighths. Okay, do you want a colored one? So can you make that so you can represent eighths? Just do it carefully 'cause you were right, you wanted to make eight parts but I think you lost track a little bit.

527 Brandon: Two...

[Brandon divides the section between 0 and 1 on the number line:]

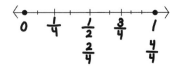

528 Ball: Okay. So, umm, this point right here [pointing to the mark] is what?

529 Brandon: It's half.

530 Ball: It's half?

531 Brandon: Mm-hmm.

532 Ball: And what's the pink one you just drew? What's that?

533 Brandon: That's half point.

534 Ball: That's a half point? So in eighths what would you write?

535 Brandon: Four-eighths.

536 Ball: Okay. So what about this one right here?

537 Brandon: This is two-eighths.

[Brandon has written:]

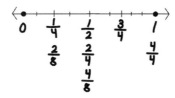

538 Ball: Okay. So now you've got two-eighths and one-fourth at the same point, right?

539 Brandon: Yeah.

Time: 00:51:58

540 Ball: So how does that go back with what we were talking about with the pictures? 'Cause here, I think you were telling me that one-fourth was more, 'cause the fourths were bigger chunks.

541 Brandon: Mm-hmm.

542 Ball: But here you've got them at the same point, so I'm curious about that.

543 Brandon: Because – 'cause you put, like – if – it depends, like – where – how – how big are the points, so if they were in fourths, then one-fourth would be bigger...

544 Ball: Mm-hmm.

545 Brandon: ...because eighths are like – fourths – some fourths are like this, but then eighths are like this.

546 Ball: But didn't you –

547 Brandon: So – but the space between them would be bigger.

548 Ball: Right. But you have that, right? See here you have one-eighth and then two-eighths. You have two-eighths right here at the same place. Is that in the wrong place, that two eighths?

549 Brandon: No. One-fourth is – one-f ... Because I put them – this is in eighths, but this – if you put them in fourths then it would be right here.

550 Ball: Okay. So are they the same or they're not the same?

551 Brandon: They're not the same.

552 Ball: And why, on this drawing, do they come out looking like they're the same? On this number line?

553 Brandon: 'Cause if you s – if you put them in fourths then this would be one ...

554 Ball: Mm-hmm.

555 Brandon: [counting the fourths along the number line] ... two, three, and then ...

556 Ball: What you have, right? You have one-fourth, two-fourths, three-fourths.

557 Brandon: All right, and then one whole ...

558 Ball: Right. So here you have – when you divide it into eighths it looks like you've got one-eighth, two-eighths – is this right? – three-eighths, and the you already wrote four-eighths –

559 Brandon: Four ...

560 Ball: – then you would have five-eighths ...

561 Brandon: Six ...

562 Ball: ... then six-eighths, and then seven eighths? Is that – is that correct?

563 Brandon: Mm-hmm.

[Ball has written:]

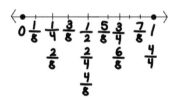

564 Ball: But then you do have – you have four-eighths at the same place as one – well what's bigger: one-half or four-eighths?

565 Brandon: I mean, the numbers are bigger, but they're both the same cause they're both the same 'cause they're both half.

566 Ball: Okay. So those you see as the same. Is two-fourths the same also?

567 Brandon: Mm-hmm.

568 Ball: Now back to this. You think these are different though –

569 Brandon: Yes –

570 Ball: – is that right?

571 Brandon: – yes because – I put – these are in fourths ...

572 Ball: Mm-hmm.

573 Brandon: ... what d – in a number line, what determines how big the fraction is that if – how big the space is.

574 Ball: Mm-hmm.

575 Brandon: 'Cause – from here [pointing to zero and referring to the distance between 0 and $\frac{1}{4}$] it's like – this is like one-fourth and two-fourths and three-fourths and one whole [counting up the number line by fourths], so the space – the space betwe – these are eighth [pointing to the distance between 0 and $\frac{1}{8}$], so – but this is one-fourth so they're in – since it's in eighths, the spaces in between it – it is smaller, so that's why one-fourth would be bigger [i.e. because the space between 0 and $\frac{1}{4}$ is bigger than the spaces between the eighths, one-fourth is bigger than two-eighths].

576 Ball: Okay. Let's – let – why don't we leave our number lines and drawings one the board and we'll switch gears a little? We've probably done enough of that for a while.

Time: 00:54:40

[Brandon and Ball: walk back to the projector.]

577 Ball: Okay. Did we get s – do we have any clean paper left?

578 Brandon: Yes.

579 Ball: Okay. Now I'm just going to show you some pictures that – some of them are easier and some of them are harder ...

580 Brandon: Mm-hmm.

581 Ball: ... and I'd like you to try to think about how much of the whole it is and what fraction you could use to express that, okay? So, how 'bout this one?

[Ball shows Brandon a figure:]

582 Ball: Let's put a piece of paper under it in case you want to write something. So the triangle – the big triangle is the whole and I'm interested in what you think about the shaded part.

583 Brandon: It's half. [He writes $\frac{1}{2}$ under the picture.]

584 Ball: Okay. And how did you decide that?

585 Brandon: 'Cause it's bo – it's evenly split –

586 Ball: Mm-hmm.

587 Brandon: – so one side is the same as this side, so ...

588 Ball: Okay. What about this one? So the whole is this whole rectangle –

[Ball shows Brandon a figure:]

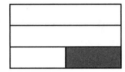

589 Brandon: Mm-hmm.

590 Ball: – okay?

591 Brandon: So would this be in – what's – is it – would this be in fourths, or ... ?
I'm not sure.

592 Ball: What are you not sure about? What is your question?

593 Brandon: Because these don't have a line in between the middle, but these two
do, so ...

594 Ball: You can do – you – you need to do something to the picture, you can.

595 Brandon: Okay.

596 Ball: If that helps you decide what you want to call that.

[Brandon draws a line through the rectangle:]

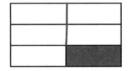

597 Ball: Okay. What did you do?

598 Brandon: I put a line though it.

599 Ball: And why – why did you want to do that?

600 Brandon: 'Cause this part doesn't have a line through it so it would – so – but
this part does, so it wouldn't make sense for just this part to have a line through
it and not the rest.

601 Ball: Okay. So now would you be able to decide how much the shaded part is?

602 Brandon: Mm-hmm.

603 Ball: What?

604 Brandon: It's one – it's one-sixth.

605 Ball: Can you write that?

Time: 00:56:36

[Brandon writes $\frac{1}{6}$.]

606 Ball: Okay. Okay, this one – this one might be a little bit trickier. Maybe not, I don't know.

[Ball shows Brandon a figure:]

607 Ball: So the whole is the whole rectangle, okay? And what you want to try to figure out is how much is shaded.

608 Brandon: Mm-hmm.

609 Ball: You can talk about it as you think about it, if you want to.

610 Brandon: I – I'm not really sure 'cause – I mean t – this – if this was a fraction [pointing to the second square of the picture], this [pointing to the last square of the picture] would be half of it 'cause it's sh – and this would be half, so . . . But this isn't shaded, so I guess this is one and –

611 Ball: Why don't you keep a little record of what you're saying? So you're saying this is one . . .

[Brandon writes:]

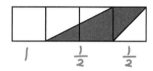

612 Brandon: Yeah. And that would be the half – that would be half.

613 Ball: Mm-hmm.

614 Brandon: And this would be half, so . . .

615 Ball: Okay. But you're – what are not sure about then?

616 Brandon: 'Cause – 'cause this isn't shaded, so – but this i – this part is shaded. This is half shaded.

617 Ball: Okay. So if you just looked at these two squares right here?

618 Brandon: Mm-hmm.

619 Ball: How much of the whole are these two? Forget about the shading. So you have this whole rectangle. What if – how much of the whole are these two together?

620 Brandon: It's two squares.

[Brandon draws:]

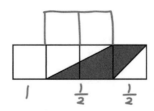

621 Ball: Mm-hmm.

622 Brandon: Okay.

623 Ball: A – and how much of the whole is that? What fraction of the whole are those two squares?

624 Brandon: Which fraction is shaded?

625 Ball: Which-uh, yeah. Like if we shaded this ... And there's your original rectangle. What fraction of the whole would be shaded?

[Ball has drawn:]

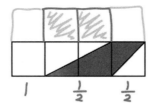

626 Brandon: Two-fourths.

Time: 00:58:27

627 Ball: Two-fourths, okay. But you just said only half of that is shaded, right? So what's half of two-fourths?

628 Brandon: Two – I mean half is two-fourths. One-fourth?

629 Ball: Okay. So here you've got one-fourth is shaded and one-fourth isn't?

630 Brandon: No.

631 Ball: No?

632 Brandon: I'm not really s – I'm not sure.

633 Ball: Okay. Why don't – should we leave that one? What about this one?

[Ball shows Brandon a figure:]

634 Brandon: I'm guessing they're six. They're six 'cause –

635 Ball: Do you want to draw some lines again?

636 Brandon: Sure.

[Brandon draws:]

637 Ball: Okay. So now would be able to say how much of the – this whole is
shaded?

638 Brandon: This part is, like – so all of these are eighths, so I think – so five
eighths?

639 Ball: Okay. Does it matter that they're not next to each other?

640 Brandon: Mm-mm

641 Ball: How would you write five-eighths?

[Brandon writes $\frac{5}{8}$.]

642 Ball: Okay [puts a new example on the projector].

643 Brandon: Those would be – that would be half 'cause there're four segments
of it –

644 Ball: There're what?

645 Brandon: There're four pieces of it and they all come together to be one.

646 Ball: Can you draw that? Can you – you want to put any lines on to show
me what you mean? Or you're saying one, two, three, four [pointing to the
different areas on the picture]

647 Brandon: Yeah. It – I mean they're different shapes, but there are four segments
of it and they come together to make one square, so –

648 Ball: Uh-huh.

649 Brandon: – but two of the segments are shaded, so it would be two-fourths, or
half.

[Brandon writes $\frac{2}{4}$ under the figure.]

650 Ball: Okay, how did you decide that? Because . . .

651 Brandon: Because since it's four of – segments then – and two are shaded so
two out of four would be half.

652 Ball: Earlier when you were folding the paper you made the big point to me that, umm, like a little kid might fold them and not mi – might not make the parts the same size.

653 Brandon: Mm-hmm.

654 Ball: Do you think these are the same size there?

655 Brandon: Mm-mm.

656 Ball: You don't think they're the same size. Which ones do you think are different?

657 Brandon: These two – I mean these two from this – from there [pointing to the two small rectangles that comprise the shaded square] or these two from this – from this... [pointing to the unshaded quadrilaterals along the bottom and right side].

658 Ball: Okay. So how come you're – how come you're going to call those each one-fourth then if they're not the same part – the same size?

659 Brandon: 'Cause one – they're all one piece but there a line that puts them into – that separates them 'cause, it's like... This is one,

[Brandon draws:]

660 Brandon: this is one,

[Brandon draws:]

661 Brandon: and then this part is...

[Brandon draws:]

662 Brandon: they're all just kind of – there's a line that separates them –

663 Ball: Mm-hmm

664 Brandon: – That puts them into fourths, but since they're different sizes they're four – they're still four segments, so – and two of them are shaded, so it's two-fourths.

665 Ball: Okay. Do we have more paper? Are we running out?

666 Brandon: Yeah.

667 Ball: We are running out?

668 Brandon: Mm-hmm. Well we still have a few.

669 Ball: Okay. We're okay then. Okay. So if I write this fraction. Here, what is that?

[Ball has written $\frac{3}{6}$.]

670 Brandon: Half. Three-sixths or half.

671 Ball: Okay. Well actually I was going to ask you about that. Earlier you said something about reducing fractions.

672 Brandon: Mm-hmm.

673 Ball: How would you reduce that fraction?

674 Brandon: Find a number that c – that goes into both the numbers evenly.

675 Ball: Mm-hmm. So can you show me?

676 Brandon: So three – I picked three. Three can go into three one, and three go – can go into six two times.

[Brandon has written $\frac{1}{2}$ directly to the right of $\frac{3}{6}$.]

677 Ball: Okay. What if somebody didn't believe you that one-half was the same amount as three sixths? Is there some way you could show them?

678 Brandon: Mm-hmm. Or you could just – if you multi – multiply these like – no, that's wrong...

679 Ball: You can scratch it out if you want.

680 Brandon: Okay.

681 Ball: Okay. So how could you – what if somebody didn't think those were the same? Are you saying that they're the same?

682 Brandon: Mm-hmm.

683 Ball: How could – what if somebody didn't believe you? 'Cause, I mean, a lot of people might say those numbers are a lot bigger.

684 Brandon: Umm. I think – I'm not s – I'm not sure. There're so many ways, but...

685 Ball: Well what's a good way, do you think?

686 Brandon: Umm. You could – I could draw – draw, umm, a picture to explain it.

687 Ball: Mm-hmm.

688 Brandon: Or something that it – that'll show that three-sixths is the same as half.

689 Ball: So what do you think – what is one way that you could do it?

[Brandon draws:]

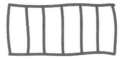

690 Ball: Okay.

691 Brandon: Those are sixths so ...

[Brandon colors the rectangle:]

692 Ball: Okay.

693 Brandon: There's the fraction for how much of it is shaded then it would be three-sixths

[Brandon writes under $\frac{3}{6}$ the rectangle.]

694 Brandon: and all together it's six, so three would be half and the other three would be half.

695 Ball: Okay. All right. So why do you call that reducing?

696 Brandon: Reducing? Umm, 'cause you – if – 'cause some numbers can't be reduced, or some fractions can't be reduced, but redu – called reduce when it's reducing when you make it smaller, but if it's reduced then it's already small.

697 Ball: Okay, but are you actually making the fractions smaller? Like is one-half smaller than three-sixths?

698 Brandon: I mean, they're both the same, like, form, but they're – they're different numbers

699 Ball: Okay. S – can you write down a fraction that can't be reduced? Can you think of one?

700 Brandon: Mm-hmm.

[Brandon writes $\frac{5}{6}$ on the overhead.]

701 Brandon: Five-sixths.

702 Ball: Okay, why can't that be reduced?

703 Brandon: Because there's no number than can into five and six evenly.

704 Ball: Mm-hmm. Umm, can you reduce fractions that are a little less familiar? Like, for example, if I wrote a number like ...

[Ball writes $\dfrac{24}{42}$.]

705 Ball: What does that fraction say?

706 Brandon: Twenty-four-s – er – uh – er – twenty-four-forty-twos?

707 Ball: Yeah. Can you reduce that?

708 Brandon: Yeah.

709 Ball: How would you do it? How could you tell so quickly that you could?

710 Brandon: Because they end and start with a number that – that – that a number can go into.

711 Ball: Okay. So what do you – what do you – what do you think you'd do to do it?

712 Brandon: One is that they're both even.

713 Ball: Okay, and why does that tell you it can be reduced?

714 Brandon: Well, if they're even, and some numbers that are odd can be reduced too, but ...

715 Ball: Okay, but if they're even you're sure they can be reduced?

716 Brandon: Yeah.

717 Ball: Why is that?

718 Brandon: 'Cause if it ends with, like, a number with two, four, six –

719 Ball: Mm-hmm –

720 Brandon: – it can be reduced to ... some could be even, but some numbers are odd that can be –

721 Ball: Okay.

[Brandon writes $\dfrac{6}{7}$.]

722 Brandon: Six-sevenths.

723 Ball: Okay. How did you do that?

724 Brandon: 'Cause I picked a number that can go into t – twenty-four and forty-two, and it –

725 Ball: What number did you pick?

726 Brandon: Six.

727 Ball: You picked six?

728 Brandon: Wait no. Actually it's ...

729 Brandon: Well actually it's ...

[Brandon corrects his work: $\dfrac{\cancel{6}^{4}}{7}$.]

730 Brandon: Four-sevenths.

731 Ball: Okay. What number did you pick?

732 Brandon: Six.

733 Ball: Okay. And what did you do?

734 Brandon: I... six can go into twenty-four four times, and six can go into seventy-two seven times.

735 Ball: Okay. All right. Umm, let's see. Are you tired? Do you want something to drink?

736 Brandon: Mm-mm.

737 Ball: No? Okay.

738 Ball: They're not being very good are they? Aren't they supposed to be quiet?

739 Brandon: They were quiet.

740 Ball: For a while. Maybe they'll be a little too quiet now. Okay. Well actually I was going to show something sort of tricky. I need another piece of paper. We're going to run out very fast. So just now you were reducing fractions...

741 Brandon: Mm-hmm.

742 Ball: ...but, umm, sometimes when I've been teaching kids, they do other things to reduce fractions and I wanted to show you something that I saw a student do, and I want to know if you think you could do it this way. So I showed him this fraction –

[Ball writes $\dfrac{13}{43}$.]

743 Ball: – and he said, "Well, I c – I know that that's – I can reduce that just by crossing out, or canceling the numbers that are in the ones place"

[Ball crosses out the threes, changing the fraction to $\dfrac{1\cancel{3}}{4\cancel{3}}$.]

744 Brandon: Mm-hmm.

745 Ball: "... and so I can just do that and it will be one fourth." Is that a correct way to reduce fractions?

746 Brandon: I can't really say if it's right or wrong cause I've never tried it.

747 Ball: Mm-hmm. What do – what would you try to – if you had to try to figure out, like, if that was a good method, or if it really was a method, what would you do to try to decide?

748 Brandon: Umm. I would see – I would see what number can go into those and see if it comes out to one fourth.

749 Ball: Mm-hmm. And does any number come to mind that you could do to divide into both of those to try to reduce it?

750 Brandon: No.

751 Ball: You can't think of one, or there isn't one, or what?

752 Ball: What're you trying?

[Brandon has written $13 \overline{)\,43}$.]

753 Brandon: I'm trying to see if thirteen can go into forty-three.

754 Ball: Can you think of a number that you can use to divide into thirteen that you could also divide into forty-three?

755 Brandon: Umm. Three?

756 Ball: Three would go into 13?

757 Brandon: No. I'm not – I don't think – I'm not sure, 'cause – no, I don't think any number can go into thirteen evenly.

758 Ball: Uh-huh. So then what does that tell you about this? [i.e. 13/43]

759 Brandon: Umm. I'm not sure. I can't find a number that'll go into thirteen and thirty-three evenly – I mean forty-three evenly.

760 Ball: Okay. So what's your view about this method right now? You're saying you don't know, right? Is that what you're saying? Or are you thinking it doesn't – it's not a good idea?

761 Brandon: I'm thinking it's not a good idea.

762 Ball: Uh-huh. Why a – why are you thinking that?

763 Brandon: 'Cause – I don't know how, but I can't – there – I don't think there's a number that can go into thirteen and forty-three evenly.

764 Ball: Okay. Have you done some adding and subtracting of fractions?

765 Brandon: Mm-hmm.

766 Ball: And you did – you showed me you were doing some dividing also, right?

767 Brandon: Mm-hmm.

768 Ball: What else? Have you multiplied fractions too?

769 Brandon: Mm-hmm.

770 Ball: You want to do a few of those now? Okay. Can I have a, umm – can you hand me one of those pens. And I think we're okay with paper. Yeah. Thank you.

771 Ball: How would you do that problem? You can use your own. Here.

[Ball has written:]

$$\frac{2}{3} + \frac{2}{3} =$$

772 Brandon: It would equal four-thirds.

[Brandon writes his answer:]

$$\frac{2}{3} + \frac{2}{3} = \frac{4}{3}$$

773 Ball: How'd you do it?

774 Brandon: You added them, but some frac – some – along the line there's going to be a problem that – where the denominators aren't the same . . .

775 Ball: Mm-hmm.

776 Brandon: . . . so you would have to make the denominators the same.

777 Ball: Uh-huh. But if they're the same, then what?

778 Brandon: Then you can add.

779 Ball: Okay. Well what if somebody said the answer to this was actually four-sixths, what would you say? Do you see how somebody might get that?

780 Brandon: Mm-hmm.

781 Ball: What would you say about that?

782 Brandon: It's not correct.

783 Ball: Why?

784 Brandon: Because, umm. 'Cause you don't – you don't add the denominator.

785 Ball: Okay.

786 Brandon: I think you just add the numerator.

787 Ball: All right. So let – want to try one where the denominators are different? Can you show me how you do that? Are you going to write one? Okay.

[Brandon writes $\frac{4}{6}$ and then $\frac{2}{3}$ directly under it.]

788 Ball: So you're doing four-sixths plus two-thirds?

789 Brandon: Mm-hmm.

790 Ball: Okay. Can we – let's – You want to write it horizontally and then show me how you would do it, or is this how you would normally write it, vertically?

791 Brandon: This way.

[Brandon writes:]

792 Ball: Okay. So what would you do?

793 Brandon: I would find a number that can go into three and six evenly.

794 Ball: Mm-hmm.

795 Brandon: The smallest number that can go . . .

796 Ball: So like what?

797 Brandon: Three.

798 Ball: Okay.

799 Brandon: Wait, no. I'm getting confused . . . like six. 'Cause si – times two and then . . . You – whatever you do, whatever you multiply to get – get to the number, you do to the numerator, so . . .

800 Ball: Okay.

801 Brandon: . . . it's times would be times two by the numerator.

[Brandon has written:]

$$\frac{4 \times 1}{6 \times 1} \quad \frac{4}{6}$$
$$+\frac{2 \times 2}{3 \times 2} \quad \frac{4}{6}$$

802 Ball: Mm-hmm.

803 Brandon: Then you would add.

[Brandon writes:]

$$\frac{4 \times 1}{6 \times 1} \quad \frac{4}{6}$$
$$+\frac{2 \times 2}{3 \times 2} \quad \frac{4}{6}$$
$$\frac{8}{6}$$

804 Brandon: But that can be – that can be put into another number.

805 Ball: Okay, what did you get for that?

806 Brandon: Eight-sixths.

807 Ball: Okay.

808 Brandon: But that's a improper fraction. When – it's when the – the numerator is bigger than the denominator. So what – I would divide six into eight.

[Brandon writes:]

$$6 \overline{\smash{)}8} \\ \underline{6} \\ 2$$
with quotient 1

809 Brandon: Two – So you would get one-and-two-sixths, or one-and-one-third.

810 Ball: Okay. And why do you divide it to find out how to write it another way?

811 Brandon: Umm.

812 Ball: You've done that actually a few times. You did that with the three-over-two a couple times, right?

813 Brandon: Mm-hmm.

814 Ball: Why are you dividing it?

815 Brandon: 'Cause, umm – 'cause there's a – a different way that the numerator can be – it can be smaller than the denominator. It can – it'll be back to a whole number and a fraction of the whole.

816 Ball: Okay. I'm going to write a addition problem, okay? How 'bout…But you prefer written this way?

[Ball writes the fraction problem for Brandon both horizontally and vertically:]

$$\frac{3}{5}+\frac{2}{3} \qquad \begin{array}{r} \frac{3}{5} \\ +\frac{2}{3} \\ \hline \end{array}$$

817 Ball: So either way. How would you do that? There the denominators aren't the same, right?

818 Brandon: Mm-hmm.

819 Ball: So what would you do in order to add those?

820 Brandon: Find the, uh, the smallest number that three and five could go into.

821 Ball: Okay, so what would that be?

822 Brandon: Fifteen.

823 Ball: All right.

[Brandon writes:]

$$\begin{array}{r} \frac{3\times3}{5\times3} \quad \frac{9}{15} \\ +\frac{2\times5}{3\times5} \quad \frac{10}{15} \\ \hline \frac{19}{15} \end{array}$$

824 Ball: Okay. So let's try a subtraction problem, okay? What if I wrote down four-and-one-third minus two? What would the answer to that be?

[Ball has written:]

$$4\frac{1}{3} - 2 =$$

825 Brandon: One-and – one-and-two-thirds?

826 Ball: How'd you decide that?

827 Brandon: 'Cause minus two...

[Brandon writes:]

$$4\tfrac{1}{3}$$
$$-2$$

828 Brandon: 'Cause, umm ... 'Cause it's – it's not whole number minus whole number so it – so it – it's a fraction, so it would... So if you minus two, you get two, but it's – it's part of one fraction so it – one-and-two-thirds – and two-thirds – the two-thirds – two-thirds plus one-third would equal three.

829 Ball: I don't understand where you got the two-thirds from. I don't see where this came from.

830 Brandon: 'Cause that plus that would equal two.

831 Ball: That – this one-and-two-thirds –

832 Brandon: That's two-third plus – yeah one-and-two-thirds plus one-and-three-thirds would equal two.

833 Ball: Okay. And then what?

834 Brandon: What?

835 Ball: And then – so then you're have two. That one – one-and-two-thirds plus one-third equals two.

836 Brandon: That would equal two, so that would equal a number so that's why I got that cause it's different.

837 Ball: What if you had a fraction that you were subtracting on the bottom as well? So what if you had four-and-one-third minus, umm, two-and-one-half? Then what would you do?

838 Brandon: Hmm. Find the number that two and third – two and three can go to, 'cause you can't – you can't subtract.

839 Ball: So what would that be?

840 Brandon: It'd be six.

841 Ball: Okay.

842 Brandon: But you don't do anything to the whole numbers.

843 Ball: Okay. So now what?

844 Brandon: And then you subtract.

845 Ball: Okay, so how would you do that?

846 Brandon: Umm. Hold on, you can't subtract yet 'cause two is – two – you can't subtract two from – you can't subtract three from two so –

847 Ball: Mm-hmm

848 Brandon: – you would cross out four and make it a three.

849 Ball: Mm-hmm.

850 Brandon: And would I turn this in – and would you turn this into a twelve?

851 Ball: How – what are you trying to decide?

852 Brandon: Umm, cross this out, make the three, and put the whole number right here.

853 Ball: Okay.

854 Brandon: And then you subtract.

855 Ball: Okay, so what would you get?

856 Brandon: You would get nine – you'd get nine ...

857 Ball: It's kind of hard to see what you've got here. You look like you've got twelve-sixths minus three-sixths, and you have three minus two. Do you want me to rewrite it?

858 Brandon: Yeah.

859 Ball: Three-and-twelve-sixths, that's what you have?

860 Brandon: Mm-hmm.

861 Ball: And you have, umm, two-and-three-sixths.

[Ball has written:]

$$3\frac{12}{6}$$
$$2\frac{3}{6}$$

862 Brandon: Hmm. That's a hard question 'cause if you subtract it, it would be nine – it would be one-and-nine-sixths.

863 Ball: Yeah.

864 Brandon: So it'll be a whole number and an improper fraction.

865 Ball: Why did you put the one here? I mean I think I have an idea about why you put the one, but I wasn – I'd like to hear you explain it to me.

866 Brandon: 'Cause –

867 Ball: Here, I mean.

868 Brandon: Here – here it – you put three, 'cause three ...

869 Ball: You crossed off the four and made it a three. Then why did you write a one there?

870 Brandon: Because I borrowed, umm, a whole from the four 'cause you can't sub – you can't subtract three from two, so I just crossed out the three, made it – crossed out the four and made it a three and put the one that was taken from the four and put it by the two.

871 Ball: Okay. So what would you say if I said to you you're actually working in sixths here, right? So a whole is six-sixths, is that right?

872 Brandon: Mm-hmm.

873 Ball: So if you have six-sixths then you wouldn't be moving a ten over there, right, you'd be moving six-sixths over there. Can you try doing it that way?

874 Ball: Alright . . . we'll make it clear. So here you have . . . Actually let's start where we were. So we had four-and-two-sixths, right?

875 Brandon: Mm-hmm.

876 Ball: And we were trying to subtract two-and-three-sixths?

877 Brandon: Mm-hmm.

[Ball has written:]

$$4\frac{2}{6}$$
$$-2\frac{3}{6}$$

878 Ball: That's after you found a common denominator. So what I'm saying is you said you were borrowing a ten, but the whole here would be six-sixths, right? So what if we try again and say you're going to make that a three and now you've got a whole that you can put together with the two-sixths, but what is that whole . . .

[Ball has written:]

879 Brandon: I would cross the two out and put a six.

880 Ball: Well you have two-sixths already and now you're having six more sixths that you got from this one. Now how many six-sixths would you have all together?

881 Brandon: Umm, six?

882 Ball: Well you have six-sixths from this whole?

883 Brandon: Mm-hmm.

884 Ball: . . . but you already had two-sixths, so how much is six-sixths plus two-sixths together?

885 Brandon: One-and-two-sixths?

886   Ball: Okay, but why don't you write it as an improper fraction because you're going to be subtracting. So how much would that be? Six-sixths plus two-sixths?

887   Brandon: Mm, it would be eight-sixths.

888   Ball: So write an eight here instead of the two.

[Brandon writes:]

889   Ball: Okay, now can you subtract?

890   Brandon: Uh-huh.

[Brandon writes:]

891   Ball: Okay. So, now walk back 'cause I kind of helped you with that one a little bit. Can you try to explain what I was trying to show you? See if you understand what I was trying to show you?

892   Brandon: 'Cause –

893   Ball: Originally what you did is – it seems like you borrowed, like, a ten, the way you normally would borrow, but since we're working with fractions, I tried to show you a different way to think about the whole. Can you try to explain it back to me?

894   Brandon: Umm, what you would do is that ...

895   Ball: Okay.

896   Brandon: ... is that you can't borrow, like ten, so you would cross, making it eight, so – wait how ...

897   Ball: Why did we get the eight from, though?

898   Brandon: I'm not sure. I'm not sure.

899   Ball: Okay. Okay. So what we were talking about in part is if you borrow when you're working with fractions, you have to think what the whole is, right?

900   Brandon: Six?

901   Ball: It's six-sixths, so what we did is we took the six-sixths from this whole and we put it together with the two-sixths we already had. Have you seen that before?

902 Brandon: Probably, but I probably got confused on it.

903 Ball: Mm-hmm.

904 Brandon: So, okay we took – what we – we took the whole from the three and we added it with the two sixths.

905 Ball: Mm-hmm.

906 Brandon: So two plus six would be eight –

907 Ball: Mm-hmm.

908 Brandon: so it would be eight-sixths.

909 Ball: Okay. How're we doing? We're almost out of time I think. You want to do one more thing, or are you tired?

910 Brandon: Oh no, we can...

911 Ball: You're not tired; you could do this all day? Yeah? Okay? Every time I kid with you they get badly behaved. Okay. Umm.

912 Ball: So I was thinking of telling you a problem, a story. Have you really ever done anything like that at all, or have you just been working with a lot of numbers and drawings?

913 Brandon: Mm-hmm.

914 Ball: Okay. So here's the story I'm going to tell you, but I'll have to tell it to you probably even more than one time. Maybe we want some paper to keep track. I'll tell you the story first. So imagine your teacher, Carol...

915 Brandon: Mm-hmm

916 Ball: Imagine she bought a cake, but she only bought half a cake. Okay, she didn't buy a whole cake, she just bought half a cake. Ever see it in the store where they sell half-cakes sometimes? She bought half a cake. Okay? And then she ate half of that. Can you make a picture to show me how much she ate?

917 Brandon: Mm-hmm.

[Brandon draws:]

918 Brandon: This would be half of the cake, so if she ate half of that, she ate half this part.

919 Ball: Okay. So how much did she eat of the whole cake? If there had been a whole cake, how much would she have eaten of the whole?

920 Brandon: One-fourth out of it.

921 Ball: One-fourth of the whole cake, and how much of her cake?

922 Brandon: Half.

923 Ball: Half of her cake. Now she was full then ...

924 Brandon: Mm-hmm.

925 Ball: ... and she decided to split the rest of her cake between you and one of
the other kids in the class, but equally. So this is the part she didn't eat right
here.

[Ball colors Brandon's drawing indicating that the shaded portion is what is left
uneaten:]

926 Ball: Okay, so how could she divide that part equally and give you each an
equal piece?

[Brandon draws:]

927 Ball: And how much would guys each be getting?

928 Brandon: What?

929 Ball: How much of the orig – how much of the whole cake would you each be
getting?

930 Brandon: Umm, a half of that piece.

931 Ball: A half of this piece is the remainder, right?

932 Brandon: Mm-hmm.

933 Ball: So how much cake is that?

934 Brandon: Like this?

935 Ball: Yeah, how much cake is that?

936 Brandon: Out of the whole – out of this –

937 Ball: Out of the whole cake.

938 Brandon: One – I mean one-third. If Carol hadn't eaten this part then it would
just be one out of third.

939 Ball: Okay. So you're getting one-third of her – the cake she bought?

940 Brandon: Yeah.

941 Ball: And the other student?

942 Brandon: Mm-hmm. But since she ate it, since she ate one-third of it then there's only two-thirds of it remaining.

943 Ball: But she ate half of it, not a third of it.

944 Brandon: Oh yeah. So I – I ate and the other friend ate one-half of – the half of her whole cake, or the whole cake

945 Ball: Right, so what's one-half of a half?

946 Brandon: I'm not sure.

947 Ball: Okay. So you were thinking of these as three equal pieces or three different pieces.

948 Brandon: Three different pieces.

949 Ball: Okay. All right. Want to try one more?

950 Brandon: Mm-hmm.

951 Ball: Okay. Somebody showed you this problem and said you shouldn't find a common denominator, or you shouldn't find an exact answer.

[Ball shows Brandon the problem:]

$$\frac{19}{21} + \frac{52}{55}$$

952 Brandon: Mm-hmm.

953 Ball: They just wanted to know about how big is the answer to that. Like is it about – is the total of those two numbers together about one-half? Is the total about one? Is the total about one-and-a-half? About two? What would you say?

954 Brandon: Umm, I –

955 Ball: Just approximately.

956 Brandon: About – about one.

957 Ball: How did you figure that out?

958 Brandon: 'Cause – wait no. Actually it would be seventy-out-of – seventy-out-of-eighty, I think. 'Cause since this number is five then I can go up so it would be sixty...

959 Ball: Mm-hmm

960 Brandon: ...and this is approximately twenty. And that is approximately twenty, and this is approximately fifty, so fifty plus twenty would be...

961 Ball: You can write it out.

[Brandon writes $\frac{70}{80}$.]

962 Brandon: Okay. So this would be seventy and this would be eighty.

963 Ball: Okay, so if you were to write a fraction that was a whole with twenty-one in the denominator, what would it be?

964 Brandon: It – to make it a whole?

965 Ball: Mm-hmm

966 Brandon: Twenty-one over twenty-one.

967 Ball: Okay. Is that pretty close to that?

968 Brandon: Mm-hmm.

969 Ball: And if you were going to make a fraction that had fifty-five in the denominator that was one, what would you write?

970 Brandon: Fifty-five.

971 Ball: Is that pretty close to that?

972 Brandon: Yeah, but – yeah. Mm-hmm.

973 Ball: So this one's pretty close to one and this one's pretty close to one. When you add them together what're you going to get?

974 Brandon: Two.

975 Ball: You're going to get pretty close to two. Do you see why?

976 Brandon: Yeah. 'Cause –

977 Ball: Can you explain it to me?

978 Brandon: Yeah, 'cause this is almost one. It's approximately and this approximately one, so ... If we were to round then it – they would be one plus one will, will equal to two.

979 Ball: Okay. So maybe we should stop. I think we've been working a long time and I wanted to ask you a couple questions. What was – of all the stuff we did, and when you look back – we have tons of paper up here now – what do you think was, umm, the most fun that we did, or was anything fun?

980 Brandon: Yeah. I – I like the number line –

981 Ball: Uh-huh.

982 Brandon: – 'cause, umm, I kn – I probably did it few time, but I don't really work with number lines so it was fun to learn to do something new.

983 Ball: Uh-huh. So that was pretty new for you?

984 Brandon: Mm-hmm.

985 Ball: And of the things we did, what was – what did you find – was there anything that you found difficult that we did?

986 Brandon: Umm, yeah. Well no, not actually. Mm-hmm.

987 Ball: No?

988 Brandon: 'Cause I – 'cause probably in the past I probably said, "I don't remember this," but sort of like if I see it then, yeah I'll probably remember or . . . from a long time ago.

989 Ball: Mm-hmm. Were there things you found that you had to think about, umm, very hard, that you hadn't been thinking about for a while?

990 Brandon: Yeah.

991 Ball: What – do you remember an example today of something you had to think really hard about?

992 Brandon: Umm. About the subtracting – about the subtracting. About, umm, for – for, umm, like if a number – if – if it's – if you have a whole number and a fraction, but the numerator of the fraction can't subtract the numerator of the other fraction, you have to borrow from the other one – from the whole.

993 Ball: Mm-hmm. And you think you had learned that before, or you don't think you've learned that before?

994 Brandon: I – yeah I learned it, but – but then I get confused with it, so I use it a different way –

995 Ball: Uh-huh.

996 Brandon: – the incorrect way.

997 Ball: And how did it feel to have all these people out there? What did you think about that?

998 Brandon: Nothing, 'cause we were basically just focusing, like on the board or on the chalkboard –

999 Ball: Mm-hmm.

1000 Brandon: – so it wasn't a big deal.

1001 Ball: Uh-huh. Except when they laughed, of course. Right? Umm, do you have any questions you wanted to ask me, or anything you wanted to, I don't know, anything? Anything about what we did or anything specific?

1002 Brandon: Mm-mm.

1003 Ball: No? Do you want to stop?

1004 Brandon: Mm-hmm.

1005 Ball: So you're – there's going to be a little presentation for you, to thank you for this, so at this point people are going to pay attention to you again. But the director of the Mathematical Sciences Research Institute – I'm going to introduce him to you and he's going to make a little presentation to you. Is that okay?

1006 Brandon: Mm-hmm.

Assessing Mathematical Proficiency
MSRI Publications
Volume **53**, 2007

Chapter 16
Reflections on an Assessment Interview:
What a Close Look at Student
Understanding Can Reveal

ALAN H. SCHOENFELD

This chapter offers a brief introduction to and commentary on some of the issues raised by Deborah Ball's interview with Brandon Peoples, which was transcribed as Chapter 15 and can be seen in its entirety at http://www.msri.org/ publications/ln/msri/2004/assessment/session5/1/index.html. The video and its transcript are fascinating, and they reward close watching and reading. What follows in the next few pages barely scratches the surface of the issues they raise.

Before getting down to substance, one must express admiration for both of the participants in the conversation transcribed in Chapter 15. Brandon, a sixth grader, is remarkably poised and articulate. He responds openly and thoughtfully to questioning from Ball on a wide range of issues, in a conversation that is casual but intense — in front of a very large audience of adults! He also shows a great deal of stamina — the interview lasted an hour and a half! Ball demonstrates extraordinary skill in relating to Brandon, and in establishing a climate in which he feels comfortable enough to discuss mathematics in public, and to reveal what he knows. She covers a huge amount of territory with subtlety and skill, examining different aspects of Brandon's knowledge of fractions, revealing connections and confusions, and spontaneously pursuing issues that open up as Brandon reveals what he knows. The interview is a tour de force, demonstrating the potential of such conversations to reveal the kinds of things that students understand.

To begin, the interview reveals the complexity of both what it means to understand fractions and what it means for a student to come to grips with fractions. Consider the range of topics that was covered, roughly in this order: conversions from fractions to decimals and percents; the algorithm for multiplying or

269

dividing one fraction by another; the meaning of fractions as parts of a whole
(and the need for the parts to be equal); equivalent fractions and what it means
for fractions to be equivalent; the role of the numerator and denominator in
determining the magnitude of a fraction; comparing magnitudes of fractions;
improper fractions; rectangular, circular, and more complex area models for
fractions; number-line representations of fractions; which fractions can be "reduced" and why; finding the area of subsets of complex geometric figures; algorithms for adding and subtracting proper and improper fractions; models for
the multiplication of fractions; approximating the result of computations with
fractions.

For the person who understands the mathematics deeply, all of these topics
are connected; all the pathways between topics tie them together neatly. For
someone learning about the material, however, things are very different. Some
connections exist, at various degrees of robustness and some are being formed;
some are missing, and some mis-connections exist as well. This is a fact of
life, and an interesting one. At its best, assessment serves to reveal this set of
invisible mental connections in the same way that first x-rays and now MRI
techniques serve to reveal that which is physiologically beneath the surface. Let
us take a brief tour of Brandon's interview, to see what is revealed in his case.
The idea is not to do an exhaustive commentary, but to highlight the kinds of
things that a sensitive assessment interview can reveal.

Broadly speaking, Brandon is quite comfortable with, and competent in, the
procedural aspects of working with fractions. At the very beginning of the interview he produces the algorithm for dividing fractions (a topic he has just
studied) and demonstrates its use. He demonstrates how to multiply fractions
and reduce the result to lowest terms. Although he does make an initial canceling
error in multiplying $\frac{2}{3} \times \frac{4}{6}$ (remember, he is a sixth grader performing in front
of an auditorium full of adults!), he notes his error, corrects it, and confidently
confirms that the answer is correct. He confidently converts simple fractions
such as $\frac{1}{2}$ and $\frac{1}{3}$ to percents and decimals. Throughout the interview he easily
generates fractions equivalent to a given fraction; and he knows how to add and
subtract mixed fractions.

At the beginning of the interview, Brandon demonstrates his understanding of
certain area models. He explains after folding a paper in half that the two parts
have to be equal in size, and goes on to show that the two halves could each be
represented by $\frac{19}{38}$ or $\frac{37}{74}$. In folding a paper into thirds, he works to make the
three parts equal. He shows that $\frac{1}{4}$ can also be written as $\frac{2}{8}$, and as 25%. He
knows that $\frac{1}{6}$ is between 16% and 17%, and justifies his claim by showing that
$6 \times 16 < 100$, while $6 \times 17 > 100$.

Asked how much paper the fraction $\frac{3}{2}$ would represent, Brandon does a long division to show that $\frac{3}{2}$ is equal to "a whole and a half." Yet he pronounces $\frac{3}{2}$ as "three twos, I think," suggesting that he has not spent very much if any time talking about improper fractions. He is able to put the fractions $\frac{1}{2}$, $\frac{2}{3}$, and $\frac{3}{4}$ in increasing order, and justify the ordering, and, with a slight bit of prompting, to show that the figure

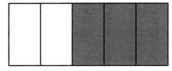

can be written as 60%, $\frac{60}{100}$, $\frac{3}{5}$, and $\frac{6}{10}$ — also dividing the figure in half horizontally to reveal six shaded pieces out of ten.

But then, life gets cognitively interesting. Ball asks (turn 267), "of these three fractions you've written [$\frac{60}{100}$, $\frac{3}{5}$, and $\frac{6}{10}$], which one's the largest?" and the following dialogue ensues:

Brandon: Three-fifths?

Ball: Why is it the largest?

Brandon: 'Cause three-fifths is like like we said earlier, it – 'cause one-hundredths are really small . . .

Ball: Mm-hmm.

Brandon: I mean it – In – I – my opinion it's not – it's not about the numerator, I think it's about the denominator.

Brandon goes on to draw a circle partitioned into five equal pieces and another into ten equal pieces, noting that the tenths are much smaller. He says,

Brandon: So – so it's six of that, even though the numer – the numerator's bigger than this numerator, it – my opinion is that the denominator – how big the denominator determines how – how big the fraction – the whole fraction is.

Moving to different numbers chosen by Ball, he indicates that $\frac{7}{8}$ is less than $\frac{1}{4}$, by placing a $\frac{7}{8}$ card to the left of the $\frac{1}{4}$ card on the board. He then goes on (turns 292–293) to draw the following two figures:

claiming that is larger than "because it's – you have fourths – I meant – I mean eighths is – eighths are a lot smaller, so seven of them would have to – you have to shade in 'cause you couldn't put seven into four." The discussion continues, with Brandon explaining that when two numbers have different denominators, the one with the smaller denominator is larger; but when two numbers have the same denominator, the one with the larger numerator is larger. Ball tries to reconcile these statements by having Brandon compare $\frac{1}{4}$, $\frac{2}{8}$, and $\frac{7}{8}$. Ball reminds Brandon that he had earlier said that $\frac{1}{4}$ and $\frac{2}{8}$ are equal. He draws pictures of those two fractions, and the following dialogue ensues:

Ball: Mm-hmm. When you look at these two pictures, which one do you think is greater: two-eighths or one fourth?

Brandon: Umm... One-fourth?

Ball: Why do you think one-fourth?

Brandon: Umm. 'Cause it has – it has bigger chunks into it to make fourths, so – but these are all, like li – sorta small, so just one out of four is bigger than two out of eight.

What this indicates is that the various pieces of the fractions puzzle have not yet fallen into place for Brandon. Although he has mastered some of the relevant algorithms and some aspects of the area model (especially with rectangles, where relative sizes can be perceived more readily), he is not confident of his (sometimes incorrect) judgments about relative sizes when it comes to circle models, despite the formal calculation that says that $\frac{1}{4}$ and $\frac{2}{8}$ are equal. And his incorrect algorithm for comparing fractions leads him to claim that $\frac{7}{8} < \frac{1}{4}$, despite the pictures he has produced that suggest the contrary. I stress that these confusions are normal and natural — they are part of coming to grips with a complex subject matter domain — and that they are shared by many students. Part of what this interview reveals is how complex it is to put all the pieces of the puzzle together, and how easy it is to overlook such difficulties, if one focuses on just the procedural aspects of understanding that comprised the first part of the interview.

On turns 332–576 of the interview, Ball conducts what is commonly called a "teaching experiment," in which she introduces Brandon to some new ideas and monitors what he learns and how he connects it with what he already knows. Brandon says that he is not familiar with the number-line representation of integers and fractions, and Ball takes him into new territory when she introduces him to it. Brandon seems confident and comfortable with labeling fractions between 0 and 1. It is interesting to see him grapple with numbers greater than 1: at first he labels the number half-way between 1 and 2 as $\frac{1}{2}$, and he needs to work to see that it should be labeled $1\frac{1}{2}$. Once he does, however, the rest of the labeling scheme seems to fall into place: he labels 1, $1\frac{1}{4}$, $1\frac{1}{2}$, $1\frac{3}{4}$, and 2. But then there is the following exchange:

Ball: Okay. All right. So, umm, can you think of any other fractions that you could put up here? Like, is there any fraction that goes between one-fourth and two-fourths, for example?

Brandon: Umm, mm-mm. No.

Ball: No? Er, have we put all the fractions up here that we can?

Brandon: Yeah.

It is worth noting that, earlier, Brandon had put the fractions $\frac{1}{2}$, $\frac{2}{3}$, and $\frac{3}{4}$ in linear order — he had observed that $\frac{1}{2} < \frac{2}{3} < \frac{3}{4}$, — so it should follow that $1\frac{1}{2} < 1\frac{2}{3} < 1\frac{3}{4}$. But what should follow (at least to the cognoscenti) is not necessarily what does follow, when one is new to a domain. This is a critically important fact about learning. At the same time, much of what Brandon understands about proper fractions does transfer well onto his developing understanding of the interval between 0 and 1; he fills in that part of the number line with no difficulty. In short, learning is complex!

Ball then (turns 506–575) revisits the issue of the relationship between $\frac{2}{8}$ and $\frac{1}{4}$ with Brandon, this time using the number line. As in Alice in Wonderland, things get curiouser and curiouser. With Brandon, Ball co-constructs a representation of some of the fractions and their equivalents between 0 and 1, inclusive:

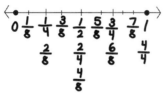

Brandon insists (turn 565) that $\frac{1}{2}$ and $\frac{4}{8}$ are the same size ("I mean, the numbers [in $\frac{4}{8}$] are bigger, but they're both the same [i.e., the fractions] cause they're both the same 'cause they're both half"). But then he goes on to assert that $\frac{1}{4}$ is larger than $\frac{2}{8}$ (turn 575):

Brandon: 'Cause – from here [pointing to zero and referring to the distance between 0 and $\frac{1}{4}$] it's like – this is like one-fourth and two-fourths and three-fourths and one whole [counting up the number line by fourths], so the space – the space betwe – these are eighth [pointing to the distance between 0 and $\frac{1}{8}$], so – but this is one-fourth so they're in – since it's in eighths, the spaces in between it – it is smaller, so that's why one-fourth would be bigger [i.e. because the space between 0 and $\frac{1}{4}$ is bigger than the spaces between the eighths, one-fourth is bigger than two-eighths].

Again, this should not come as a tremendous surprise. To whose who have the big picture, Brandon's reasoning is inconsistent: he is not applying the same logic to the relationship between $\frac{1}{2}$ and $\frac{4}{8}$ and the relationship between $\frac{1}{4}$ and $\frac{2}{8}$. But Brandon does not yet have the big picture; he is in process of constructing it. As he does, he does not have the bird's eye view that enables him to see contradictions, or the tight network of relationships that would constrain him to rethink his judgment. As he sees it, the facts about the two relationships are independent. Hence for him there is no contradiction.

After this discussion, Ball turns once again to area models (turn 576). Given a triangle divided in half, Brandon has no trouble calling each piece a half. Given the figure

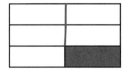

he tentatively identifies the shaded area as one fourth, but given the opportunity to "do something to the picture," he draws a line through the rectangle as below, and calls the shaded area one sixth.

This appears to indicate solid mastery of the "equal parts" part of the definition of fractions. Another more complex figure,

causes him some difficulty, but I suspect that one would cause a lot of people difficulty. On the other hand, he is able to modify the figure on the left below to the figure on the right, and say correctly that the shaded area is $\frac{5}{8}$ of the whole.

So far, so good: Brandon seems to have part-whole down pretty well. But the next figure,

,

throws him for a loop. Possibly because the shaded and unshaded pieces are so different in shape, he loses track of their relative size and claims "two of the segments are shaded, so it would be two fourths, or half." This claim resists a fairly strong examination by Ball.

In the discussion that follows, Brandon correctly produces a rectangular area representation for the fraction $\frac{3}{6}$, "reduces" the fraction to $\frac{1}{2}$, and provides cogent descriptions of what it means to reduce a fraction and when a fraction cannot be reduced. He has thus shown that he is on solid ground with regard to the straightforward representation of some simple areas, but that the knowledge is not yet robust.

At this point (turn 764) the conversation turns to adding and subtracting proper fractions, improper fractions, and mixed numbers. Brandon demonstrates clear mastery of the addition algorithm. Subtraction proves more complex, and Ball once again provides a tutorial — once again revealing the complexities of the learning process, as Brandon works to connect what he already knows (e.g., subtraction of integers and the conversion of integers into fractions) with the new context that calls for their use (e.g., computing $4\frac{2}{6} - 2\frac{3}{6}$). This conversation continues through turn 893.

The next part of their conversation reveals how, when some things are familiar, a student can produce correct answers and seem to have deep understanding; but that going beyond the familiar can reveal the fragility of the underlying knowledge. Ball asks Brandon to draw a picture of half of half a cake, and to say how much of the cake that is. Brandon produces the following picture and says that the area in question (which Ball colors in) is one-fourth of the original cake.

So far, so good. But when Ball asks Brandon, "Okay, so how could she divide that part equally and give you each an equal piece?," he draws the correct figure:

but says that the remaining part is one-third of the whole cake. It is difficult to know why he did this, although one can speculate. First, he must be tired by now; second, the whole cake is not there for him to see; third, the combination of those two factors might cause him to focus just on the picture and forget the "equal size" criterion for fraction definition. The part he is interested in is one part of three in the diagram he sees, and is thus labeled as one-third.

In the final mathematical segment of their conversation (turns 949–978), Ball
asks Brandon to estimate the sum

$$\frac{19}{22} + \frac{52}{55}.$$

Brandon — who, earlier, had demonstrated his competence in using the standard
algorithm to add fractions with more manageable denominators — now employs
the frequently used (and incorrect) procedure

$$\frac{a}{c} + \frac{c}{d} \;\rightarrow\; \frac{a+b}{c+d}$$

to arrive at the approximate sum

$$\frac{20+50}{20+60} = \frac{70}{80} \approx 1.$$

At this point Ball leads Brandon to the observation that each of the two fractions
in the original sum is nearly 1 in value, so that their sum is close to 2.

Discussion

Beyond admiring Brandon for his intelligence, bravery, and stamina, and Ball
for her skill as interviewer, there are at least two major points to take away from
their exchange. But first one must stress that the point of their exchange, and of
this chapter, is not to evaluate Brandon. Rather, it is to see what one can learn
from their conversation.

One point that comes through with great force is the complexity of what
it means to learn and understand a topic such as fractions. Knowing is not
a zero-one valued variable. The interview reveals that Brandon knows some
things in some mathematical contexts, but not in others; that in some places
his knowledge is robust and that in others it is shaky; that some connections
are strong and others not; and that he (as can everyone) can have in his mind
pieces of information that, when put side by side, can be seen as contradicting
each other. When he is on solid ground, for example when he is working with
rectangular figures as models of area, he makes certain that all the pieces of a
figure are the same size before counting them as n-ths; see turns 588–604 and
635–641. However, when the figures get complex or he gets tired, he loses sight
of this constraint: see turns 643–647 and 916–942. Similarly, Brandon correctly
uses the standard algorithm to calculate the sum of two fractions when the denominators are relatively small and the task is familiar; but when confronted
with an estimation task using unfamiliar denominators he makes the common
error of adding the fractions by adding numerators and denominators. In some
contexts and with some representations he asserts confidently that $\frac{2}{8}$ is equal to

$\frac{1}{4}$; but in other contexts or using other representations, he will assert that $\frac{2}{8}$ is less than $\frac{1}{4}$. And, he will assert that $\frac{7}{8}$ is less than $\frac{1}{4}$, justifying his statement by saying that whenever fraction A has a larger denominator than fraction B, that A is smaller in value than B.

The point is not that Brandon is confused, or that he does not understand. The point is that he is building a complex network of understandings — we see him doing so in interaction with Ball — and that some partial understandings and misunderstandings are natural and come with the territory. Anyone who thinks that understanding is simple does not understand understanding. That is why assessment is such a subtle art.

The second point to be observed is there are significant differences in the potential, cost, and utility of different kinds of assessments. It should be clear that it would be impossible to reveal the complexity of Brandon's understanding of fractions using a typical paper-and-pencil test. Simple multiple-choice tests can be useful for accountability purposes, providing a rough accounting of what students know at various points during their academic careers; but they are not really useful for diagnostic purposes, or fine-grained enough to support teachers' decision-making in the classroom. More complex "essay questions" of the type discussed in Chapter 14 of this volume provide a much richer picture of the various aspects of student understanding of fractions and can be used for accountability purposes and to support instruction. But these too are incomplete, as Ball's interview with Brandon demonstrates. The more that teachers can "get inside their students' heads" in an ongoing way, in the way that Ball interacted with Brandon, the more they will be able to tailor their instruction to students' needs. To the degree that we can foster such inclinations and skills in all teachers, and add the diagnostic interview to their toolkits (in addition to more formally structured assessments) the richer the possibilities for classroom instruction. This is not to suggest that teachers should do separate 90-minute interviews with each of their students. The idea is that ordinary classroom interactions provide significant opportunities for noticing what students understand — and for asking probing questions, if teachers are prepared to take advantage of these opportunities. The more teachers know what students know, the more they will be able to build on their strengths and to address their needs.

Section 6
The Importance of Societal Context

To invert Vince Lombardi: context isn't the only thing, it's everything. As the six chapters in this section show, it is sometimes the case that the best way to look inward is to begin by looking outward. Just as fish are said to be unaware of the water in which they swim, those embedded in a cultural context may be unaware of just how much of what they do is socially and culturally determined, and not simply "the way it is" or "the way it must be."

In Chapter 17, Michèle Artigue takes readers on a tour of mathematics assessments in France. As she does, it becomes clear that some aspects of assessment are universal and that some are very particular to local circumstances. The broad goals of mathematics assessment discussed by Artigue are parallel to those discussed elsewhere in this volume. Assessments should, she writes, reflect our deeply held mathematical values; they should help teachers to know what their students are understanding, in order to enhance diagnosis and remediation; they should provide the government with a general picture of the country based on representative samples of pupils. Note that different assessments in the U.S. serve these functions. For example, the National Assessment of Educational Progress (NAEP) provides some "benchmarking" of American student performance. However, NAEP provides no direct feedback to teachers or students. Numerous high-stakes tests at the state level — in some cases, at every grade from second grade on — are highly consequential for students but have no diagnostic value. (No other nation tests students as much and with as little useful feedback as the U.S. It is the case that local assessments such as those discussed in Chapters 10, 12, and 14 do provide diagnostic information at the student, classroom, and district levels. However, they are not used on a larger scale.)

Artigue also discusses the fact that, in France, assessments given before the end of high school are explicitly conceptualized as a mechanism for systemic change and for teacher professional development. The impact of such "levers for change" in a coherently organized system — in France the Ministry of Education essentially determines curricula nationwide — is very different than the impact in a distributed assessment system such as in the United States, where each of the fifty states has its own assessment standards and the nation's 15,000 relatively independent school districts can set about meeting their state's standards with a fair degree of latitude.

In Chapter 18, Mark Wilson and Claus Carstensen take us to the shores of science learning and of assessment design more generally; then they bring us back to mathematics. Originally the Berkeley Evaluation and Assessment Research Assessment system was designed to provide diagnostic and conceptual information about students' understanding of science concepts. The system is grounded in the "assessment triangle" put forth by the National Research Council (2001) in *Knowing What Students Know*, and in four building blocks for constructing an assessment system: (1) examining the growth of knowledge over time (a "developmental perspective"), (2) seeking a close match between instruction and assessment, (3) generating high-quality evidence, that (4) yields information that is useful to teachers and students, and can feed back into instruction. (In Chapter 10, we have seen examples of how (3) and (4) might happen.) Wilson and Carstensen describe the recent application of the system to mathematics assessments.

Chapters 19 and 20 explore issues of mathematics assessment, teaching, and learning for students whose first language is not English. To make the extreme case related to Lily Wong Fillmore's argument in Chapter 19: how well would you do on a mathematics assessment administered in Swahili, and what would your responses reveal about your understandings of mathematics? Also to the point, how long would it take you to get up to speed if you were being taught mathematics in Swahili? To varying degrees, this is the dilemma being faced by the myriad students in U.S. schools for whom English is not yet a comfortable language. In Chapter 20, Judit Moschkovich addresses some related themes. If a student's mathematical understandings are assessed in a foreign tongue, that student's competencies may not be revealed. That is, the person being assessed may well know some things, but he or she may be unable to express them in the language used for the assessment. Note that in this case, providing remediation on content related to the assessment is precisely the wrong thing to do! In addition, Moschkovich makes a convincing case that diagnosis and remediation on the basis of vocabulary is far too narrow. She discusses the linguistic resources that students bring to their mathematical encounters, and the ways that we can recognize and capitalize on them.

The final two chapters in this volume, by Elizabeth Taleporos and Elizabeth Stage, focus on external, systemic realities — on the ways that political contexts shape assessment and its impact. In Chapter 21, Taleporos describes some of the history of assessment and its impact in New York City over the past few decades. Unsurprisingly, people who get mixed messages about what is expected of them get confused — and the confusion breeds problems. Taleporos describes such conflicts in New York in the 1980s, where statewide assessments focused on minimal competency while city-wide assessments focused on the percentage of students who performed above average on a norm-referenced nation-wide test. Chaos resulted. Later on, when testing in the system got more consistent, the situation improved. (There are lessons to be learned here, given the plethora of testing requirements catalyzed by the No Child Left Behind Act.) In Chapter 22, Stage reviews some of the history of California's statewide mathematics assessments. Stage describes the origins of the California Assessment Program, a low-stakes testing regime that provided teachers and school districts with useful information about what students were learning in their classrooms and schools. Political pressures raised the stakes, and with higher stakes came greater statistical constraints and more narrowly defined items. These are very consequential changes: as the title of Stage's chapter suggests, the items that teachers see as exemplifying the test have a significant impact in shaping what they teach. This is not always a good thing.

When it comes to assessment, then, we are living the ancient curse: these are indeed interesting times. There is much to learn if we heed the lessons of the chapters in this section.

Assessing Mathematical Proficiency
MSRI Publications
Volume **53**, 2007

Chapter 17
Assessment in France

MICHÈLE ARTIGUE

Introduction

This chapter provides an overview of the methods currently used in France to assess students' mathematics learning, of the changes occurring in this area, and the rationale for them. As in any country, the methods of assessment used in France are part of the country's general educational culture. In order to understand them, it is helpful to know about the characteristics of this culture. I briefly describe these cultural characteristics, before discussing the methods of assessment. These include both internal assessments done by teachers in classrooms and external assessments.[1] In this chapter, I will focus on the external assessments, in particular, on two different kinds of assessment: the *Baccalauréat*, which is the national examination at the end of high school, and the national diagnostic assessment at the beginning of middle school.[2] Then, after briefly discussing current research on alternative modes of assessment relying on technology, I end this chapter with some general comments on assessment issues.

Some Characteristics of the French Educational System

Students in France begin secondary education at age eleven, after two to three years of kindergarten[3] and five years of elementary school. Secondary education

---

[1] As in any country, the situation is in flux; by the time this volume appears, some things will have changed. Readers who speak French can find current information at the Web sites listed below.

[2] Updated statistics from the *Baccalauréat* can be found on the Web site of the French Ministry of Education: http://www.education.gouv.fr/stateval/. The diagnostic assessments discussed in this paper are accessible on the Web site http://cisad.adc.education.fr/eval/.

[3] In France, the term equivalent to kindergarten, *maternelle*, also designates preschool. Available to children aged from two to five, it is not compulsory — yet almost 100% of children aged three to five attend it. It is regulated by the French Department of Education.

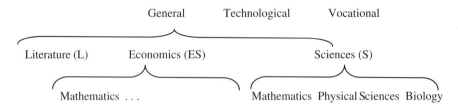

Figure 1. Differentiation in high school. Top row: grade 10 (streams). Middle row: grade 11 (orientations). Bottom row: grade 12 (specialties).

is comprised of two parts: first, a four-year program called *collège* (which I will call "middle school"), and then the *lycée* (which I will call "high school") beginning at grade 10. Compulsory education ends at age sixteen, and thus includes the first year of high school. Prior to high school, the curriculum is the same for all students.[4] In high school, there are three main streams: the general and the technological streams which are three-year programs, and the vocational which includes two successive two-year programs. At present, approximately half of the high school population is in a technological or vocational stream. Each stream is further differentiated. For the general stream for instance, differentiation occurs at grade 11. Then students can choose between three different orientations: Literature (L), Economic Sciences (ES) and Sciences (S); at grade 12 (the last year of high school), they have also to choose a speciality. In S for instance, this can be mathematics, physical sciences, or biology (see Figure 1). The different high school programs all end with a national examination: the *Baccalauréat* (mentioned above) which allows students to enter post-secondary education. However, some post-secondary programs such as the CPGE (specific classes preparing students for national competitions for the most prestigious higher institutions such as the *Ecole Polytechnique* and the *Ecoles Normales Supérieures*, as well as for many engineering and business schools) select their students on the basis of their academic results in high school, before knowing their results for the *Baccalauréat*.

France has a national curriculum. Until 2005, the 21-member national council for curriculum (CNP) designed the general curricular organization and prepared specific guidelines for each discipline; group of experts created the syllabuses for each discipline. These syllabuses were then submitted for approval to the CNP and to various other authorities. As of 2005, the CNP was replaced by the 9-member Haut Conseil de l'Education (HCE), which has less content-specific expertise.

The mathematics curriculum is integrated in the sense that chapters on algebra and geometry alternate in textbooks. Synthetic geometry still occupies a rather

---

[4]Specific instititutions exist for students with disabilities.

important place and, in it, transformations play an essential role. Due to history, and to the introduction of the metric system during the French Revolution, in the curriculum for elementary school and even in the first years of middle school, numbers with finite decimal representations play a more important role than rational numbers in general, and fraction representations are not really used in elementary school. Mathematics occupies about 5.5 hours per week in elementary school and about 3.5 in middle school. Time allotment varies in high school according to the different streams and orientations. Table 1 shows the respective timetables for the general stream in high school. Note that mathematics is optional for the last year in L (this is the only class where it is optional in secondary schools).

Grade and Orientation	Timetable						Workload (range)
	C	M	AI	TD	O	S	
10	3 hr	1 hr	1 hr				4–5 hrs
11 L	1 hr			1 hr	3 hr		2–5 hrs
11 ES	2.5 hr			0.5 hr	2 hr		3–5 hrs
11 S	4 hr			1 hr			5 hrs
12 L					3 hr		0–3 hrs
12 ES	4 hr					2 hr	4–6 hrs
12 S	4.5 hr			1 hr		2 hr	5.5–7.5 hrs

Table 1. Current mathematics allotment for the general stream. C means "Cours"; M, "Modules"; TD, "Travaux dirigés"; O, "Option"; S, "Spécialité" and AI, "Aide individualisée." "Cours" denotes whole-class activities, "Modules" and "Travaux dirigés," half-class activities, "Option," optional courses, "Specialité," the compulsory speciality courses, "Aide individualisée," remedial activities organized for groups of at most eight students.

In high school, the syllabus also varies according to the different streams intended to serve the diversity of students' interests and needs.[5] For instance, the L syllabus for grade 11, called *Informatique et Mathématiques*, focuses on mathematics used in a visible way in society: tables of numbers, percentages, some statistical parameters, and graphical representations; it encourages the systematic use of spreadsheets. The ES syllabus includes more statistics than the others and an option for the speciality in grade 12 includes graph theory, which

---

[5]The official descriptions are accessible on the Web site of the Ministry of Education (www.education.gouv.fr) in French. For overviews in English, the reader can consult [Artigue 2003] or the description of the French system prepared by the CFEM, which is the French subcommission of the International Commission on Mathematical Instruction, for ICME-10, which is accessible on the Web site of the CFEM, http://www.cfem.asso.fr/syseden.html.

does not appear in the other syllabuses. In the S syllabus calculus plays a larger role and the mathematics speciality includes number theory, a topic that does not occur in others.

As mentioned above, mathematics learning in France is assessed in different ways, both inside and outside the classroom. High school teachers are in charge of formative and summative assessments for their courses. The results of these assessments serve to decide if the students will have to repeat courses or not, and to decide which high school programs they will take. Beyond that, there are external modes of assessment. These include some traditional examinations such as the *Brevet des Collèges* at the end of middle school, the *Certificat d'Aptitude Professionnelle* and the *Brevet d'Enseignement Professionnel* at the end of the first two-year program in vocational high schools, and of course the *Baccalauréat*. About fifteen years ago, diagnostic tests taken at the beginning of the academic year were introduced for some students. In this chapter, I will focus on the external assessments, beginning with the traditional examinations.

Traditional Examinations: The Case of the *Baccalauréat*

The first examination students take is the *Brevet des Collèges*. This is an examination taken at the end of middle school that combines written tests in French, Mathematics, Geography, and History, and the results of the internal evaluation of the students in grades 8 and 9. This examination is not compulsory and results do not affect students' choice of high school programs. In 2003, it was taken by about 700,000 students and the overall success rate was 78.3% (82% for girls and 75% for boys).

The most important examination is the *Baccalauréat*. This is a national examination at the end of high school, and it is necessary to pass it to enter post-secondary education. About 69% of the population now attend the last year of high school (75.4% for girls and 63% for boys) and take the *Baccalauréat*. In 2002, the 628,875 candidates for the *Baccalauréat* were distributed as follows: general *Baccalauréat* 52.2%, technological *Baccalauréat* 29.3%, vocational *Baccalauréat* 18.5%.

Overall pass rates for the different streams in 2003 are given in Table 2.

In the same year, in the general stream, pass rates for the different orientations were: 82.2% in L, 79.4% in ES, and 80.1% in S. Girls represented 45.6% of the new *bacheliers* in S, 65.7% in ES, and 83.6% in L. Pass rates increased sharply over the past twenty years (in 1980, only 34% of an age-cohort attended the last year of high school) but they have now stabilized.

The duration and the content of the mathematics part of the examination is different in the different streams and orientations, as is the weight of mathematics in the global mean which decides of the attribution of the *Baccalauréat*.

Stream	Girls	Boys	Total
General	82.3%	77.5%	80.3%
Technological	79.8%	73.8%	76.8%
Vocational	78.6%	75.2%	76.6%
Total	81.0%	75.8%	78.6%

Table 2. 2003 pass rate for the *Baccalauréat*.

Table 3 gives the corresponding statistics for the general stream. As can be seen, even in the S orientation, the weight of mathematics is only about 20%.

Traditionally, the mathematics test consists of two exercises and one problem with several parts, the whole covering important parts of the syllabus. Exercises may involve different topics: number theory (in S with mathematics speciality), complex numbers, two- and three-dimensional geometry, probability and statistics, sequences, graph theory (in ES with mathematics speciality). One of the exercises is different for those having choosen the mathematics speciality in ES or S. The problem itself generally deals with functions and calculus, and in ES it is often motivated by an economics context.

In order to give the reader a precise idea of what such a test might be, I have translated the 2003 test given to the students with orientation S who specialized in mathematics[6] (see Appendix 1). This test follows the traditional structure but was considered very difficult by students and teachers. On the one hand, it was the first test aligned with the new high school curriculum,[7] and on the other hand while globally obeying the traditional test structure, it did not respect some traditional standards. Thus, it is a good illustration of what is standard in such a test and what is not. The examination was the source of passionate

	Written test duration in hours	Coefficient
L	1.5 (taken at grade 11)	2 out of 38
L	3 (if option taken in grade 12)	4 out of 34
ES	3	5 out of 37 (+2 if speciality)
S	4	7 out of 38 (+2 if speciality)

Table 3. Weight of mathematics in the G stream

---

[6]The test, referred to as the *metropolitan* test, is given in continental France. The tests given in the overseas territories and in foreign centers did not generate the same discussions.

[7]The new high school curriculum was introduced in grade 10 in 2000–2001, and thus reached grade 12 in 2002–2003.

discussions all over the country and in the media. Indeed, the Minister of Education himself had to make a statement on national TV. The importance given to the *Baccalauréat* in French culture, the fact that the scientific orientation is still considered the elite orientation certainly contributed to the strength of the reactions.

Analyzing the metropolitan test for the S orientation in 2003. The first exercise has a traditional form. Complex numbers are used for proving geometrical properties: showing that a given triangle is a right isosceles triangle and that given points lie on the same line. A geometrical transformation is involved. In this case, it is a rotation (it could have been a similarity) and students have to express it as a complex transformation. Students have to use different representations for complex numbers (cartesian and exponential) and to express the image of a set of complex numbers (a circle, in this case) under the transformation. Such an exercise cannot be solved without a reasonable knowledge of complex numbers and connections between complex numbers and geometry, but these are standard expectations. Moreover, students are carefully guided by a lot of subquestions, can check their answers by reasoning on a geometrical figure, and the computations are not technically very complex. This is also standard.

Exercise 2 is specific to the mathematics speciality[8] and connects three-dimensional geometry and number theory. The latter is not so traditional because three-dimensional geometry is not frequently examined in the *Baccalauréat*. Many teachers do not feel comfortable with it or do not like it, and when they lack time, they often skip this part of the course in high school. Students are thus less prepared to solve problems in three-dimensional geometry. Moreover the test questions cannot be considered routine. This is especially the case for Question 2 whose form is not common and whose difficulty could have been easily limited by saying that the planes are parallel to coordinate planes instead of parallel to a coordinate axis. The last part including number theory is not completely new in spirit because number theory has sometimes been linked to a geometrical context in such *Baccalauréat* exercises, but generally it has been related to two- rather than three-dimensional geometry. What is new is the expected reasoning based on computations using congruences (which have just entered the curriculum), the autonomy given to the students in selecting a solution method for solving Question 3 (which is nonroutine), and the fact that solving of Question 4 involves an infinite descent process or reasoning by

---

[8]The second exercise for the students who had not chosen the mathematics speciality dealt also with three-dimensional geometry. Students were asked to study different properties of a tetrahedron OABCD whose faces OAB, OAC, and OBC were right triangles (with right angle at O), and verifying that OA = OB = OC. This exercise, which used knowledge of three-dimensional geometry taught two years before in grade 10, was more criticized than the speciality exercise.

contradiction and minimality. These are not familiar forms of reasoning for the students and the question format does not guide them.

What was in fact most criticized on this test was not Exercise 2, even if it was considered rather difficult, but the problem. At first look, this problem could be considered standard — it deals with calculus, has a long multipart text, and students are guided by a lot of questions and hints. Nevertheless, it is far from standard. First of all, it reflects some new aspects of the curriculum, especially the emphasis on relationships between different scientific disciplines and on modeling activities. The problem tests students' knowledge of calculus as usual, but by asking them to use this knowledge in the analysis of a situation in biology. Two competing models are given for representing the increase in a population of bacteria, and students after studying these two models are asked to judge their adequacy to represent some empirical data. This by itself would have been enough to make this problem unfamiliar because students would not find anything similar in the *Annales du Baccalauréat* traditionally used to prepare for the examination. This unfamiliarity, nevertheless, was not the only cause of the violent reaction the problem generated. What was strongly criticized was the way the problem was set up and made more complicated by the introduction from the beginning of many parameters (in the French secondary curriculum, the use of parameters is very limited and students are not used to carrying out symbolic computations with many parameters). Moreover, the problem did not follow the usual tradition that the beginning of each part is straightforward and that the difficulty increases progressively. For instance, Question 1a in Part B asked the students to make a change of unknown in a differential equation. This is a complex task which requires having an elaborated conceptualization of the idea of function. For many students this was undoubtedly the first time they encountered such a task (the study of differential equations is only a small part of the syllabus, and students only have to know how to solve very simple linear differential equations with constant coefficients). Question 2d was also criticized, this time because it did not make sense in this situation, and seemed to have been introduced only to check that students could relate mean and integral, and were able to calculate the antiderivative of the function g.

Teachers and students are used to the fact that items in the *Baccalauréat* are long, occupying several pages, and that they cover a wide part of the syllabus. They are also used to items where the autonomy of the student is rather limited. The problems posed are not generally easy to solve but many subquestions and intermediate answers help the students, enabling them to continue even if they have not succeeded at solving a particular question. The 2003 test items for the orientation S, while it did have many of these characteristics, was much more cognitively demanding. This, together with the existence of the new syllabus for the test, deeply destabilized teachers and students. This example raises unavoidable questions about what is really assessed or can be really assessed through this kind of examination, and about the distinction between what is routine and what is nonroutine — a distinction which is necessarily fuzzy and dependent on institutions and educational cultures.

Current debates and changes. The *Baccalauréat* appears today to be a very burdensome and costly system of evaluation with both positive and negative effects. Its national and external character is generally seen as positive, together with the fact that the desire to pass the exam motivates students to work. But its existence makes the academic year one month shorter because high school teachers have to administer the tests, correct the tests, and organize oral examinations for those whose score lies between 8 and 10. Because of the test's national character, a minor local incident may have profound effects. Moreover, mathematics teaching in grade 12 is, in many places, too strongly oriented toward preparation for the test, and the types of knowledge that this kind of test cannot assess tend to be neglected. Nevertheless, changing the *Baccalauréat* appears very difficult. The *Baccalauréat* is like a monument, a myth, which is part of the culture. Each time the possibility of change is mentioned, impassioned reactions come from all over. Commissions have been created, different possibilities have been discussed (and some locally tested): replacing the present structure by four or five exercises in order to avoid the constraints resulting from a long problem, giving more autonomy to the students with shorter texts and favoring more reflective work, combining written tests with other forms of assessment . . . but, no substantial changes have occurred.

The events of 2003 did have some consequences. The Ministry of Education decided to prepare secondary education for some unavoidable changes. The General Inspectorate, an institutional body of general inspectors that is responsible for the *Baccalauréat*, was asked to prepare exercises aligned with the new curriculum, which were more diverse in both form and content, and better aligned with the role played by technology in mathematics education today. A first set was put on-line in December 2003[9] and has been discussed in regional meetings with inspectors and in teacher training sessions. Translations of two of its exercises appear in Appendix 2.

As of 2004 the *Baccalauréat* assessments have been effectively structured around four to five exercises. In addition, in 2005, new items in which students are asked to prove results that are from the syllabus were introduced. These items, known as ROC (Restitution Organisée de Connaissance) items, have been

---

[9] See www.education.gouv.fr

controversial. Examples can be downloaded from the Ministry Web site, http://eduscol.education.fr/D0056/accessujets.htm.

(In 2005, the Minister of Education tried to introduce some substantial changes in the *Baccalauréat* (for example, assessing some disciplines through the results of internal assessment as is the case for the *Brevet des Collèges*) but he had to give up these plans because of strong student opposition. Students argued that the *Baccalauréat* would lose its national character and that the diplomas would be given different value depending on which high schools students came from.)

Further, a change in assessment methods resulted from the introduction of the TPE (supervised project work). This new didactic organization was introduced in grade 11 in 2000–2001 and extended to grade 12 the following year. Working in small groups over one semester, students work on a collective product, starting from varied resources, on a subject chosen by them from a list of national topics. The TPE must involve at least two disciplines, including one required for the students' high school program. The students' project is supervised by teachers from the disciplines concerned. Two hours per week are reserved for the TPE in the students' timetable.

This new didactic organization has several aims:

• providing students with the opportunity to develop a multidisciplinary approach to questions which are not just school questions;
• helping them to mobilize their academic knowledge in such a context;
• broadening their intellectual curiosity;
• developing their autonomy;
• helping them to acquire methods and the competencies required for working in groups;
• developing the abilities necessary for an effective search, selection, and critical analysis of documentary resources;
• and, finally, establishing more open relationships between teachers and students.

This work and the scientific competencies it allows students to build cannot be assessed by the standard examination. In fact, its evaluation takes into account the production of the group of students as well as their written and oral presentations. As of 2005, the TPE only exists in grade 11. Students can choose to have their TPE evaluation taken into account as part of their *Baccalauréat* evaluation. As this chapter goes to press, new forms of assessment for the *Baccalauréat,* which are more experimental and require intensive use of technology, are being explored.

As explained above, even if among the different methods of assessment, external examinations are considered most important, there are other ways to assess students' mathematical knowledge. In the next section quite different assessment methods are discussed: the diagnostic tests which were introduced about fifteen years ago.

Diagnostic Assessment: The Case of the Grade 6 Test

Diagnostic tests at the beginning of the academic year were introduced at the start of middle school and high school about fifteen years ago. For years, they were compulsory in grade 3 and grade 6; in 2002, they were extended to grade 7. They concern both French and Mathematics.

The diagnostic assessment, carried out by the DEP (Direction of Evaluation and the Future) at the Ministry of Education, has two main functions:

- providing the Ministry with a general picture of the country based on representative samples of pupils. The pupils' scores, analyzed according to different socio-demographic variables such as age, sex, nationality, parents' profession, are published every year and are also accessible on-line[10];
- providing teachers with a tool allowing them to better identify and answer the mathematical needs of the different categories of their pupils.

But the educational institutions would like it to be also:

- a way for accompanying curricular changes and influencing teachers' practices;
- a tool to help the regional academic institutions appreciate local needs for teacher training and to develop appropriate training programs.

In the following, I will focus on the diagnostic assessment administered when students enter secondary school in grade 6. The assessment items change every year but keep the same structure and form. The assessment is a collection of short exercises (in 2002, 39 exercises out of 77 questions) mixing multiple-choice and constructed-response items (some appear in Appendix 3). Thus they are more similar in format to U.S. tests than are the traditional French examinations. Items are designed according to a matrix of general competencies and mathematical domains. Table 4 gives the test structure and results for 2002.

Note that the standard deviation is rather high. The top ten percent of the students have a success rate of 91.9% while the bottom ten percent only reach 28%. There are also important differences in scores between middle schools when classified by social characteristics: the mean score of the middle schools located in ZEP (priority area for education) is only 52%. We also note that girls outperform boys in French, but not in Mathematics (66.3% versus 63%).

The test is administered during three ordinary 50-minute classroom sessions during the second and third weeks of the academic year. Middle schools receive

---

[10] See www.education.gouv.fr/Evace26

Domains	number of items	mean	median	standard deviation
Numeration and number notation	17	65.2	70.6	21.8
Operations	18	69.9	72.2	21.5
Numerical problems	6	65.2	66.7	29.6
Geometrical tasks	20	61.6	65	21.8
Data processing	16	63.3	68.7	23.3
Total score		65	67.5	19.6
Competencies				
Seek information, interpret, reformulate	8	61.7	62.5	24.9
Analyze a situation, organize a process	29	64.3	65.5	22
Produce an answer, justify it	10	57.7	60	24.9
Apply a technique	7	78	85.7	19.8
Directly use a piece of knowledge	23	66.1	69.6	22.1

Table 4. Mathematics test structure and 2002 scores.

from the Ministry of Education as many booklets with the different exercises as they have pupils, and specific booklets for the teachers. The latter explain the general structure of the test, test administration conditions (for instance, the amount of time given to pupils for solving specific exercises or groups of exercises in order to increase uniformity in testing conditions), the aim of each exercise and the competencies it is intended to test, how to score the answers, and what kind of help can be provided to the pupils who fail.

The teachers supervise the administration of the test, and then enter their students' answers into a document using specific software. The documents are then sent to the DEP which selects representative samples and produces an overall statistical analysis. By using the software, teachers can access the results of their pupils. They can also use it to obtain different statistics, and to identify groups of students according to characteristics of their answers. Middle schools are also asked to meet with the pupils' parents in order to present to them the results of the evaluation.

The test does not attempt to cover the whole elementary syllabus. Teachers are invited to use the on-line database for previous tests to find complementary items if they want to have a more accurate vision of the state of their class when beginning particular chapters. Moreover, having in view formative rather than summative assessment, the test designers include tasks associated with competencies that students are not expected to have fully mastered by the end

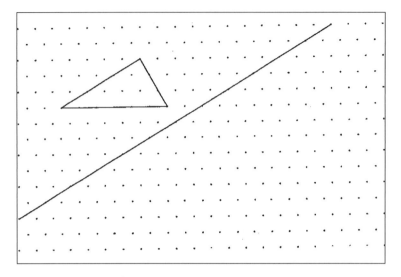

Figure 2. Illustration of task asking for reflection of a triangle.

of elementary school. This is, for instance, the case for the task illustrated in Figure 2 which asks for the construction of the reflection of a triangle.

In elementary school, students become familiar with tasks that involve draw- ing reflections of figures reflected across a horizontal or vertical axis on rect- angular grids. For this, perception is very helpful. In contrast, the task on the diagnostic test uses a grid of points that form a triangular network and the axis of reflection is neither horizontal nor vertical (see Figure 2). Drawing the image requires more than perception, students have to use the mathematical properties of reflection in an analytical way. Didactic research shows that different kinds of errors can be expected in such a context, for instance, mixtures between reflection and translation. Some of these are given in the teachers' booklets in order to enable teachers become aware of them. Such tasks can thus con- tribute to showing the teachers the new competencies which must be gained in grade 6 with respect to reflection. The same can be said for tasks involving the invariance of area under decomposition and recomposition as in the exercise shown in Figure 3; and also for tasks asking for justifications about the nature of quadrilaterals or for tasks involving decimals (see Appendix 3).

There is no doubt that this diagnostic assessment offers a view of assessment which is very different from that described for the *Baccalauréat*. The tools that were developed for it in the last fifteen years, the collected data and their statistical analysis are clearly useful for understanding and adapting instruction to pupils' mathematical knowledge upon entering middle school. As mentioned above, information from diagnostic tests can also help smooth the transition between the cultures of elementary and secondary school, by pointing out some

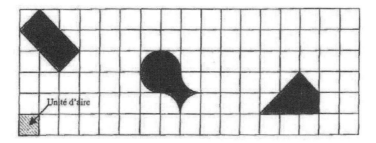

Figure 3. Expressing areas of figures with respect to a given unit of area.

necessary evolutions in the institutional relationship to mathematics, and by making middle school teachers sensitive to the fact that many mathematical notions introduced at elementary school are still under construction. At a more global level, the results of the diagnostic test display what is hidden by a score given by a single number: 65% which can be considered acceptable because the scope of the test is not limited to notions that must be fully mastered at the end of elementary school. The test results also display differences between performance in different areas and competencies (see Table 4), but, most importantly, differences between the top ten and the bottom ten percent of students, between middle schools in low-income areas and others. Even at the beginning of middle school, strong educational inequalities exist in France and must be addressed. The information I collected when preparing this chapter left me with the feeling that the diagnostic test is not used as well as it could be, and that we could benefit more from the investment made in its development. It is the case that teacher training sessions have been regularly organized with both elementary teachers and secondary teachers in order to use the results of the diagnostic test to promote reflection on the transition from elementary to secondary school; and, to help secondary teachers reconsider classroom management and remediation. Nonetheless, these sessions do not seem to have seriously influenced teachers' practices.

Understanding the precise effects of the different methods of assessment on the learning and teaching of mathematics, overcoming the limitations of most current assessment tools, and using the information they give us in order to improve mathematics teaching and learning is not easy. It surely requires research. In the next section, I describe briefly a research project that I am involved in, whose aim is to develop diagnostic and remedial tools in a specific domain — algebra — by relying on the new possibilities offered by technology.

Looking Towards the Future: The LINGOT Project

The LINGOT project is a multidisciplinary effort involving researchers in mathematics education, artificial intelligence, and cognitive ergonomy.[11] Its origin was a multidimensional model for considering competence in elementary algebra developed by Grugeon in her doctoral thesis [Grugeon 1995]. To this multidimensional model was attached a set of 20 tasks which allowed the exploration of the students' algebraic competence according to the various dimensions of the model. One ambition of this model and of the associated set of tasks is to allow the identification of lines of coherence in students' algebraic behavior, which are then used in order to design appropriate didactic strategies. The empirical part of the doctoral thesis showed how productive this model was in identifying such lines of coherence and helping students who had difficulty with algebra.

But the diagnosis was burdensome and rather sophisticated. Collaboration with researchers in artificial intelligence offered new perspectives. Building a computer version of the set of tasks, automatically coding responses to multiple-choice tasks, and designing ways to help teachers code other responses could certainly lighten the diagnostic process. This was the goal of the PEPITE project.

The LINGOT project currently under development is a third phase [Delozanne et al. 2003]. One aim is to transform the diagnostic tool provided by PEPITE into a flexible one which can be adapted by the teachers for their particular aims without losing the automated diagnosis. Another aim is to reduce the number of tasks necessary for diagnosis by relying on a dynamic diagnosis using Bayesian models. As requested by the teachers who piloted PEPITE, the LINGOT project intends to link the diagnosis to didactic strategies by associating to it potential learning trajectories. A final aim is to understand how teachers can appropriate such tools — which are at variance with their usual diagnostic methods — and transform these into professional instruments. Of course, in the project all these dimensions mutually intertwine. We are far from the end of this project, and the difficulties we encounter often seem barely surmountable, but at the same time, I have the feeling that it helps us to better understand the cognitive and institutional complexity of assessment issues.

Concluding Comments and Questions

In this chapter, I have briefly presented some of the methods used in France to assess students' mathematical knowledge, focusing on external examinations,

---

[11] For a description of the PEPITE and LINGOT projects see http://pepite.univ-lemans.fr/. There is a link to a program description for English speakers.

but balancing the contribution between two very different types of assessment: (1) the traditional examination whose paradigmatic example is the *Baccalauréat*, and (2) the diagnostic test whose paradigmatic example is the test taken by pupils on entering middle school. Of course, these two types of assessment do not have the same function. The *Baccalauréat* is a summative evaluation, and the grade 6 test is a diagnostic evaluation which is situated within a formative perspective. Due to its summative character and its institutional importance — it is a condition for entrance into post-secondary education and gives students the right to enter a university — the *Baccalauréat* strongly influences teaching in the last year of high school. The grade 6 test does not have a similar influence on teaching in grade 5, the last year of elementary school. These two tests raise common issues which are general issues relevant for any form of assessment. I will mainly focus on two of these general issues:

- How can one make assessment reflect fundamental characteristics of mathematical knowledge such as the following: the diversity of its facets on the one hand, the fact that it is to a large extent contextualized knowledge on the other?
- How can one make assessment reflect the values we want to transmit in mathematics education?

Students' mathematical knowledge is not easy to understand and is inherently multidimensional. Researchers have built different models in order to describe this multidimensionality; for example, the model created by Kilpatrick was mentioned several times during the MSRI conference [NRC 2001]. Some dimensions seem *a priori* easier to understand than others: for instance, investigating technical competencies seems easier than investigating mathematical attitude or creativity. But even technical competencies are not so easy to determine. Small variations in the labelling of the tasks, or in their didactic variables can generate large differences in students' responses. The boundary between what is routine and what is not is fuzzy and depends on each student's experience, because knowledge emerges from mathematical practices. Moreover, an important part of this knowledge remains tightly attached to the precise context of these practices. What takes the status of what we call in French *savoir* (knowledge) in order to differentiate it from *connaissance* (acquaintance), and is thus partially decontextualized is only a small part of what we know [Brousseau 1997]. What can be discovered through an assessment instrument is thus strongly dependent on the nature of the instrument, and on the way this instrument is situated with respect to the mathematical practices of the student who takes this test. What a test reveals is also very partial, a small part of what might be examined. The characterization of stuent competencies is created by the instrument. Another instrument that appears very similar might give very different results.

The theoretical model of Grugeon mentioned previously had four main dimensions: formation and treatment of algebraic objects, conversions between the language of algebraic representation and other semiotic registers (see, e.g., [Artigue 1999; Duval 1996]), the tool dimension of algebra, and the nature of the argumentation used in algebra. To each dimension were attached several subdimensions and for each of these, different criteria. In order to determine coherences in the algebraic functioning of the students, it was decided to test each criterion at least five times in different contexts. This resulted in an complex interrelated set of tasks: an approximately two-hour test was necessary just to get a workable account of the students' relationship with elementary algebra! This is the reason why we turn today towards technology, with the expectation that, thanks to the potential it offers, we could design a dynamic test where the tasks given to the students would be, after a first exploration of their cognitive state, carefully chosen in order to test and improve provisional models of their understanding of algebra, up to a reasonably stable level. Anyone who attended the workshop certainly remembers Deborah Ball's one and a half hour interview of the grade 6 boy (see Chapter 15), and can easily imagine how difficult it would be gain comparable insight into this boy's understanding of fractions using a written test, and how very different the images of his understanding might be if he were given different written tests. Despite the length of the interview, Ball mainly approached the object rather than the tool dimension of fractions, making the image we got rather partial.

Up to now, in this conclusion, I have focused on the difficulties we have approaching students' mathematical knowledge due to the diversity of its forms, due to its dependence on particular practices and contexts. This certainly must make us modest when approaching assessment issues. This also shows the importance of research. Thanks to educational research on algebra and fractions — not to mention other domains — we have at our disposal ways to approach the complex structure of students' mathematical knowledge in particular areas, for drawing inferences from what we observe, and for synthesizing the partial results we obtain in a productive way.

Nevertheless, despite the limitations of the enterprise that have been indicated, assessment is necessary to the life of educational systems, to the point that any form of knowledge which is not assessed or cannot be assessed lacks legitimacy. Approaching assessment within this perspective changes the general frame. From a cognitive frame we pass to an anthropological one, considering the role assessment plays in a human organization and questioning its ecology. We come thus to the second question: how to make assessment reflect the values we want to transmit through mathematics education? At the workshop, no one could escape the feeling that, in many U.S. states, the high-stakes tests that have been introduced do not reflect these values. Their negative effects were stressed by many speakers: as a result of testing the teaching enterprise was reduced to test preparation, and whatever the quality of these tests, this reduction had dramatic effects, For example, innovative experiments could no longer survive because they followed a completely different approach. On the other hand, those involved in the design of the tests showed the complexity of the work that is being carried out, expressing themselves in terms of validity, reliability, measurement, and scientific rigor. It was quite interesting to try to understand these discussions from the viewpoint of another culture with a different experience of assessment. In regard to the values that mathematics education intended to transmit, there is no distance at all: the discourse was the current discourse of democratic and developed countries who have to face immigration and increasing social inequalities.

Conversely, the educational culture with regard to assessment is rather different. In France, as described above, a national external examination at the end of high school is part of the culture. This examination, as I have tried to show, has to be aligned with the curriculum and also to respect a certain level of mathematical performance. Making these change is not easy. Many questions can be considered as routine questions. This is not, *per se*, a problem, because one of the goals of this examination is to check that some tasks have really become routine and that students have mastered the techniques they have been taught for solving them. What has been more closely scrutinized is the balance between routine and nonroutine tasks. I may be wrong but I do not have the feeling that the scientific rigor of the assessment is a core question. What seems to be important is to have — a utopia — all the students treated the same way thanks to the same national scale. That is, equity issues are paramount. And, for the government, what is important is to mantain the current pass rates on the examination. Nevertheless, the question of the influence of the *Baccalauréat* on teaching practices is also recognized as an important issue. Even on an examination with constructed-response questions, where students are asked to justify their answers, teaching practices excessively oriented towards its preparation limit the kind of tasks given to students, and above all, tend to reduce the diversity of students' modes of expression. For example, standard sentences are learned for justifying a given property. This has negative effects on the image of mathematics that students develop and does not help them make sense of mathematical practices — but for teachers and students this formulaic approach acts as an institutional norm. Some recent changes in the calculus syllabus have been introduced explicitly in order to avoid what have become template questions but there is no doubt that schools will quickly adapt to these changes.

Limiting the unavoidable negative effects from assessment and enhancing potential positive effects is not easy for educational systems. Doing so requires a better understanding of the effects of assessment and of the mechanisms that produce them, and it requires adequate systems of regulations grounded in those understandings. Further research is necessary in order to inform political decisions.

Appendix 1: Baccalauréat Test for Orientation S with Mathematics Speciality Given in Continental France in 2003

Exercise 1 (4 points)

In the complex plane with an orthonormal coordinate system (with unit length 2 cm), consider the points A, B and C whose respective affixes[12] are $a = 2$, $b = 1 - i$, and $c = 1 + i$.

1. (1 point)

   a. Place the points A, B and C on a figure.

   b. Compute $(c - a)/(b - a)$. Deduce from this computation that ABC is a right isosceles triangle.

2. (1 point)

   a. r is the rotation with center A such that $r(B) = C$. Determine the angle of r and compute the affix d of point $D = r(C)$.

   b. Let Γ be the circle with diameter BC.
      Determine and construct the image Γ' of Γ under r.

3. (2 points)
   Let M be a point of Γ with affix z, distinct from C, and let M' with affix z' be its image under r.

   a. Show that there exists a real number θ belonging to $[0, \pi/2) \cup (\pi/2, 2\pi)$ such that $z = 1 + e^{i\theta}$.

   b. Express z' as a function of θ.

   c. Show that $(z' - c)/(z - c)$ is a real number. Deduce from this that the points C, M, and M' are on the same line.

   d. Place on the figure the point M with affix $1 + e^{i2\pi/3}$ and construct its image M' under r.

Exercise 2 (speciality exercise, 5 points)

Questions 3 and 4 are independent of Questions 1 and 2, except that the equation Γ given in Question 1c occurs in Question 4.

---

[12] A point in the complex plane with cartesian coordinates (a, b) is said to have affix $a + bi$.

1. (1.5 points)

The space is equipped with an orthonormal system of coordinates.

a. Show that the planes P and Q whose respective equations are:

$$x + \sqrt{3}\,y - 2z = 0 \quad \text{and} \quad 2x - z = 0$$

are not parallel.

b. Give a parametric representation for the line Δ, the intersection of the planes P and Q.

c. Consider the cone of revolution Γ whose axis is the x-axis and has the line Δ as a generator.
Prove that the equation of Γ is $y^2 + z^2 = 7x^2$.

2. (1 point)

Two intersections of Γ with planes, each plane parallel to a coordinate axis, are represented below.

In each case, determine a possible equation of the plane, and carefully justify your answer.

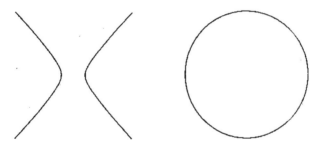

3. (1.5 points)

a. Show that the equation $x^2 \equiv 3 \pmod{7}$, whose unknown x is an integer, does not have solutions.

b. Prove the following property:
For every integer a and b, if 7 divides $a^2 + b^2$, then 7 divides a and 7 divides b.

4. (1 point)

a. Let a, b and c be integers different from 0. Prove the following property:
if point A with coordinates (a, b, c) is on the cone Γ, then a, b, and c are multiples of 7.

b. Deduce from it that the only point of Γ whose coordinates are integers is the summit of this cone.

Problem (11 points)

Let N_0 be the number of bacteria introduced in a growth medium at time $t = 0$ (N_0 is a positive real number in millions of bacteria).

This problem aims to study two growth models for this bacterial population:

A first model for a short period following the introduction of the bacteria into the growth medium (Part A).

A second model which can be used over a longer period (Part B).

Part A (2 points). In the time following the introduction of the bacteria into the growth medium, assume that the rate of increase of the bacteria is proportional to the number of bacteria.

In this first model, denote by $f(t)$ the number of bacteria at time t (expressed in millions of bacteria). The function f is thus a solution of the differential equation $y' = ay$ (where a is a positive real number depending on the experimental conditions).

1. Solve this differential equation with the initial condition $f(0) = N_0$.
2. T is the time of doubling of the population. Prove that, for every positive number t:
$$f(t) = N_0 2^{(t/2T)}$$

Part B (3 + 4 points). The medium being limited (in volume, in nutritive elements ...), the number of bacteria cannot indefinitely increase in an exponential way. The previous model cannot thus be applied over a long period. In order to take these observations into account, the increase in the population of bacteria is represented as follows.

Let $g(t)$ be the number of bacteria at time t (expressed in millions of bacteria), g is a strictly increasing function which admits a derivative on $[0, \infty)$, and satisfies on this interval the relationship:

(E) $$g'(t) = ag(t)\left[1 - g(t)/M\right]$$

where M is a positive constant which depends on the experimental conditions and a is the real number given in Part A.

1. a. Prove that if g is a strictly positive function satisfying (E) then the function $1/g$ is a solution of the differential equation (E'): $y' + ay = a/M$.

   b. Solve (E').

   c. Prove that if h is a strictly positive solution of (E') $1/h$ satisfies (E).

2. In the following, it is assumed that, for every positive real number t, we have $g(t) = M/(1 + Ce^{-at})$, where C is a constant, strictly larger than 1, dependent on the experimental conditions.

a. Find the limit of g at ∞ and prove that, for every nonnegative real number, the two inequalities $0 < g(t) < M$ hold.

b. Study the variations of g (it is possible to use the relation (E)).
Prove that there exists a unique real number t_0 such that $g(t_0) = M/2$.

c. Prove that $g'' = a(1 - 2g/M)g'$. Study the sign of g''. Deduce from this that the rate of increase of the number of bacteria is decreasing after the time t_0 defined above. Express t_0 as a function of a and C.

d. Knowing that the number of bacteria at time t is $g(t)$, calculate the mean of the number of bacteria between time 0 and time t_0, as a function of M and C.

Part C (2 points)

1. The table below shows that the curve representing f passes through the points with coordinates $(0, 1)$ and $(0.5, 2)$.
Deduce from that the values of N_0, T and a.

2. Knowing that $g(0) = N_0$ and that $M = 100N_0$, prove that for every positive real number the following equality holds:

$$g(t) = 100/(1 + 99 \times 4^{-t}).$$

3. Draw on the attached sheet [see graph below] the curve Γ representing g, the asymptote to Γ and the point of Γ with abscissa t_0.

4. Under what conditions does the first model seem to you consistent with the observations made?

t (in h)	0	0.5	1	1.5	2	3	4	5	6
Millions of bacteria	1.0	2.0	3.9	7.9	14.5	37.9	70.4	90.1	98

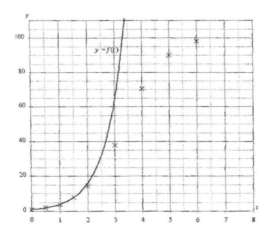

Appendix 2: New Proposals For Assessment at the Baccalauréat

Exercise 6 The plane is equipped with an orthonormal system of coordinates. Let f be the function defined on $\mathbb{R}$ by

$$f(x) = e^{2x}/2 - 2.1e^x + 1.1x + 1.6.$$

1. Graph the curve representing f in the window $-5 \le x \le 4$, $-4 \le y \le 4$. Reproduce the curve on your assessment sheet.
2. From this graphic representation, what do you conjecture:

   a. About the variation of the function f (where f is positive or negative, increases or decreases).

   b. About the number of solutions of the equation $f(x) = 0$.

3. Now we want to study the function f.

   a. Solve in $\mathbb{R}$ the inequality $e^{2x} - 2.1e^x + 1.1 \ge a$.

   b. Study the variation of f.

   c. Deduce from that study the number of solutions of the equation $f(x) = 0$.

4. If you want to draw the curve of f on the interval $[-0.05, 0.15]$ and to visualize the results of Question 3, what extreme values for y can be chosen for the calculator window?

Exercise 7 Consider a sequence $\{u_n\}$ with $u_n \ge 0$, and the sequence $\{v_n\}$ defined by

$$v_n = u_n/(1 + u_n).$$

Are the following propositions true or false? Justify your answer in each case.

1. For every n we have $0 \le v_n \le 1$.
2. If the sequence $\{u_n\}$ is convergent, so is the sequence $\{v_n\}$.
3. If the sequence $\{u_n\}$ is increasing, so is the sequence $\{v_n\}$.
4. If the sequence $\{v_n\}$ is convergent, so is the sequence $\{u_n\}$.

Exercise 20 The different questions are independent. All the answers have to be carefully justified.

1. In each of the following cases, give a function f satisfying the given properties. An expression for f must be given.

   a. f is defined on $\mathbb{R}$ by $f(x) = ae^{2x} + be^x + c$, the limit of f at ∞ is ∞, and the equation $f(x) = 0$ has two solutions, 0 and $\ln 2$.

   b. f is defined on $[0, \infty)$, $f(2) = 4$ and, for every positive[13] real numbers x and y, we have $f(xy) = f(x) + f(y)$.

---

[13] Here and below "positive" is used with its usual English meaning, so in this case $x > 0$ and $y > 0$.

c. f is a polynomial function of degree at least 2 and the mean value of f on $[-2, 2]$ is 0.

2. Let g be a function that is defined and differentiable with derivative g' that is continuous on $[-1, 1]$. The curve representing g is drawn below.
Are the following assertions consistent with this graph?

a. $\int_{-1}^{1} g'(x)\, dx = 0$

b. $\int_{-1}^{1} g(x)\, dx > -\frac{1}{2}$

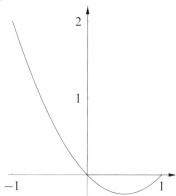

Exercise 30 Strange numbers!
Numbers such as 1, 11, 111, 1111, etc. are called rep-units. They are written only with the digit 1. They have a lot of properties which intrigue mathematicians. This exercise has the aim of revealing some of these.

For a positive integer k, the rep-unit written with k 1's is denoted N_k:

$$N_1 = 1, \quad N_2 = 11, \ldots$$

1. Give two prime numbers less than 10 that never appear in the decomposition in prime factors of a rep-unit. Justify your answer.
2. Give the decomposition in prime factors of N_3, N_4, N_5.
3. Let n be an integer strictly larger than 1. Suppose that the decimal expression of n^2 ends in 1.

a. Prove that the decimal expression of n ends in 1 or in 9.

b. Prove that there exists an integer m such that n can be written in the form $10m + 1$ or $10m - 1$.

c. Deduce from this that $n^2 \equiv 1 \pmod{20}$.

4. a. Let $k \geq 2$. What is the remainder in the division of N_k by 20?

b. Deduce from this that a rep-unit different from 1 cannot be a square.

Appendix 3: 2002 Grade 6 Diagnostic Test Exercises with Allocated Time and Pass Rates

NUMBERS AND OPERATIONS

Exercise 1 (2.5 min)   *Mental calculation*

In box a, write the result of: 198 plus 10   84.1%
In box b, write the result of: 123 plus 2 tens   73.8%
In box c, write the result of: 37 divided by 10   56.0%
In box d, write the result of: 7 multiplied by 10,000   87.9%
In box e, write the result of: 405 minus 10   80.8%

Exercise 5 (2 min)   86.5%

A bus started from the school at 8:30 a.m. and arrives at the museum at 9:15 a.m. How long did it take to travel?

Exercise 12 (2 min)   72.4%

Order the following numbers from smallest to largest.

$$2 \quad 2.02 \quad 22.2 \quad 22.02 \quad 20.02 \quad 0.22$$

$$\ldots\ldots < \ldots\ldots < \ldots\ldots < \ldots\ldots < \ldots\ldots < \ldots\ldots$$

Exercise 13 (2 min)   57.6%

Here are five numbers ordered from smallest to largest.
Write the number 3.1 in the correct place.

$$\ldots\ldots 2.93 \ldots\ldots 3 \ldots\ldots 3.07 \ldots\ldots 3.15 \ldots\ldots 3.4$$

Exercise 18 (3 min)

Calculate:

a) $58 - (8 + 22) =$ 73.7%
b) $5 \times (9 - 6) =$ 82.0%
c) $(7 + 13) \times 3 =$ 86.6%
d) $15 \div (2 + 3) =$ 68.2%

Exercise 21 (3 min)

Write the calculations in the boxes provided: [boxes omitted]

  a. $8.32 + 15.87$ 84.6%   b. $15.672 + 352.21$ 79.2%

PROBLEMS

Exercise 10 (3 min)   48.0%

A school has two classes. In the school there are 26 girls. In the first class, there are 12 girls and 11 boys. In the second class, there are 27 pupils. How many boys are there in the second class?

Exercise 31 (4 min 30 sec)

A pack of mineral water is made of 6 bottles of 1.5 liters.
A shopkeeper arranges 25 such packs.

a. Circle the calculation that yields the number of bottles arranged. 81.4%

$25 + 6$	25×6	1.5×6
$25 - 6$	25×1.5	$6 - 1.5$

b. The shopkeeper made the calculation $(25 \times 6) \times 1.5$.
The result of this calculation is 225.
What quantity did the shopkeeper want to know?   59.1%

Exercise 14 (4 min)

The following graph represents the increase in the world population since 1850 (in billions).

a. In what year did the world population reach 2 billion?   78.4%
b. What was the world population in 2000?   86.1%
c. How many years were necessary to go from 1 to 2 billion?   40.1%
d. Approximately what was the world population in 1950?   53.2%

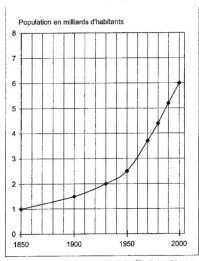

D'après «Histoire-Géographie 6ème» - Hachette Education

GEOMETRY

Exercise 24 (4 min)   95.7%

Below, on the left, is a figure made of a square and a circle.
Somebody began to reproduce this figure. Two sides of the square are already
drawn. Complete the drawing.

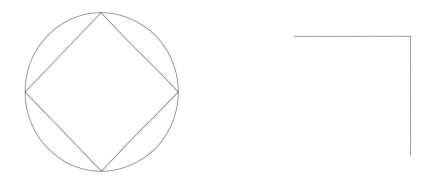

Exercise 6 (5 min)

Here are three figures.

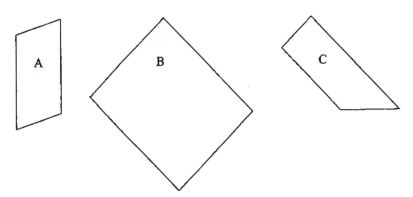

Fill in the table.

Figure	Is it a rectangle ? (Circle the answer.)	Explain how you found it.
A	YES NO	42.9%
B	YES NO	37.2% + 45.9%
C	YES NO	39.9% + 47.9%

References

[Artigue 1999] M. Artigue, "The teaching and learning of mathematics at the university level: Crucial questions for contemporary research in education", *Notices Amer. Math. Soc.* **46**:11 (1999), 1377–1385.

[Artigue 2003] M. Artigue, "The teaching of mathematics at high school level in France. Designing and implementing the necessary evolutions", pp. 9–34 in *Proceedings of the Third Mediterranean Conference on Mathematics Education* (Athens, 2003), edited by A. Gagatsis and S. Papastavridis, Athens: Hellenic Mathematical Society, 2003.

[Brousseau 1997] G. Brousseau, *Theory of didactical situations in mathematics: Didactique des mathématiques, 1970-1990*, edited by N. Balacheff et al., Mathematics education library **19**, Dordrecht: Kluwer, 1997.

[Delozanne et al. 2003] E. Delozanne, D. Prévit, B. Grugeon, and P. Jacoboni, "Supporting teachers when diagnosing their students in algebra", pp. 461–470 in *Advanced technologies in mathematics education: Supplementary proceedings of the 11th International Conference on Artificial Intelligence in Education* (AI-ED2003), edited by E. Delozanne and K. Stacey, Amsterdam: IOS Press, 2003.

[Duval 1996] R. Duval, *Semiosis et pensée humaine*, Paris: Peter Lang, 1996.

[Grugeon 1995] G. Grugeon, *Étude des rapports institutionnels et des rapports personnels des élèves à l'algèbre élémentaire dans la transition entre deux cycles d'enseignement: BEP et Première G*, thèse de doctorat: Université Paris 7, 1995.

[NRC 2001] National Research Council (Mathematics Learning Study: Center for Education, Division of Behavioral and Social Sciences and Education), *Adding it up: Helping children learn mathematics*, edited by J. Kilpatrick et al., Washington, DC: National Academy Press, 2001.

Assessing Mathematical Proficiency
MSRI Publications
Volume **53**, 2007

Chapter 18
Assessment to Improve Learning in Mathematics:
The BEAR Assessment System

MARK WILSON AND CLAUS CARSTENSEN

Introduction

The Berkeley Evaluation and Assessment Research (BEAR) Center has for the last several years been involved in the development of an assessment system, which we call the BEAR Assessment System. The system consists of four principles, each associated with a practical "building block" [Wilson 2005] as well as an activity that helps integrate the four parts together (see the section starting on p. 325). Its original deployment was as a curriculum-embedded system in science [Wilson et al. 2000], but it has clear and logical extensions to other contexts such as in higher education [Wilson and Scalise 2006], in large-scale assessment [Wilson 2005]; and in disciplinary areas, such as chemistry [Claesgens et al. 2002], and the focus of this chapter, mathematics.

In this paper, the four principles of the BEAR Assessment System are discussed, and their application to large-scale assessment is described using an example based on a German assessment of mathematical literacy used in conjunction with the Program for the International Student Assessment [PISA 2005a]; see also Chapter 7, this volume). The BEAR Assessment System is based on a conception of a tight inter-relationship between classroom-level and large-scale assessment [Wilson 2004a; Wilson and Draney 2004]. Hence, in the process of discussing this large-scale application, some arguments and examples will be directed towards classroom-level applications, or, more accurately, towards the common framework that binds the two together [Wilson 2004b].

The Assessment Triangle and the BEAR Approach

Three broad elements on which every assessment should rest are described by the Assessment Triangle from the National Research Council's report *Knowing What Students Know* [NRC 2001] shown in Figure 1.

According to *Knowing What Students Know*, an effective assessment design requires:

- *a model of student cognition and learning* in the field of study;
- well-designed and tested assessment questions and tasks, often called *items*;
- and ways to make *inferences about student competence* for the *particular context of use*. (p. 296)

These elements are of course inextricably linked, and reflect concerns similar to those addressed in the conception of constructive alignment [Biggs 1999], regarding the desirability of achieving goodness-of-fit among learning outcomes, instructional approach, and assessment.

Models of student learning should specify the most important aspects of student achievement to assess, and they provide clues about the types of tasks that will elicit evidence and the types of inferences that can connect observations to learning models and ideas about cognition. To collect responses that serve as high-quality evidence, items themselves need to be systematically developed with both the learning model and the character of subsequent inferences in mind, and they need to be trialed, and the results of the trials systematically examined. Finally, the nature of inferences desired provides the "why" of it all—if we don't know what we want to do with the assessment information, then we can't figure out what the student model or the items should be. Of course, context determines many specifics of the assessment.

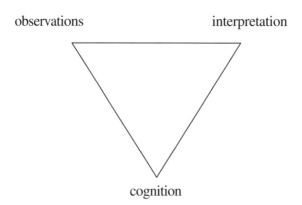

Figure 1. The *Knowing What Students Know* assessment triangle.

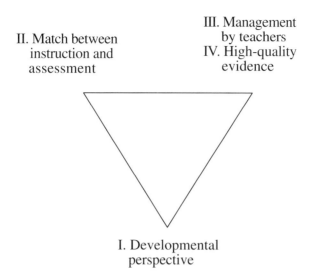

Figure 2. The principles of the BEAR assessment system.

The BEAR Assessment System is based on the idea that good assessment addresses these considerations through four principles: (1) a developmental perspective, (2) a match between instruction and assessment, (3) the generating of high-quality evidence, and (4) management by instructors to allow appropriate feedback, feed-forward, and follow-up. Connections between these principles and the assessment triangle are illustrated in Figure 2. See [Wilson 2005] for a detailed account of an instrument development process based on these principles. Next, we discuss each of these principles and their implementation.

Principle 1: Developmental Perspective

A "developmental perspective" regarding student learning means assessing the development of student understanding of particular concepts and skills over time, as opposed to, for instance, making a single measurement at some final or supposedly significant time point. Criteria for developmental perspectives have been challenging goals for educators for many years. What to assess and how to assess it, whether to focus on generalized learning goals or domain-specific knowledge, and the implications of a variety of teaching and learning theories all impact what approaches might best inform developmental assessment. From Bruner's nine tenets of hermeneutic learning [Bruner 1996] to considerations of empirical, constructivist, and sociocultural schools of thought [Olson and Torrance 1996] to the recent National Research Council report *How People Learn* [NRC 2000], broad sweeps of what might be considered in a developmental perspective have been posited and discussed. Cognitive taxonomies such as Bloom's Taxonomy of Educational Objectives [1956], Haladyna's Cognitive Operations Dimensions [1994] and the Structure of the Observed Learning Outcome (SOLO) Taxonomy [Biggs and Collis 1982] are among many attempts to concretely identify generalizable frameworks. One issue is that as learning situations vary, and their goals and philosophical underpinnings take different forms, a "one-size-fits-all" development assessment approach rarely satisfies course needs. Much of the strength of the BEAR Assessment System comes in providing tools to model many different kinds of learning theories and learning domains. What is to be measured and how it is to be valued in each BEAR assessment application is drawn from the expertise and learning theories of the teachers and/or curriculum developers involved in the developmental process.

Building block 1: Progress variables. Progress variables [Masters et al. 1990; Wilson 1990] embody the first of the four principles: that of a developmental perspective on assessment of student achievement and growth. The four building blocks and their relationship to the assessment triangle are shown in Figure 3. The term "variable" is derived from the measurement concept of focusing on one characteristic to be measured at a time. A progress variable is a well-thought-out and researched ordering of qualitatively different levels of performance. Thus, a variable defines what is to be measured or assessed in terms general enough to be interpretable at different points in a curriculum but specific enough to guide the development of the other curriculum and assessment components. When the goals of the instruction are linked to the set of variables, then the set of variables also define what is to be taught. Progress variables are one model of

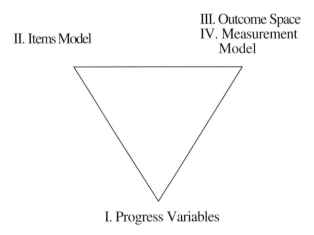

Figure 3. The building blocks of the BEAR assessment system.

how assessments can be connected with instruction and accountability. Progress variables provide a way for large-scale assessments to be linked in a principled way to what students are learning in classrooms, while remaining independent of the content of a specific curriculum.

The approach assumes that, within a given curriculum, student performance on progress variables can be traced over the course of the year, facilitating a more developmental perspective on student learning. Assessing the growth of students' understanding of particular concepts and skills requires a model of how student learning develops over a set period of (instructional) time. A growth perspective helps one to move away from "one-shot" testing situations, and away from cross-sectional approaches to defining student performance, toward an approach that focuses on the process of learning and on an individual's progress through that process. Clear definitions of what students are expected to learn, and a theoretical framework of how that learning is expected to unfold, as the student progresses through the instructional material, are necessary to establish the construct validity of an assessment system.

Explicitly aligning the instruction and assessment addresses the issue of the content validity[1] of the assessment system as well. Traditional testing practices — in standardized tests as well as in teacher-made tests — have long been criticized for oversampling items that assess only basic levels of knowledge of content and ignore more complex levels of understanding. Relying on progress variables to determine what skills are to be assessed means that assessments focus on what is important, not what is easy to assess. Again, this reinforces the central instructional objectives of a course. Resnick and Resnick [1992, p. 59] have argued: "Assessments must be designed so that when teachers do the natural thing — that is, prepare their students to perform well — they will exercise the kinds of abilities and develop the kinds of skill and knowledge that are the real goals of educational reform." Variables that embody the aims of instruction (e.g., "standards") can guide assessment to do just what the Resnicks were demanding. In a large-scale assessment, the notion of a progress variable will be more useful to the parties involved than simple number-correct scores or standings relative to some norming population, providing the diagnostic information so often requested (see also Chapters 10, 12, 14, 21 and 22 of this volume.)

The idea of using variables (note that, for the sake of brevity, I will refer to these as "variables") also offers the possibility of gaining significant *efficiency* in assessment: Although each new curriculum prides itself on bringing something new to the subject matter, in truth, most curricula are composed of a common

---

[1]Content validity is evidence that the content of an assessment is a good representation of the construct it is intended to cover (see [Wilson 2005, Chapter 8]).

stock of content. And, as the influence of national and state standards increases, this will become more true, and also easier to codify. Thus, we might expect innovative curricula to have one, or perhaps even two progress variables that do not overlap with typical curricula, but the remainder will form a fairly stable set that will be common across many curricula.

Progress variables are derived in part from research into the underlying cognitive structure of the domain and in part from professional opinion about what constitutes higher and lower levels of performance or competence, but are also informed by empirical research into how students respond to instruction or perform in practice [NRC 2001]. To more clearly understand what a progress variable is, let us consider an example.

The example explored in this chapter is a test of mathematics competency taken from one booklet of a German mathematics test administered to a random subsample of the German PISA sample of 15-year-old students in the 2003 administration [PISA 2004]. The test was developed under the same general guidelines as the PISA mathematics test (see Chapter 7 in this volume), where Mathematical Literacy is a "described variable" (i.e., the PISA jargon for progress variable) with several successive levels of sophistication in performing mathematical tasks [PISA 2005a; 2005b]. These levels are as follows:

PISA Levels of Mathematical Literacy

VI. At Level VI students can conceptualize, generalize, and utilize information based on their investigations and modeling of complex problem situations. They can link different information sources and representations and flexibly translate among them. Students at this level are capable of advanced mathematical thinking and reasoning. These students can apply their insight and understandings along with a mastery of symbolic and formal mathematical operations and relationships to develop new approaches and strategies for attacking novel situations. Students at this level can formulate and precisely communicate their actions and reflections regarding their findings, interpretations, arguments, and the appropriateness of these to the original situations.

V. At Level V students can develop and work with models for complex situations, identifying constraints and specifying assumptions. They can select, compare, and evaluate appropriate problem-solving strategies for dealing with complex problems related to these models. Students at this level can work strategically using broad, well-developed thinking and reasoning skills, appropriate linked representations, symbolic and formal characterizations, and insight pertaining to these situations. They can reflect on their actions and formulate and communicate their interpretations and reasoning.

IV. At Level IV students can work effectively with explicit models for complex concrete situations that may involve constraints or call for making assumptions. They can select and integrate different representations, including symbolic, linking them directly to aspects of real-world situations. Students at this level can utilize well-developed skills and reason flexibly, with some insight, in these contexts. They can construct and communicate explanations and arguments based on their interpretations, arguments, and actions.

III. At Level III students can execute clearly described procedures, including those that require sequential decisions. They can select and apply simple problem-solving strategies. Students at this level can interpret and use representations based on different information sources and reason directly from them. They can develop short communications reporting their interpretations, results and reasoning.

II. At Level II students can interpret and recognize situations in contexts that require no more than direct inference. They can extract relevant information from a single source and make use of a single representational mode. Students at this level can employ basic algorithms, formulae, procedures, or conventions. They are capable of direct reasoning and making literal interpretations of the results.

I. At Level I students can answer questions involving familiar contexts where all relevant information is present and the questions are clearly defined. They are able to identify information and to carry out routine procedures according to direct instructions in explicit situations. They can perform actions that are obvious and follow immediately from the given stimuli.

The levels shown above were derived from a multistep process [PISA 2005b] as follows: (a) Mathematics curriculum experts identified possible subscales in the domain of mathematics, (b) PISA items were mapped onto each subscale, (c) a skills audit of each item in each subscale was carried out on the basis of a detailed expert analysis, (d) field test data were analyzed to yield item locations on subscales, (e) the information from the two previous steps was combined. In this last step, the ordering of the items was linked with the descriptions of associated knowledge and skills, giving a hierarchy of knowledge and skills that defined possible values of the progress variable. This results in natural clusters of skills, which provides a basis for understanding and describing the progress variable. The results of this last step were also validated with later empirical data, and by using a validation process involving experts. Note that this method of developing a progress variable is much less precise than the approaches described in the references above (e.g., [Wilson et al. 2000; Wilson and Scalise 2006], and will thus usually result in a progress variable that is much broader in its content.

Principle 2: Match Between Instruction and Assessment

The match between instruction and assessment in the BEAR Assessment System is established and maintained through two major parts of the system: progress variables, described above, and assessment tasks or activities, described in this section. The main motivation for the progress variables so far developed is that they serve as a framework for the assessments and a method of making measurement possible. However, this second principle makes clear that the framework for the assessments and the framework for the curriculum and instruction must be one and the same. This is not to imply that the needs of assessment must drive the curriculum, nor that the curriculum description will entirely determine the assessment, but rather that the two, assessment and instruction, must be in step — they must both be designed to accomplish the same thing, the aims of learning, whatever those aims are determined to be.

Using progress variables to structure both instruction and assessment is one way to make sure that the two are in alignment, at least at the planning level. In order to make this alignment concrete, however, the match must also exist at the level of classroom interaction and that is where the nature of the assessment tasks becomes crucial. Assessment tasks need to reflect the range and styles of the instructional practices in the curriculum. They must have a place in the "rhythm" of the instruction, occurring at places where it makes instructional sense to include them, usually where instructors need to see how much progress their students have made on a specific topic. See [Minstrell 1998] for an insightful account of such occasions.

One good way to achieve this is to develop both the instructional materials and the assessment tasks at the same time — adapting good instructional sequences to produce assessable responses and developing assessments into full-blown instructional activities. Doing so brings the richness and vibrancy of curriculum development into assessment, and also brings the discipline and hard-headedness of explaining assessment data into the design of instruction.

By developing assessment tasks as part of curriculum materials, they can be made directly relevant to instruction. Assessment can become indistinguishable from other instructional activities, without precluding the generation of high-quality, comparative, and defensible assessment data on individual students and classes.

Building block 2: The items design. The items design governs the match between classroom instruction and the various types of assessment. The critical element to ensure this in the BEAR assessment system is that each assessment task is matched to at least one variable.

A variety of different task types may be used in an assessment system, based on the requirements of the particular situation. There has always been a tension in assessment situations between the use of multiple-choice items, which are perceived to contribute to more reliable assessment, and other, alternative forms of assessment, which are perceived to contribute to the validity of a testing situation. The BEAR Assessment System includes designs that use different item types to resolve this tension.

When using this assessment system within a curriculum, a particularly effective mode of assessment is what we call *embedded assessment*. By this we mean that opportunities to assess student progress and performance are integrated into the instructional materials and are virtually indistinguishable from the day-to-day classroom activities. We found it useful to think of the metaphor of a stream of instructional activity and student learning, with the teacher dipping into the stream of learning from time to time to evaluate student progress and performance. In this model or metaphor, assessment becomes *part* of the teaching and learning process, and we can think of it being assessment for learning [Black et al. 2003]. If assessment is also a learning event, then it does not take unnecessary time away from instruction, *and* the number of assessment tasks can be more efficiently increased in order to improve the reliability of the results [Linn and Baker 1996]. But, for assessment to become fully and meaningfully embedded in the teaching and learning process, the assessment must be linked to a specific curriculum, i.e. it must be curriculum dependent, not curriculum independent as must be the case in many high-stakes testing situations [Wolf and Reardon 1996].

In embedded assessment in classrooms, there will be a variety of different types of assessment tasks, just as there is variety in the instructional tasks. These may include individual and group "challenges," data processing questions, questions following student readings, and even instruction/assessment events such as "town meetings." Such tasks may be constructed-response, requiring students to fully explain their responses in order to achieve a high score, or they may be multiple choice, freeing teachers from having to laboriously hand score all of the student work [Briggs et al. 2006].

There are many variations in the way that progress variables can be made concrete in practice, from using different assessment modes (multiple choice, performance assessment, mixed modes, etc.), to variations in the frequency of assessing students (once a week, once a month, etc.), to variations in the use of embedding of assessments (all assessments embedded, some assessments in a more traditional testing format, etc.).

In large-scale testing situations, the basis on which the mix of assessment modes is decided may be somewhat different from that in embedded assessment contexts. Many large-scale tests are subject to tight constraints both in terms of the time available for testing, and in terms of the financial resources available for scoring. Thus, although performance assessments are valued because of their perceived high validity, it may not be possible to collect enough information through performance assessments alone to accurately estimate each examinee's proficiency level; multiple-choice items, which require less time to answer and which may be scored by machine rather than by human raters, may be used to increase the reliability of the large-scale test.

Returning to the German Mathematical Literacy example, the test booklet contained 64 dichotomous items; 18 of these items were selected for this example. Examples of these items are the tasks Function, Rectangle and Difference, shown on the next page. Each item was constructed according to Topic Areas and the Types of Mathematical Modeling required. The Topic Areas were: Arithmetic, Algebra, and Geometry. The Modeling Types were: Technical Processing, Numerical Modeling, and Abstract Modeling. The Technical Processing dimension requires students to carry out operations that have been rehearsed such as computing numerical results using standard procedures — see, for example, the item Function. Numerical Modeling requires the students to construct solutions for problems with given numbers in one or more steps — see the item Rectangle. In contrast, Abstract Modeling requires students to formulate rules in a more general way, for example by giving an equation or by describing a general solution in some way — see the item Difference. Because the collection of items follows an experimental design, the responses may also be considered data from a psychological experiment. The experimental design has two factors, Topic Area and Modeling Type. In sum, the selected set of items has a 3×3 design with two observations of each pair of conditions, resulting in 18 items in total.

Principle 3: Management by Teachers

For information from the assessment tasks and the BEAR analysis to be useful to instructors and students, it must be couched in terms that are directly related to the instructional goals associated with the progress variables. Constructed response tasks, if used, must be quickly, readily, and reliably scorable. The categories into which the scores are sorted must be readily interpreted in an educational setting, whether it is within a classroom, by a parent, or in a policy-analysis setting. The requirement for transparency in the relationship between scores and actual student responses to an item leads to the third building block.

Example tasks from the German Mathematical Literacy booklet

(Copyright German PISA Consortium)

Function (a technical processing item in algebra)

Consider the function given by the equation $y = 2x - 1$. Fill in the missing values.

x	-2	-1	0		3	$\cdots$		
y						$\cdots$	19	

Rectangles (a numerical modeling item in algebra)

Around a small rectangle a second one is drawn. A third rectangle is drawn around the two and so on. The distance between the sides is always 1 cm.

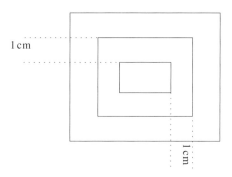

By how much do the length, width and perimeter increase from rectangle to rectangle?

The length increases by ___ cm.   The width increases by ___ cm.

The perimeter increases by ___ cm.

Difference (an abstract modeling item in algebra)

Put the digits 3, 6, 1, 9, 4, 7 in the boxes so that the difference between the two three-digit numbers is maximized. (Each digit may be used only once.)

first number ☐ ☐ ☐

second number ☐ ☐ ☐

Building block 3: The outcome space. The outcome space is the set of outcomes into which student performances are categorized for all the items associated with a particular progress variable. In practice, these are presented as scoring guides for student responses to assessment tasks. This is the primary means by which the essential element of teacher professional judgment is implemented in the BEAR Assessment System. These are supplemented by "exemplars": examples of student work at every scoring level for every task and variable combination, and "blueprints," which provide the teachers with a layout showing opportune times in the curriculum to assess the students on the different progress variables.

For the information from assessment opportunities to be useful to teachers, it must be couched in terms that are directly interpretable with respect to the instructional goals associated with the progress variables. Moreover, this must be done in a way that is intellectually and practically efficient. Scoring guides have been designed to meet these two criteria. A scoring guide serves as a operational definition for a progress variable by describing the performance criteria necessary to achieve each score level of the variable.

The scoring guides are meant to help make the performance criteria for the assessments clear and explicit (or "transparent and open" to use Glaser's [1990] terms) — not only to the teachers but also to the students and parents, administrators, or other "consumers" of assessment results. In fact, we strongly recommend to teachers that they share the scoring guides with administrators, parents and students, as a way of helping them understand what types of cognitive performance were expected and to model the desired processes.

In addition, students appreciate the use of scoring guides in the classroom. In a series of interviews with students in a Kentucky middle school that was using the BEAR Assessment System (reported in [Roberts and Sipusic 1999]), the students spontaneously expressed to us their feeling that, sometimes for the first time, they understood what it was that their teachers expected of them, and felt they knew what was required to get a high score. The teachers of these students found that the students were often willing to redo their work in order to merit a higher score.

Traditional multiple-choice items are, of course, based on an implicit scoring guide — one option is correct, the others all incorrect. Alternative types of multiple-choice items can be constructed that are explicitly based on the levels of a construct map [Briggs et al. 2006], and thus allow a stronger interpretation of the test results. For the German Mathematical Literacy example, the items are all traditional multiple choice — their development did not involve the explicit construction of an outcome space.

Principle 4: High-Quality Evidence

Technical issues of reliability and validity, fairness, consistency, and bias can quickly sink any attempt to measure values of a progress variable as described above, or even to develop a reasonable framework that can be supported by evidence. To ensure comparability of results across time and context, procedures are needed to (a) examine the coherence of information gathered using different formats, (b) map student performances onto the progress variables, (c) describe the structural elements of the accountability system — tasks and raters — in terms of the progress variables, and (d) establish uniform levels of system functioning, in terms of quality control indices such as reliability. Although this type of discussion can become very technical to consider, it is sufficient to keep in mind that the traditional elements of assessment standardization, such as validity/reliability studies and bias/equity studies, must be carried out to satisfy quality control and ensure that evidence can be relied upon.

Building block 4: Wright maps. Wright maps represent the principle of high-quality evidence. Progress maps are graphical and empirical representations of a progress variable, showing how it unfolds or evolves in terms of increasingly sophisticated student performances. They are derived from empirical analyses of student data on sets of assessment tasks. Maps are based on an ordering of these assessment tasks from relatively easy tasks to more difficult and complex ones. A key feature of these maps is that both students and tasks can be located on the same scale, giving student proficiency the possibility of substantive interpretation, in terms of what the student knows and can do and where the student is having difficulty. The maps can be used to interpret the progress of one particular student, or the pattern of achievement of groups of students, ranging from classes to nations.

Wright maps can be very useful in large-scale assessments, providing information that is not readily available through numerical score averages and other traditional summary information — they are used extensively, for example, in reporting on the PISA assessments [PISA 2005a]. A Wright map illustrating the estimates for the Rasch model is shown in Figure 4. On this map, an "X" represents a group of students, all at the same estimated achievement level. The logits (on the left-hand side) are the units of the Wright map — they are related to the probability of a student succeeding at an item, and are specifically the log of the odds of that occurring. The symbols "T," "N" and "A" each represent a Technical Processing, Numerical Modeling, and Abstract Modeling item, with the Topic Area indicated by the column headings above. Where a student is located near an item, this indicates that there is approximately a 50% chance of the student getting the item correct. Where the student is above the item, the

```
logits   students                      Topic Areas
_____        _____
   3
                    X  Arithmetic  Geometry        Algebra      Levels
                    X  _____
                   XX
                  XXX
                   XX
   2               XX                                              .
                  XXX                                              .
                  XXX                                              |
                 XXXX                                             
                 XXXX                                             
              XXXXXXXX                                A           
   1           XXXXXXX                                            V
               XXXXXXX                                            |
               XXXXXXX                                A N         |
             XXXXXXXXX                    A A         N          
              XXXXXXXX |N                                         |
            XXXXXXXXXX                                           IV
   0           XXXXXXX |N                  N N                   -
               XXXXXXX                      T         T          |
               XXXXXXX                      T         T          
               XXXXXXX |A                                        
             XXXXXXXXX |A                                        
            XXXXXXXXXX                                           III
  -1           XXXXXXX                                           
               XXXXXXX |T                                        |
                 XXXXX |T                                        
                 XXXX                                            |
                 XXXX                                            
                 XXXX                                            .
  -2               XX                                             .
                   XX
                   XX
                    X
                    X
                    X
                    X
  -3
```

Figure 4. A Wright map of the mathematical literacy variable.

chance is greater than 50%, and the further it is above, the greater the chance. Where the student is lower than the item, the chance is less than 50%, and the further it is below, the lesser the chance. Thus this map illustrates the description of the Mathematical Literacy progress variable in terms of the Levels from page 316 as well as the Topic Areas and the Modeling Types in the items design. The Topic Areas reflect the earlier placement of Arithmetic in the curriculum than Geometry and Algebra. The ordering of Modeling Types is generally consistent with what one might expect from the definitions of the Levels, except for the Arithmetic Abstract Modeling items, which seem to be somewhat easier than expected. This is a topic that deserves a follow-up investigation.

We typically use a multi-dimensional Rasch modeling approach to calibrate the maps for use in the BEAR Assessment System (see [Adams et al. 1997] for the specifics of this model). These maps have at least two advantages over the traditional method of reporting student performance as total scores or percentages: First, it allows teachers to interpret a student's proficiency in terms of average or typical performance on representative assessment activities; and second, it takes into consideration the relative difficulties of the tasks involved in assessing student proficiency.

Once constructed, maps can be used to record and track student progress and to illustrate the skills that students have mastered and those that the students are working on. By placing students' performance on the continuum defined by the map, teachers, administrators, and the public can interpret student progress with respect to the standards that are inherent in the progress variables. Wright maps can come in many forms, and have many uses in classroom and other educational contexts. In order to make the maps flexible and convenient enough for use by teachers and administrators, we have also developed software for teachers to use to generate the maps. This software, which we call *GradeMap* [Kennedy et al. 2005], allows consumers to enter the scores given to the students on assessments, and then map the performance of groups of students, either at a particular time or over a period of time.

Bringing It All Together: Performance Standard Setting

The final ingredient in the BEAR Assessment System is the means by which the four building blocks discussed thus far are brought together into a coherent system — in the case of large-scale assessment, by standard setting. We have developed a standard-setting procedure, called "construct mapping" [Wilson and Draney 2002], that allows the standard-setting committee members to use the item response map as a model of what a student at a given level knows and can do. The map is represented in a piece of software [Hoskens and Wilson 1999] that allows standard-setting committee members to find out about the details of student performance at any given proficiency level, and to assist them in deciding where the cutoffs between performance levels should be.

An example showing a section of such an item map is given in Figure 5. This illustration is from a somewhat more complicated example than the German Mathematical Literacy Test, involving both multiple-choice items and two items that required written responses (WR1 and WR2) and which were scored into five ordered categories [Wilson 2005]. The column on the far left contains a numerical scale that allows the selection and examination of a given point on the map, and the selection of the eventual cut scores for the performance levels. This scale is a transformation of the original logit scale, designed to

Scale	Multiple Choice		P	WR 1		P	WR 2		P
620									
610									
600									
590									
580									
570	37		.30				2.3		.26
560	15		.34						
550									
540	28 39		.38						
530	27		.41						
520	19 38		.45						
510									
500	34 43 45 48		.50	1.3		.40			
490	17 18 20 40 50		.53						
480	4 31		.56						
470	11 32 33 44 47		.59						
460	5 9 12 46		.61						
450	3 6 7 10 16 29		.64						
440	36		.67						
430	8 14 22 23 26 35		.69						
420	13 24 25		.71						
410	41 42		.73						
400	1 21 30 49		.76						
390									
380							2.2		.56
370	2		.82						
360				1.2		.40			
350									

Figure 5. A screen-shot of a cut-point setting map.

have a mean of 500, and to range from approximately 0 to 1000. (This choice of scale is a somewhat arbitrary, but designed to avoid negative numbers and small decimals, which some members of standard-setting committees find annoying.) The next two columns contain the location of the multiple-choice items (labeled by number of appearance on the examination), and the probability that a person at the selected point would get each item correct (in this case, a person at 500 on the scale — represented by the shaded band across the map). The next two sets of columns display the thresholds for the two written-response items — for example, the threshold levels for scores of 2 and 3 on written-response item 1 are represented by 1.2 and 1.3, respectively (although each item is scored on a scale of 1 to 5 on this particular examination, only the part of the scale where a person would be most likely to get a score of 2 or 3 on either item is shown) — and the probability that a person at 500 on the scale would score at that particular score level on each item. The software also displays, for a person at the selected point on the logit scale, the expected score total on the multiple-choice section (Figure 5 does not show this part of the display), and the expected score on each of the written response items.

In order to set the cut points, the committee first acquaints itself with the test materials. The meaning of the various parts of the map is then explained, and the committee members and the operators of the program spend time with the software familiarizing themselves with points on the scale.

The display of multiple-choice item locations in ascending difficulty, next to the written-response thresholds, helps to characterize the scale in terms of what increasing proficiency "looks like" in the pool of test-takers. For example, if a committee were considering 500 as a cut point between performance levels, it could note that 500 is a point at which items like 34, 43, 45, and 48 are expected to be chosen correctly about 50% of the time, a harder item like 37 is expected to be chosen correctly about 30%, and easier items like 2 are expected to be chosen correctly 80% of the time. The set of multiple-choice items, sorted so they are in order of ascending difficulty, is available to the committee so that the members can relate these probabilities to their understanding of the items. The committee could also note that a student at that point (i.e., 500), would be equally likely to score a 2 or a 3 on the first written-response item (40% each) and more likely to score a 2 than a 3 on the second (56% vs. 26%). Examples of student work at these levels would be available to the committee for consideration of the interpretation of these scores. Committee members can examine the responses of selected examinees to both the multiple-choice and written-response items, chart their locations on the map, and judge their levels.

The committee then, through a consensus-building process, sets up cut points on this map, using the item response calibrations to allow interpretation in terms of predicted responses to both multiple-choice items and open-ended constructed-response items. Locations of an individual student's scores and distributions of the scaled values of the progress variable are also available for interpretative purposes. This procedure allows both criterion-referenced and norm-referenced interpretations of cut scores.

Use of the maps available from the item response modeling approach not only allows the committees to interpret cut-offs in a criterion-referenced way, it also allows maintenance of similar standards from year to year by equating of the item response scales. This can be readily accomplished by using linking items on successive tests to keep the waves of data on the same scale — hence the cut-offs set one year can be maintained in following years.

Discussion

A central tenet of the assessment reforms of recent years ("authentic," "performance," etc.) has been the WYTIWYG principle — "What you test is what you get." This principle has led the way for assessment reform at the state or district level nationwide. The assumption behind this principle is that assessment reforms will not only affect assessments *per se*, but these effects will trickle down into the curriculum and instruction that students receive in their daily work in classrooms. Hence, when one looks to the curricula that students are experiencing, one would expect to see such effects, and, in particular, one would expect to see these effects even more strongly in the cutting-edge curricula that central research agencies such as the U.S. National Science Foundation (NSF) sponsor. Thus it is troubling to find that this does not seem to be the case: An NSF review of new middle school science curricula [NSF 1997] found only one where the assessment itself reflected the recent developments in assessment. For that one (the *IEY Assessment System* — see [Wilson et al. 2000]), it was found that the reformed assessment did indeed seem to have the desired sorts of effects [Wilson and Sloane 2000], but for the other curricula no such effects were possible, because the assessment reforms have not, in general, made it into them.

We have demonstrated a way in which large-scale assessments can be more carefully linked to what students are learning. The key here is the use of progress variables to provide a common conceptual framework across curricula. Variables developed and used in the ways we have described here can mediate between the level of detail that is present in the content of specific curricula and the necessarily more vague contents of standards documents. This idea of a "crosswalk between standards and assessments" has also been suggested by Eva Baker of the Center for Research on Evaluation, Standards, and Student Testing [Land 1997, p. 6]. These variables also create a "conceptual basis" for relating a curriculum to standards documents, to other curricula, and to assessments that are not specifically related to that curriculum.

With the assessments to be used across curricula structured by progress variables, the problem of item development is lessened — ideas and contexts for assessment tasks may be adapted to serve multiple curricula that share progress variables. The cumulative nature of the curricula is expressed through (a) the increasing difficulty of assessments and (b) the increasing sophistication needed to gain higher scores using the assessment scoring guides. Having the same underlying structure makes clear to teachers, policy-makers, and parents what is the ultimate purpose of each instructional activity and each assessment, and also makes easier the diagnostic interpretation of student responses to the assessments.

The idea of a progress variable is not radically new — it has grown out of the traditional approach to test content — most tests have a "blueprint" or plan that assigns items to particular categories, and hence, justifies why certain items are there, and others aren't. The concept of a progress variable goes beyond this by looking more deeply into why we use certain assessments when we do (i.e., by linking them to growth through the curriculum), and by calibrating the assessments with empirical information.

Although the ideas inherent in components of the BEAR Assessment System are not unique, the combination of these particular ideas and techniques into a usable system does represent a new step in assessment development. The implications for this effort for other large-scale tests, for curricula, and for assessment reform on a broader level, need to be explored and tested through other related efforts. We hope our efforts and experiences will encourage increased discussion and experimentation of the use of state of the art assessment procedures across a broad range of contexts from classroom practice to large-scale assessments.

References

[Adams et al. 1997] R. J. Adams, M. Wilson, and W.-C. Wang, "The multidimensional random coefficients multinomial logit model", *Applied Psychological Measurement* **21**:1 (1997), 1–23.

[Biggs 1999] J. B. Biggs, *Teaching for quality learning at university*, Buckingham: SRHE and Open University Press, 1999.

[Biggs and Collis 1982] J. B. Biggs and K. F. Collis, *Evaluating the quality of learning: The SOLO taxonomy*, New York: Academic Press, 1982.

[Black et al. 2003] P. Black, C. Harrison, C. Lee, B. Marshall, and D. Wiliam, *Assessment for learning*, London: Open University Press, 2003.

[Bloom 1956] B. S. Bloom (editor), *Taxonomy of educational objectives: The classification of educational goals: Handbook I, cognitive domain*, New York and Toronto: Longmans, Green, 1956.

[Briggs et al. 2006] D. Briggs, A. Alonzo, C. Schwab, and M. Wilson, "Diagnostic assessment with ordered multiple-choice items", *Educational Assessment* **11**:1 (2006), 33–63.

[Bruner 1996] J. Bruner, *The culture of education*, Cambridge, MA: Harvard University Press, 1996.

[Claesgens et al. 2002] J. Claesgens, K. Scalise, K. Draney, M. Wilson, and A. Stacy, "Perspectives of chemists: A framework to promote conceptual understanding of chemistry", paper presented at the annual meeting of the American Educational Research Association, New Orleans, April 2002.

[Glaser 1990] R. Glaser, *Testing and assessment: O tempora! O mores!*, Pittsburgh: Learning Research and Development Center, University of Pittsburgh, 1990.

[Haladyna 1994] T. M. Haladyna, "Cognitive taxonomies", pp. 104–110 in *Developing and validating multiple-choice test items*, edited by T. M. Haladyna, Hillsdale, NJ: Lawrence Erlbaum Associates, 1994.

[Hoskens and Wilson 1999] M. Hoskens and M. Wilson, *StandardMap* [Computer program], Berkeley, CA: University of California, 1999.

[Kennedy et al. 2005] C. A. Kennedy, M. Wilson, and K. Draney, *GradeMap 4.1* [Computer program], Berkeley, California: Berkeley Evaluation and Assessment Center, University of California, 2005.

[Land 1997] R. Land, "Moving up to complex assessment systems", *Evaluation Comment* **7**:1 (1997), 1–21.

[Linn and Baker 1996] R. Linn and E. Baker, "Can performance-based student assessments be psychometrically sound?", pp. 84–103 in *Performance-based student assessment: Challenges and possibilities. Ninety-fifth Yearbook of the National Society for the Study of Education*, edited by J. B. Baron and D. P. Wolf, Chicago: University of Chicago Press, 1996.

[Masters et al. 1990] G. N. Masters, R. A. Adams, and M. Wilson, "Charting student progress", pp. 628–634 in *International encyclopedia of education: Research and studies*, vol. 2 (Supplementary), edited by T. Husen and T. N. Postlethwaite, Oxford and New York: Pergamon, 1990.

[Minstrell 1998] J. Minstrell, "Student thinking and related instruction: Creating a facet-based learning environment", paper presented at the meeting of the Committee on Foundations of Assessment, Woods Hole, MA, October 1998.

[NRC 2000] National Research Council (Committee on Developments in the Science of Learning, Commission on Behavioral and Social Sciences and Education), *How people learn: Brain, mind, experience, and school*, expanded ed., edited by J. D. Bransford et al., Washington, DC: National Academy Press, 2000.

[NRC 2001] National Research Council Committee on the Foundations of Assessment, Board on Testing and Assessment, Center for Education, Division of Behavioral and Social Sciences and Education), *Knowing what students know: The science and design of educational assessment*, edited by J. Pelligrino et al., Washington, DC: National Academy Press, 2001.

[NSF 1997] National Science Foundation, *Review of instructional materials for middle school science*, Arlington, VA: Author, 1997.

[Olson and Torrance 1996] D. R. Olson and N. Torrance (editors), *Handbook of education and human development: New models of learning, teaching and schooling*, Oxford: Blackwell, 1996.

[PISA 2004] PISA, *PISA 2003: Der Bildungsstand der Jugendlichen in Deutschland: Ergebnisse des zweiten internationalen Vergleichs*, Münster: Waxmann, 2004.

[PISA 2005a] Programme for International Student Assessment, "Learning for tomorrow's world: First results from PISA 2003", Technical report, Paris: Organisation for Economic Co-operation and Development, 2005.

[PISA 2005b] Programme for International Student Assessment, "PISA 2003 Technical Report", Technical report, Paris: Organisation for Economic Co-operation and Development, 2005.

[Resnick and Resnick 1992] L. B. Resnick and D. P. Resnick, "Assessing the thinking curriculum: New tools for educational reform", pp. 37–76 in *Changing assessments*, edited by B. R. Gifford and M. C. O'Connor, Boston: Kluwer, 1992.

[Roberts and Sipusic 1999] L. Roberts (producer) and M. Sipusic (director), "Moderation in all things: A class act" [Film], available from the Berkeley Evaluation and Assessment Center, Graduate School of Education, University of California, Berkeley, CA 94720–1670, 1999.

[Wilson 1990] M. Wilson, "Measurement of developmental levels", pp. 628–634 in *International encyclopedia of education: Research and studies*, vol. Supplementary vol. 2, edited by T. Husen and T. N. Postlethwaite, Oxford: Pergamon Press, 1990.

[Wilson 2004a] M. Wilson, "A perspective on current trends in assessment and accountability: Degrees of coherence", pp. 272–283 in [Wilson 2004b], 2004.

[Wilson 2004b] M. Wilson (editor), *Towards coherence between classroom assessment and accountability: One hundred and third yearbook of the National Society for the Study of Education*, part 2, Chicago: University of Chicago Press, 2004.

[Wilson 2005] M. Wilson, *Constructing measures: An item response modeling approach*, Mahwah, NJ: Lawrence Erlbaum Associates, 2005.

[Wilson and Draney 2002] M. Wilson and K. Draney, "A technique for setting standards and maintaining them over time", pp. 325–332 in *Measurement and multivariate analysis*, edited by S. Nishisato et al., Tokyo: Springer-Verlag, 2002.

[Wilson and Draney 2004] M. Wilson and K. Draney, "Some links between large-scale and classroom assessments: The case of the BEAR Assessment System", pp. 132–154 in [Wilson 2004b], 2004.

[Wilson and Scalise 2006] M. Wilson and K. Scalise, "Assessment to improve learning in higher education: The BEAR Assessment System", *Higher Education* **52**:4 (2006), 635–663.

[Wilson and Sloane 2000] M. Wilson and K. Sloane, "From principles to practice: An embedded assessment system", *Applied Measurement in Education* **13**:2 (2000), 181–208.

[Wilson et al. 2000] M. Wilson, L. Roberts, K. Draney, and K. Sloane, *SEPUP assessment resources handbook*, Berkeley, CA: Berkeley Evaluation and Assessment Research Center, University of California, 2000.

[Wolf and Reardon 1996] D. P. Wolf and S. Reardon, "Access to excellence through new forms of student assessment", pp. 52–83 in *Performance-based student assessment: Challenges and possibilities. Ninety-fifth yearbook of the National Society for the Study of Education*, edited by J. B. Baron and D. P. Wolf, Chicago: University of Chicago Press, 1996.

Chapter 19
English Learners and Mathematics Learning: Language Issues to Consider

LILY WONG FILLMORE

The Problem

TUTOR: "OK — let's begin with some of the terms you are going to need for this kind of math."

STUDENT: "Whadyu mean 'terms'?"

TUTOR: "You don't know the word, 'term'?"

STUDENT: "Nuh-uh."

TUTOR: "Have you ever heard it before?"

STUDENT: "No — huh-uh."

An unpromising start for the tutoring session: the student, a "former" English learner, needs help with math to pass the California High School Exit Examination (CAHSEE). He has failed the math part of the test twice already, and will not graduate from high school if he does not improve his performance substantially this year.[1] But as we see in the vignette, not only does this student have to learn more math this year — he also has to gain a much better understanding of the language used in math.

Although "Freddy" is no longer categorized as an English learner at school, his English remains quite limited. When he began school 12 years earlier, he was a Spanish monolingual. Now, at age 17, he speaks English mostly, although Spanish is still spoken in the home. Despite having been in a program described

---

[1] A summary of the results for the 2004 administration of the CAHSEE [California 2004b] indicates that of the nearly half million tenth-grade students who took the test in 2004, 74% passed the math part of the test. English-language learners performed 25% lower, with a 49% passing rate. In the English language arts part of the test, 75% of all students passed, while the passage rate for English-language learners was 39%, or 36% lower than all students.

as "bilingual" during his elementary school years, his schooling has been entirely in English from the time he entered school. The "bilingual" program provided him and his Spanish-speaking classmates one Spanish language session each week. Over the years, he has learned enough English to qualify as a "fluent English-speaker."[2] His English is adequate for social purposes, allowing him to communicate easily enough with peers and teachers. However, he lacks the vocabulary and grammatical resources to make much sense of the materials he reads in school, or to participate in the instructional discourse that takes place in his classes. Like many former English learners in California, Freddy has language problems at school, but he is no longer included in the official record of English learners kept by the California State Department of Education.

Officially, there are 1.6 million English learners in California's public schools. They constitute 25% of the student population, grades K through 12, for whom language is a clear barrier to the school's curriculum. Unofficially, there are many others—students like Freddy—who have learned enough English through schooling to be classified as English speakers but whose problems in school nonetheless can be traced to language difficulties. Added to the students who are still classified as English learners, they comprise some 2.25 million students in the state whose progress in school is impeded because they do not fully understand the language in which instruction is presented.

Would Freddy have been helped if this session had been conducted in Spanish, his primary language? The answer is no — not anymore. It would have helped when he was younger, but now, at age 17, it makes no difference. Having been schooled exclusively in English, Freddy no longer understands or speaks Spanish as well as he does English. He was no more familiar with Spanish 'término,' than he was with English 'term.'

Had he heard of expressions that included the word, 'term' — for example, "term paper," "terms of endearment," and "term limit"?

"Well, no, not really."

How important is the word 'term' itself? It was easy enough to tell him that in this context, 'term' means 'word' — words that will be used in discussing the mathematical concepts and operations that he would be learning about. His lack of familiarity with such a word, however, is telling. He knows and can use everyday words, but he lacks familiarity with words that figure in academic discourse — the kind of language used in discussing academic subjects such as mathematics, science, and history. But the requirements of language go beyond vocabulary; they include familiarity with distinctive grammatical structures and

---

[2]English learners are designated "fluent English proficient" when they achieve a certain score on a test of English proficiency administered by their school districts. The test California schools must use for this purpose is the California English Language Development Test.

rhetorical devices that add up to a different register of the language that is used in academic settings. Clearly, Freddy needed help with academic English as much as he did with the mathematics concepts he needed to learn. The problem, however, is that while he recognized and acknowledged his need for help with math ("Me and math jus' don't get along"), he was quite oblivious to his need for help in English. In this, he is not unlike the many mathematics teachers he has had over the years. He was asked:

> "Math makes use of many special terms, such as *probability, radius,* and *factor.* Did your teachers explain what those terms meant?"

> "Well, yeah — but I still don't know what they mean."

Evidently, the explanations that teachers have offered have not been meaningful or memorable enough, because Freddy could not say what any of these terms meant, nor would he venture any guesses. It was not surprising then that he would have difficulty dealing with the CAHSEE *Mathematics Study Guide* that the California State Department of Education makes available to students preparing to take the test. The Guide offers Spanish equivalents for key terms in the expectation that students whose primary language is Spanish would find it easier to understand the concepts if they were identified in Spanish. Consider for example, the following definitional paragraph from the glossary of the 2004 CAHSEE *Study Guide* [California 2004a, p. 83]:

> The **probability** of an event's happening is a number from 0 to 1, which measures the chance of that event happening. The probability of most events is a value between 0 (impossible) and 1 (certain). A probability can be written as a fraction, as a decimal or as a percentage. Spanish word with the same meaning as *probability:* probabilidad.

Language in Learning

Freddy could read the passage, but did he understand it? What does it mean to say, "the probability of most events is a value between 0 (impossible) and 1 (certain)"?[3] What could it possibly mean to say, "a probability can be written as a fraction, as a decimal or as a percentage"? His efforts to paraphrase what he had read revealed that he understood virtually nothing of this definition, nor was

---

[3] The adequacy of the definitions provided in the *Study Guide* is not an issue in this discussion — nevertheless, it should be noted that this is not the clearest explanation of the concepts of probability, or of dependent and independent events! For example, it does not make any appeal to the ordinary use of "probable," "probably," or "probability" in terms of chance or likelihood. The technical term is defined as a number, or a numerical value, which is correct, but except for the phrase that it "measures the chance" there is no exploitation of the everyday concept.

he helped by the inclusion of the Spanish equivalent, *probabilidad*, a word that was entirely unfamiliar to him. '*Probabilidad*' is as much a technical term in Spanish as 'probability' is in English. His teachers used some Spanish in school through the second grade, but mathematics, even then, was taught in English. There was little likelihood that he would have learned mathematical terminology in Spanish even in school. He was asked to read on, in the hope that this text from the *Study Guide*'s glossary would help clarify the concept of probability.

> **Independent events, dependent events:** These terms are used when fig-
> uring probabilities. In probability, an event is a particular happening that
> may or may not occur. Some examples of events are: "A fair coin will come
> up heads on the next flip," and "Rain will fall in Oakland tomorrow," and
> "Trudy Trimble will win next week's California lottery."
>
> One event is said to be independent of another if the first event can
> occur with absolutely no effect on the probability of the second event's
> happening. For example, suppose you are going to flip a fair coin two times
> and on the first flip it comes up heads. On the second flip, the probability of
> the coin coming up heads is still 50%. Each flip of the coin is *independent*
> of all other flips.
>
> But some events are *dependent*; that is, the probability of one event
> depends on whether the other event occurs. For example, suppose you
> are randomly choosing two marbles, one after another, from a bag that
> contains three blue marbles and three red marbles. On your first draw, you
> have a 50% chance of drawing a blue marble. But on your second draw, the
> probability of drawing a blue marble depends on which color you pulled
> out on the first draw. The probability of getting a blue marble on your
> second draw is *dependent* upon the result of the first draw. Spanish words
> with the same meaning as *independent events, dependent events*: eventos
> independientes, eventos dependientes. [California 2004a, p. 81]

When Freddy was done reading this passage, he was even less sure that he understood what "probability" meant than he was before reading it. Added to his confusion was what this word had to do with coins, fair or not, or with rain in Oakland. The *Study Guide* is meant only as a review of materials students have already covered in coursework at school. The textbook used in Freddy's ninth grade mathematics class [Fendel et al. 2003] covered probability over 30 days of games and experimentation, and was as clear as a textbook could be on the subject. Nevertheless, he could not remember anything of the concepts he learned in that course, nor could he relate what he found in the *Study Guide* to that experience. What was his problem? Was he simply intellectually incompetent, or was there something wrong with the textbook? Is he a hopeless case? Should he be graduating from high school if he can't pass a test like the CAHSEE, which is not really all that difficult?

I will argue that the problem lies neither in Freddy nor in the textbook. Freddy is no different from the many other students in the state who know English, but not the kind of English used in academic discourse, and especially in instructional texts. And there are problems with this textbook, to be sure, but not more so than with other texts addressed to students of Freddy's age. The passage is not so different from other instances of expository texts meant to convey information about subject matter such as probability and statistics for high school students. The problem is that students like Freddy — in fact, any student who has not learned the academic register used in written texts and in instructional discourse — need help learning it or they will find language to be an insuperable barrier to learning, whatever the subject matter. In mathematics education, it is a truism that the learner's language does not figure importantly, because "math has its own universal language; it is the same no matter what language is spoken by the students," as I am often reminded when I argue that mathematics teachers need to help their students who are English learners deal with the language in which math instruction is presented.

But is the language of mathematics universal? If students are unable to understand the words or the structures used in texts like the *Study Guide* we have been considering, could they understand the explanations offered in them? Would they be able to understand a teacher's explication of these ideas? Would they be able to solve the problems they encountered in high-stakes tests like the CAHSEE, which are often presented as word problems — simpler than the examples we have looked at from the CAHSEE *Study Guide*, but equally challenging linguistically? Consider, for example, the following released item from the 2004 CAHSEE [California 2004b, p. 25]:

	Highway		
	A	**B**	**C**
1	**A1**	**B1**	**C1**
Road **2**	**A2**	**B2**	**C2**
3	**A3**	**B3**	**C3**
4	**A4**	**B4**	**C4**

To get home from work, Curtis must get on one of the three highways that leave the city. He then has a choice of four different roads that lead to his house. In the diagram [to the left], each letter represents a highway, and each number represents a road.

If Curtis randomly chooses a route to travel home, what is the probability that he will travel Highway B and Road 4?

A. $1/16$ B. $1/12$ C. $1/4$ D. $1/3$

A reader who understands the language and recognizes the format in which mathematics problems are presented might not have difficulty with this one, although a student trained to visualize the situation described in a word problem may experience time-wasting cognitive strain imagining a choice of three highways, each of which offers exits onto any of the same four roads. A student like Freddy, however, had no idea what was being asked for because he understood neither the concept of probability nor the reasoning called for in asking about the likelihood of an event, given the conditions presented in the problem. He had no clue as to how he might respond. And so he took a guess, having no idea as to why it might or might not be the correct answer. Would understanding the problem have helped? I will argue that no understanding of the problem was possible without an understanding of the concepts that figured in this problem, and there was no way Freddy could have understood those concepts without the language in which they can be explicated and discussed whether by a teacher in class or in the textbooks in class. Clearly, students like Freddy, a second language speaker of English, need greater familiarity with the language of textbooks and instructional discourse in order to deal with school subjects like mathematics. But in what ways is such language different from everyday English?

Academic English

There are various characterizations of academic English. The one I operate with is this: It is extended, reasoned discourse — it is more precise than ordinary spoken language in reference, and uses grammatical devices that allow speakers and writers to pack as much information as necessary for interpretation into coherent and logical sequences, resulting in greater structural complexity than one finds in ordinary conversational language. The differences stem from the functions such language has and the contexts in which it is used. It is often used in situations where there is an audience or a reader that may or may not have the prior knowledge or the background necessary to understand what is being talked or written about, and where the context itself offers little support for the interpretation of the text or talk. Even when using illustrations, charts, or diagrams, speakers or writers must rely on language to convey what they have to say to their audience or readers.

In the case of speakers, as opposed to writers, for example teachers explaining the concept of probability to high school students, there is the possibility of assessing the effectiveness of what they are saying by monitoring the students' faces while speaking. That kind of feedback enables speakers to rephrase, redirect, or otherwise tailor the discourse to facilitate successful communication.

The writer, however, does not have access to that kind of immediate feedback. In preparing an academic text of any sort (this chapter, for example), the writer has to build into the text itself, as much information as needed to communicate his or her message to anyone who is a likely reader of the text, given its purpose and intended readership. The writer can make some assumptions about what the intended readers are likely to know; such presumed shared background relieves him or her of having to build in background information, or to go into lengthy explanations. Anything that cannot be assumed must be explicated and contextualized.

In the case of this paper — one discussing language issues in mathematics education — the readers who are mathematics educators or mathematicians will not need to have any of the math concepts explicated (for example, no explanation is needed for "probability," a concept with which the readers are without question more familiar than the writer is). On the other hand, the grammatical and lexical features that get in the way of students understanding the language used in math texts or instruction may or may not be as familiar to such readers; hence, greater care needs to be taken in building into the discussion, explanations the writer hopes will allow the reader to follow what she is trying to communicate.

This is what gives academic discourse, whether written or spoken, its special character. Some other notable features of academic English can be seen in the two passages from the CAHSEE Study Guide [California 2004a] which were discussed above. Consider the sentence:

In probability, an event is a particular happening that may or may not occur.

The need for specificity in academic discourse results in definitional clarity — in fields of study such as mathematics, ordinary words such as "event" sometimes have special meaning, and in such cases, they are stipulated as we see above: "an event is *a particular happening that may or may not occur*." Notice that the definition is more than a lexical equivalent: it is a noun phrase[4] stipulating the conditions that must be satisfied for "a particular happening" to qualify as "an event" in this usage. Academic language makes heavy use of complex noun phrases as we see in this definition. This noun phrase contains a relative clause ("*that may or may not occur*"), which is itself a sentence modifying '*happening*.'

Another feature of academic language is exemplified in the following excerpt from the CAHSEE *Study Guide* passage explaining what a dependent event is [California 2004b, Appendix, Math Vocabulary, p. 81]:

---

[4] A phrasal unit headed by a noun and which functions as a grammatical unit: some examples of noun phrases, from simple to complex: "Language," "the study of language," "the language on which you will be tested," "languages that are related to Latin."

(But) some events are *dependent*; that is, the probability of one event depends on whether the other event occurs. For example, suppose you are randomly choosing two marbles, one after another, from a bag that contains three blue marbles and three red marbles. On your first draw, you have a 50% chance of drawing a blue marble. But on your second draw, the probability of drawing a blue marble depends on which color you pulled out on the first draw. The probability of getting a blue marble on your second draw is *dependent* upon the result of the first draw.

Here we see the use of the hypothetical introduced by "suppose": readers familiar with this construction know that they should be going into the hypothetical mode: what follows is a description of a set of actions that add up to a situation against which the rest of the text is to be understood. The actions described are complicated by the qualifier, 'randomly,' a word we recognize as a technical term in this context, but is not explained here. Reader are assumed to understand the conditions that the use of this word places on the actions being described: "Suppose you are randomly choosing two marbles, one after another, from a bag that contains three blue marbles and three red ones."

The question here is what do students like Freddy find difficult about such language? His ability to make sense of the definition contained in this passage depends on his ability to, in some sense, visualize the situation described given the description, understand the ways in which it is constrained, and then extrapolate the distinction between events that are said to be dependent from ones that are independent. On that hinges his ability to make sense of test items like the CAHSEE released problem discussed earlier: "If Curtis randomly chooses a route to travel home, what is the probability that he will travel Highway B and Road 4?" To do the reasoning called for, he has to understand each part of the problem as it is described, and to know what it means to choose a particular combination from multiple possibilities.

It is clear that Freddy needs a lot of help not only with mathematics, but with the language in which it is communicated as well. Without a massive effort from his tutors and from himself, he will do no better on the high school exit examination than he has in the past. He is discouraged, but he keeps working because he wants to learn and to complete high school.

What English Learners Need

Working with Freddy, one discovers what happens when language differences are not considered in learning. How did Freddy end up so limited in his command of English? How did he become so unprepared to handle even the limited mathematics tested in the high school exit examination?[5] Freddy is not unique. There are hundreds of thousands of other former English learners in high schools across the country with problems not unlike Freddy's. California happens to have the largest concentration of such students in its schools, where one in four students begins school with little or no English. California's solution to the problem has been to require that they be instructed entirely in English, by voter mandate.[6] But what happens when children try to learn complex subject matter — and mathematics is one such subject — in language they do not fully understand because they are in the process of learning it? Is it possible to learn mathematics without understanding the language in which it is being taught?

It may not be difficult at least during the early years of school for children to participate in instructional activities even though they do not fully understand the language in which instruction is given. It becomes problematic, once children reach the third or fourth grade where the concepts they are taught tend to be more complex and abstract, requiring more explanation, and therefore a greater understanding of the language used for explanation. If they are learning the language of instruction rapidly enough and are given ample support for dealing with the subject matter, their progress in mastering the subject matter will not be greatly hindered by their incomplete knowledge of the language.

But what of the many students who do not find it so easy to learn the language of instruction? In second language learning, much depends on access, opportunity, and support. What children learn depends on who is providing access to the language and support for learning it. If the source is mostly from other students who are also language learners, the version they end up with is a learner's incomplete version. If the source is mostly from teachers who are using the language for instructional purposes, and the students are encouraged to participate in the instructional discourse using that language, it is possible for them to acquire the academic register.

Freddy's lack of familiarity with academic English suggests that he did not have many such opportunities. In fact, tracing back over his school records, one sees that his problems with mathematics began early on, back in the early grades of elementary school, when he did not understand what it was he was supposed to have been learning. We can't know how he would have fared educationally had he been educated in a language he already knew and understood rather than in one he was trying to learn as a second language. It seems evident that he would have found it easier to learn what he was supposed to be learning in school,

---

[5]There is little in the mathematics section of the CAHSEE that is not covered in ninth grade mathematics.

[6]California voters passed Proposition 227 in 1998, greatly restricting the use of bilingual instruction in schools except under special circumstances. Voters in Arizona followed suit in 2000, with the passage of Proposition 203, and Massachusetts voters passed legislation in 2002 that altogether bans bilingual education in that state.

had he been instructed in Spanish. There is some intriguing evidence from comparisons of the academic performance of immigrant students who received some formal education before coming to the United States and that of non-immigrant students of the same ethnic, socio-economic status, and linguistic backgrounds, with the immigrants outperforming the non-immigrant students [Berg and Kain 2003; Thomas and Collier 1997]. Cummins [1981] found that the older the immigrant students were at the time they arrived in Canada the better they performed in Canadian schools, both in measures of second language learning and in academic learning.

Ideally, Freddy (and students like him) would learn subject matter like mathematics in his primary language in elementary school, while receiving the instructional support required for mastering English. With a firm grounding in the foundational concepts and operations in math, he could easily enough transfer what he knows to mathematics classes taught in English when he was ready for that. Much depends on whether students like Freddy receive instructional support to develop the academic register for mathematics in school. In an on-going experimental program, Boston University researchers, Suzanne Chapin and Catherine O'Connor [2004] have been demonstrating what a difference such support can make both in the mathematics and English language development of English learners in Chelsea, Massachusetts middle schools.

Working with mathematics teachers in the Chelsea schools, the researchers designed a mathematics program to promote the language and mathematical skills that are typical of higher performing students. The mathematics program met both the state of Massachusetts's mathematics standards and the standards adopted by the National Council of Teachers of Mathematics. Project Challenge, as the program was called, recruited at-risk minority, economically disadvantaged middle school students who had potential and an interest in mathematics. Most of these children, while average on all indicators used by the project for selecting students, had significant gaps in their mathematical and linguistic knowledge. For example, many of these inner-city students had difficulty dealing with fractions, and they also had difficulty applying the mathematics they did know to activities outside of school, such as in sports. Two-thirds of the students recruited for the program were English learners, and entered the program with limited English proficiency. Most of the English learners were Latinos, and the rest were members of other under-achieving minority groups.

During the first four years of the project, the researchers and teachers worked with nearly 400 students, from fourth through seventh grade. During those years, the children were presented with a challenging math curriculum designed to develop their understanding of mathematics, and to improve their problem-solving abilities in dealing with complex mathematics problems. They received an hour each day of instructional support from their teachers to verbalize their evolving understanding of the concepts and operations they were learning. The emphasis was on getting the children to take responsibility for understanding and explaining their reasoning in solving mathematics problems, all in the language they were just learning. The focus of this instruction was on explanation, generalization, and justification of mathematical concepts and procedures. The teachers taught the children to explain, defend, and discuss their mathematical thinking with their classmates. By 2002, the team had outstanding results to report: using measures such as the mathematics subtests of the California Achievement Test, and the Massachusetts Comprehensive Assessment System, Project Challenge students were posting mean scores in the 87th percentile, outperforming children in the state of Massachusetts as a whole, and even those in higher performing districts in the state.

This is the kind of instructional support English learners like Freddy have needed. What it takes is attention to language in mathematics instruction, and indeed across the curriculum. It begins with educators recognizing that the difficulty that English learners have in subjects like mathematics goes beyond the subject matter itself.

How meaningful is any assessment of Freddy's mathematical competence, whether in the classroom or on a formal test such as the CAHSEE, given his present linguistic limitations? As Alan Schoenfeld points out (personal communication, August 16, 2005): "you can't know what he knows mathematically unless he is given an opportunity to show it, which means being given problems whose statements he can understand." Freddy may be capable of doing the mathematics he is expected to handle, if he is able to get past the linguistic demands of the problems. Hence attending to such demands must be an essential component of assessment and of instruction.

References

[Berg and Kain 2003] N. Berg and J. F. Kain, "The academic performance of immigrant and native-born schoolchildren in Texas", Political Economy Work Paper 10/03, Dallas: University of Texas, 2003.

[California 2004a] California Department of Education, *Preparing for the California High School Exit Examination: A mathematics study guide*, Sacramento: Author, 2004. Available at http://www.cde.ca.gov/ta/tg/hs/mathguide.asp. Retrieved 8 Dec 2004.

[California 2004b] J. O'Connell, "O'Connell releases 2004 STAR and CAHSEE results", News Release #04-72, Sacramento: California Department of Education, 2004. Available at http://www.cde.ca.gov/nr/ne/yr04/yr04rel72.asp. Retrieved 27 Jan 2007.

[Chapin and O'Connor 2004] S. H. Chapin and C. O'Connor, "Report on Project Challenge: Identifying and developing talent in mathematics within low income urban schools", Report No. 1, Boston, MA: Boston University School of Education Research, Boston University, 2004.

[Cummins 1981] J. Cummins, "Age on arrival and immigrant second language learning in Canada: A reassessment", *Applied Linguistics* **1** (1981), 132–149.

[Fendel et al. 2003] D. M. Fendel, D. Resek, L. Alper, and S. Fraser, *Interactive Mathematics Program: Integrated high school mathematics, Year I*, Berkeley, CA: Key Curriculum Press, 2003.

[Thomas and Collier 1997] W. P. Thomas and V. P. Collier, *School effectiveness for language minority students*, Washington, DC: National Clearinghouse for Bilingual Education, 1997.

Chapter 20
Beyond Words to Mathematical Content:
Assessing English Learners
in the Mathematics Classroom

JUDIT MOSCHKOVICH

Introduction

In the U.S., approximately 4.5 million (9.3%) students enrolled in K–12 public schools are labeled English learners [NCES 2002]. In California, during the 2000–2001 school year 1.5 million (25%) K–12 public school students were labeled as having limited English language skills [Tafoya 2002]. For numerous reasons, the instructional needs of this large population warrant serious consideration.

Assessment is particularly important for English learners because there is a history of inadequate assessment of this student population. LaCelle-Peterson and Rivera [1994] write that English learners "historically have suffered from disproportionate assignment to lower curriculum tracks on the basis of inappropriate assessment and as a result, from over referral to special education [Cummins 1984; Durán 1989; Ortiz and Wilkinson 1990; Wilkinson and Ortiz 1986]."

Previous work in assessment has described practices that can improve the accuracy of assessment for this population [LaCelle-Peterson and Rivera 1994]. Assessment activities should match the language of assessment with language of instruction and "include measures of content knowledge assessed through

This work was supported in part by faculty research grants from the University of California at Santa Cruz Senate, a grant from the National Science Foundation to the author (No. REC-0096065, Mathematical Discourse in Bilingual Settings) and a grant from NSF to the Center for the Mathematics Education of Latinos (CEMELA, No. ESI-0424983). Any opinions, findings, and conclusions or recommendations expressed in this material are those of the author and do not necessarily reflect the views of the National Science Foundation.

the medium of the language or languages in which the material was taught."
Assessments should be flexible in terms of modes (oral and written) and length
of time for completing tasks. Assessments should track content learning through
oral reports and other presentations rather than relying only on written or one-
time assessments. When students are learning a second language, they are
able to display content knowledge more easily by showing and telling, rather
than through reading text or choosing from verbal options on a multiple-choice
test. Therefore, discussions with a student or observations of hands-on work
will provide more accurate assessment data than written assessments. Lastly,
evaluation should be clear as to the degree to which "fluency of expression, as
distinct from substantive content" is being evaluated. This last recommendation
raises an interesting and difficult question for assessing English learners' mathematical proficiency. For classroom assessments that are based on mathematical
discussions, how can we evaluate *content* knowledge as distinct from fluency of
expression in English? The next section presents two examples of how assessment during classroom discussions can, in fact, focus on mathematical content
rather than on fluency of expression in English.

Moving from Assessing Words to Assessing Mathematical Content

Example 1: Describing a pattern. The first example [Moschkovich 2002] is
from a classroom of sixth-to-eighth grade students in a summer mathematics
course. The students constructed rectangles with the same area but different
perimeters and looked for a pattern to relate the dimensions and the perimeters
of their rectangles. Below is a problem similar to the one students were working
on:

1. Look for all the rectangles with area 36 (length and width are integers). Write
   down the dimensions.
2. Calculate the perimeter for each rectangle.
3. Describe a pattern relating the perimeter and the dimensions.

In this classroom, there was one bilingual teacher and one monolingual teacher.
A group of four students were videotaped as they talked with each other and
with the bilingual teacher (mostly in Spanish). As the four students attempted
to describe the pattern in their group, they searched for the word for rectangle in
Spanish. The students produced several suggestions, including ángulo (*angle*),
triángulo (*triangle*), rángulos, and rangulos. Although these students attempted
to find a term to refer to the rectangles neither the teacher nor the other students
provided the correct word, rectángulo (*rectangle*), in Spanish.

   Following the small-group discussion, a teacher — a monolingual English
speaker — asked several questions from the front of the class. In response, Ali-

cia, one of the students in this small group, attempted to describe a relationship between the length of the sides of a rectangle and its perimeter.

TEACHER: [Speaking from the front of the class] Somebody describe what they saw as a comparison between what the picture looked like and what the perimeter was? ...

ALICIA: The longer the ah, ... the longer [traces the shape of a long rectangle with her hands several times] the ah, ... the longer the, rángulo,[1] you know the more the perimeter, the higher the pcrimeter is.

If assessment of this student's mathematical knowledge were to focus on her failed attempt to use the right word, "rectangle," and her lack of fluency in expressing her ideas in English, we would miss the mathematical content and

---

[1] Although the word does not exist in Spanish, it might be best translated as "rangle," perhaps a shortening of the word "rectángulo."

competence in her description. Focusing on the missing vocabulary word would not do justice to how this student successfully communicated a mathematical description. If we were to focus only on Alicia's inaccurate use of the term "rángulo," we might miss how her statement reflects important mathematical ideas. If we move from a focus on words and English fluency, then we can begin to assess this student's mathematical competence. This move is important for assessment because it shifts the focus from a perceived deficiency in the student that needs to be corrected (not using the word "rectangle") to competencies that are already evident and can be refined through instruction. This move also shifts our attention from words to mathematical ideas, as expressed not only through words but also other modes. This shift to considering multiple modes of expression is particularly important to assess the competencies of students who are learning English.

What competencies in mathematical practices did Alicia display? Alicia described a pattern, a paradigmatic practice in mathematics, so much so that mathematics is often defined as "the science of patterns" [Devlin 1998, p. 3]. And Alicia described that pattern correctly. The rectangle with area 36 that has the greatest perimeter is the rectangle with the longest possible length, 36, and shortest possible width, 1 (if the dimensions are integers). As the length gets longer, say in comparing a rectangle of length 12, width 3, and perimeter 30 with a rectangle of perimeter 74, the perimeter does in fact become greater. Although Alicia was not fluent in expressing these ideas, she did appropriately (in the right place, at the right time, and in the right way) describe this pattern.

What language resources did Alicia use to communicate mathematical ideas? She used her first language as a resource for describing a pattern. She interjected an invented Spanish word into her statement. Even though the word that she used for rectangle does not exist in either Spanish or English, it is quite clear from looking at the situation that Alicia was referring to a rectangle. What modes of expression other than language did this student use? Alicia used gestures to illustrate what she meant and she referred to the concrete objects in front of her, the drawings of rectangles, to clarify her description. It is clear from her gestures that even though she did not use the words "length" or "width," she was referring to the length of the side of a rectangle that was parallel to the floor.

Seeing mathematical competence as more than words shifts assessment from focusing on fluency in English expression to focusing on mathematical practices such as describing patterns, generalizing, and abstracting. Shifting the focus of assessment to mathematical ideas, practices, competencies, and multiple modes of expression has important implications for how assessment informs instruction. Certainly, Alicia needs to learn the word for rectangle, ideally in both English and Spanish, but future instruction should not stop there. Rather than only correcting her use of the word "rángulo" and recommending that she learn vocabulary, future instruction should also build on Alicia's use of gestures, objects, and description of a pattern.

Example 2: Describing parallel lines. The lesson excerpt presented below [Moschkovich 1999] comes from a third-grade bilingual California classroom where there are thirty-three students who have been identified (in a local assessment) as Limited English Proficiency. In general, this teacher introduces students to topics in Spanish and later conducts lessons in English. The students have been working on a unit on two-dimensional geometric figures. For a few weeks, instruction has included technical vocabulary such as *radius*, *diameter*, *congruent*, *hypotenuse* and the names of different quadrilaterals in both Spanish and English. Students have been talking about shapes and the teacher has asked them to point, touch, and identify different shapes. The teacher identified this lesson as an English as a Second Language mathematics lesson, one where students would be using English in the context of folding and cutting to make tangram pieces (see figure on the right).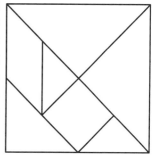

1. TEACHER: Today we are going to have a very special lesson in which you really gonna have to listen. You're going to put on your best, best listening ears because I'm only going to speak in English. Nothing else. Only English. Let's see how much we remembered from Monday. Hold up your rectangles …high as you can. (Students hold up rectangles) Good, now. Who can describe a rectangle? Eric, can you describe it [a rectangle]? Can you tell me about it?

2. ERIC: A rectangle has …two …short sides, and two …long sides.

3. TEACHER: Two short sides and two long sides. Can somebody tell me something else about this rectangle, if somebody didn't know what it looked like, what, what …how would you say it.

4. JULIAN: Paralela [holding up a rectangle, voice trails off].

5. TEACHER: It's parallel. Very interesting word. Parallel. Wow! Pretty interesting word, isn't it? Parallel. Can you describe what that is?

6. JULIAN: Never get together. They never get together [runs his finger over the top side of the rectangle].

7. TEACHER: What never gets together?

8. JULIAN: The paralela … they … when they go, they go higher [runs two fingers parallel to each other first along the top and base of the rectangle and then continues along those lines], they never get together.

9. ANTONIO: Yeah!

10. TEACHER: Very interesting. The rectangle then has sides that will never meet. Those sides will be parallel. Good work. Excellent work.

Assessing the mathematical content knowledge in Julian's contributions to this discussion is certainly a complex endeavor. Julian's utterances in turns 4, 6, and 8 are difficult both to hear and interpret. He uttered the word "paralela" in a halting manner, sounding unsure of the choice of word or of its pronunciation. His voice trailed off and it is difficult to tell whether he said "paralelo" or "paralela." His pronunciation of this word can be interpreted as a mix of English and Spanish or as just the Spanish word, whatever the vowel at the end. The grammatical structure of the utterance in line 8 is intriguing. The apparently singular "paralela" is preceded by the word "the", which can be either plural or singular, and then followed with a plural "when they go higher." What stands out clearly is that Julian made several attempts to communicate a mathematical idea in his second language.

What competencies in mathematical practices did Julian display? Julian was participating in three central mathematical practices, abstracting, generalizing, and imagining. He was describing an *abstract* property of parallel lines and making a generalization saying that parallel lines will *never* meet. He was also imagining what happens when the parallel sides of a rectangle are extended. What language resources did Julian use to communicate this mathematical idea? He used colloquial expressions such as "go higher" and "get together" rather than the formal terms "extended" or "meet."[2] What modes of expression other than language did Julian use? He used gestures and objects in his description, running his fingers along the parallel sides of a paper rectangle.

And lastly, how did the teacher respond to Julian's contributions? The teacher seems to move past Julian's confusing uses of the word "paralela" to focus on the mathematical content of Julian's contribution. He did not correct Julian's English, but instead asked questions to probe what the student meant. This response is significant in that it represents a stance towards student contributions and assessment during mathematical discussion: listen to students and try to figure out what they are saying. When teaching English learners, this means moving beyond vocabulary, pronunciation, or grammatical errors to listen for the mathematical content in student contributions. (For a discussion of the tensions between these two, see [Adler 1998].)

---

[2]It is important to note that the question of whether mathematical ideas are as clear when expressed in colloquial terms as when expressed in more formal language is highly contested and not yet, by any means, settled. For a discussion of this issue, see Tim Rowland's book [2000] *The Pragmatics of Mathematics Education: Vagueness in Mathematical Discourse.*

Summary

If classroom assessment only focuses on what mathematical words English learners know or don't know, they will always seem deficient because they are, in fact, learning a second language. If teachers perceive English learners as deficient and only assess and correct their vocabulary use, there is little room for addressing these students' mathematical ideas, building on them, and connecting these ideas to the discipline. English learners thus run the risk of being caught in a repeated cycle of remedial instruction that does not focus on mathematical content.

The two examples in this chapter show that English learners can and do participate in discussions where they grapple with important mathematical content, even if they do not always use the right words and even if they do not express themselves in a fluent manner. One of the goals of mathematics assessment for English learners should be to assess students' mathematical ideas, regardless of their proficiency or fluency in expressing their ideas in English.

Teachers can move towards this goal by learning to recognize the multiple mathematical practices, language resources, and non-language modes that students use to express mathematical ideas. Assessments of how English learners communicate mathematically need to consider more than students' use of vocabulary. Assessments should include how students participate in mathematical practices such as making comparisons, describing patterns, abstracting, generalizing, explaining conclusions, specifying claims, and using mathematical representations. Assessments should also consider how students use language resources such as colloquial expressions and their first language to communicate mathematical ideas. Lastly, assessment should include how students use modes of expression other than language to communicate. Classroom assessments that include the use of gestures, concrete objects, the student's first language, and colloquial expressions as legitimate resources for communicating mathematical ideas can support students in both displaying competencies and in further learning to communicate mathematically.

The two examples above also show that assessing the mathematical content of student contributions is a complex task. Three questions that are useful for moving beyond words and uncovering students' mathematical competencies are:

- What competencies in mathematical practices (describing patterns, abstracting, generalizing, etc.) do students display?
- What language resources do students use to communicate mathematical ideas?
- What modes of expression other than language do students use?

Assessment is certainly a complex task, perhaps especially when working with students who are learning English. It is not possible to decide whether an utterance reflects a student's conceptual understanding, a student's proficiency in expressing their ideas in English, or a combination of mathematical understanding and English proficiency. If the goal is to assess students' mathematical content knowledge, it is important to listen past English fluency and focus on the content: only then is it possible to hear students' mathematical ideas.

References

[Adler 1998] J. Adler, "A language of teaching dilemmas: Unlocking the complex multilingual secondary mathematics classroom", *For the Learning of Mathematics* **18**:1 (1998), 24–33.

[Cummins 1984] J. Cummins, *Bilingualism and special education: Issues in assessment and pedagogy*, Austin, TX: Pro-Ed, 1984.

[Devlin 1998] K. Devlin, *The language of mathematics*, New York: Freeman, 1998.

[Durán 1989] R. P. Durán, "Testing of linguistic minorities", pp. 573–587 in *Educational measurement*, 3rd ed., edited by R. L. Linn, New York: Macmillan, 1989.

[LaCelle-Peterson and Rivera 1994] M. LaCelle-Peterson and C. Rivera, "Is it real for all kids? A framework for equitable assessment policies for English language learners", *Harvard Educational Review* **64**:1 (1994), 55–75.

[Moschkovich 1999] J. N. Moschkovich, "Supporting the participation of English language learners in mathematical discussions", *For the Learning of Mathematics* **19**:1 (1999), 11–19.

[Moschkovich 2002] J. N. Moschkovich, "A situated and sociocultural perspective on bilingual mathematics learners", *Mathematical Thinking and Learning* **4**:2–3 (2002), 189–212. Special issue on diversity, equity, and mathematical learning, edited by N. Nassir and P. Cobb.

[NCES 2002] National Center for Education Statistics, "Public school student, staff, and graduate counts by state: School year 2000-01", Technical report, Washington DC: NCES, 2002. NCES Publication 2002-348.

[Ortiz and Wilkinson 1990] A. Ortiz and C. Wilkinson, "Assessment and intervention model for the bilingual exceptional student", *Teacher Education and Special Education* **Winter** (1990), 35–42.

[Rowland 2000] T. Rowland, *The pragmatics of mathematics education: Vagueness in mathematical discourse*, London: Falmer Press, 2000.

[Tafoya 2002] S. Tafoya, *The linguistic landscape of California schools*, San Francisco, CA: Public Policy Institute of California, 2002. California Counts: Population Trends and Profiles, Vol. 3, No. 4.

[Wilkinson and Ortiz 1986] C. Wilkinson and A. Ortiz, *Characteristics of limited English proficient learning disabled Hispanic students at initial assessment and at reevaluation*, Austin, TX: Handicapped Minority Research Institute on Language Proficiency, 1986.

Assessing Mathematical Proficiency
MSRI Publications
Volume **53**, 2007

Chapter 21
Assessment in the Real World:
The Case of New York City

ELIZABETH TALEPOROS

This chapter is devoted to a story from a city whose districts have faced and continue to grapple with huge challenges in helping teachers enable students from diverse backgrounds to achieve at high levels. My story is about one of the largest school systems in the United States and how it weathered storms of assessment and accountability. It begins in the hot political climate of the late 1960s.

New York City communities were clamoring for more control over their local schools. Turmoil over local control of schools in East Harlem and Ocean-Hill Brownsville sparked legislative action that resulted in the creation of thirty-two community school districts, each reflecting the unique local needs, interests, and cultural concerns of the neighborhoods that comprised them. The once all-powerful Central City Board of Education was replaced with local control, giving the districts the power to implement curriculum as it reflected local culture, and make decisions about how schools were to deliver education to neighborhood students.

The Central Board remained with a much narrower set of responsibilities, primarily focused on policy and accountability. It was responsible for setting standards, providing guidelines and support for instruction, and providing the public with test scores to reflect how successful schools were in meeting the standards. As part of the 1969 Decentralization Law, the Board was to hire a chancellor, whose responsibilities included monitoring achievement of schools in the newly created community school districts. The law required the publication of school rank by reading achievement.

Although prior city-wide testing existed, this requirement gave birth to the high-stakes city-wide testing program. Its primary mandate was reading achievement. The mathematics community advocated for city-wide assessment of mathematics also, and mathematics was added to the city-wide testing program. As it evolved, great controversy ensued about what grades and types of students were to be included, and the instructional implications of a common testing program without a common curriculum.

Later, the state testing program was implemented at grades 3, 6, 8, and high school. Its primary mandate was to assure minimal competency by the time students were ready to graduate. At high school, certain students also took a more challenging Regents examination, reflecting what has been characterized as a bifurcated system, with different academic curricula and assessments being applied to students in different tracks.

The city-wide testing program, mostly focused on elementary and middle school grades, used only norm-referenced tests. Each year, schools were ranked by the percentage of students scoring at or above grade level on these tests. Progress was tracked by this measure alone. Reading and mathematics achievement were thus monitored by the Central Board, and by real estate agents, public policy analysts, newspaper reporters, and all types of constituents from government and public advocacy groups.

The different approaches of the city and state programs led to great confusion, primarily about emphasis and curriculum. Teachers were responding to their local boards' requirements, and two sets of assessment mandates. One focused on achievement over a low bar of minimal competency; the other focused on achievement at or above a national median, where by definition half of the scores were below average. These differences perplexed teachers whose work was being assessed by different kinds of tests, with the success being evaluated with entirely different types of metrics.

In the 1990s, the city and the state began to come closer together, in the spirit of the then-burgeoning movement for school reform and high-stakes high-standards accountability. In the city, a group of people began to re-examine the city-wide program. Could we alter it, so that testing was a worthwhile experience for students and supported good instruction at the same time? We began to conceive of a mathematics performance assessment, where students were required to solve problems and explain how they got their answers, for the 75,000 students in the city. Certainly, we were concerned about a number of issues associated with scoring: getting reliable and valid scores, and providing professional development so that teachers could score the tests. Despite these concerns, we became convinced that this process was a way to turn instruction from the drill and kill approach to that of problem solving, reasoning, and communication.

We began by simply posing three problems to all seventh graders, asking them to solve the problems, describe their approach, and communicate how they reached their conclusions. We developed a general 6-point rubric and classified scores as high, medium, or low proficiency. Each score point was described in the general rubric, with specific examples cited for the given problem. Every middle school mathematics teacher was trained in scoring in which each potential score point was illustrated by a few papers gathered during pilot testing a variety of ways students could achieve the score.

We made a policy decision to give the classification "low" to responses that gave the correct answer alone, with no explanation. This focus on different ways to solve problems, with communication about the method chosen and the steps taken, had an enormous impact on teachers. The aspects of performance that were valued changed. Students had to focus on their reasoning and communication skills.

Later, the program expanded to include fifth as well as seventh grade. We developed a method to scale the Performance Assessment in Mathematics (PAM) scores with the regular multiple-choice norm-referenced scores. To build on the design established in mathematics, the city developed a reading test for grades 4 and 6 to make the assessment program comprehensive and inclusive of depth of thinking dimensions as well as the traditional sampling from the breadth of the content domains.

The state program was then modeled in the same fashion for grades 4 and 8, eventually displacing the city program and expanding the topics tested. The reading test expanded to include greater emphasis on listening skills and writing. The mathematics test expanded to include more problems, with a significant increase in the challenge they provided to students. Now, with the No Child Left Behind requirements, the state program holds great promise for teaching and learning in New York. The evolution that produced this balanced and deeper approach to assessment was slow and incremental. But its impact on instruction has been profound, and it is hoped that teachers and students will continue to benefit in meeting the continuing challenges.

Our story does not end here. The structure of the educational system in New York has changed again. There is now a newly reorganized Department of Education, which, some say, has taken more centralized authority and responsibility. The political pressures remain, and have in some ways even increased. But fortunately, the city's educational system is still struggling to balance the external political agenda with the work of a group of talented educators whose passion for standards-based curriculum, instructional approaches, and assessment is indeed to be recognized and applauded. The story of New York City is, and will undoubtedly remain, the continuing story of a work in progress.

Assessing Mathematical Proficiency
MSRI Publications
Volume **53**, 2007

Chapter 22
Perspectives on State Assessments in California:
What You Release is What Teachers Get

ELIZABETH K. STAGE

In the 1970s, the California Legislature determined that California's curriculum was not well served by national standardized tests and developed its own testing program, the California Assessment Program (CAP). Folklore gives several rationales for this action: urban superintendents wanted to conceal their students' low achievement, which would be revealed with the publication of national norms; consciousness that the California population was more diverse than the U.S. population; or awareness that California's curriculum differed from the national composite used for national tests.

California has had state-wide frameworks to guide district curricula for many years. This story begins with the 1980 addendum to the 1975 framework [California 1982], a small volume that declared "problem solving" to be the umbrella for all of the framework's curricular strands (number, algebra, geometry, statistics, etc.). The task for CAP was to provide the state and districts with information about performance on these strands. It used matrix sampling and item response theory to provide detailed analysis, on the basis of which instructional improvements could be made. CAP was not designed to yield individual student scores; these could continue to come from standardized tests so that parents, teachers, principals, and superintendents could answer the question, "How does this student, this class, this school, or this district stack up in comparison with national norms?"

CAP was designed to provide scores on mathematical topics. It used matrix sampling to make sure that enough kids took items such as "whole number division" or "similar figures" to yield a reliable and valid score. That meant that people could see quite obviously that scores on the "number" component of the exam were relatively good, while other strands had weak performances.

This was similar to California results on the National Assessment of Educational Progress and many other measures. But, CAP showed information at that level of detail for districts and schools, as well for the state as a whole. That way teachers could focus on areas of poor performance because they were getting information in a form they could use. "I can work on division. That is something that I can target."

The way in which I entered this conversation was with respect to gender differences. The pattern of CAP results, performance on the number strand that was strong relative to performance on the other strands, was even more pronounced for girls. (There wasn't very good tracking of ethnic data at the time.) Professional development programs like EQUALS at the Lawrence Hall of Science focused attention on giving girls opportunities to learn geometry. Over time, the gap narrowed. In Lee Shulman's formulation, CAP was a low-stakes, high-yield assessment. CAP yielded information that teachers could use to target instruction.

Work through this item before you read further.

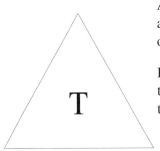

A piece of cardboard shaped like an equilateral triangle with a side 6 cm is rolled to the right a number of times.

If the triangle stops so that the letter T is again in the upright position, which one of the following distances could it have rolled?

a. 24 cm   b. 60 cm   c. 30 cm   d. 90 cm

A typical multiple-choice item on most standardized tests is allotted a response time of 45 seconds to a maximum of 1 minute. The item above, categorized "extended multiple choice," was allotted as much as 3 minutes because, at the time, in 1988, it was unconventional and it required some thought. In other words, it doesn't say, "What's the perimeter of the triangle?" and "Which one of the numbers is divisible by 18?," both of which suggest approaches to solving the problem. Another strategy is to mentally imagine the triangle rolling over, "Cabump, cabump, cabump, 18; cabump, cabump, cabump, 36"; until you get a match with one of the possible answers. Because CAP was low stakes, its designers could fool around with format and use the multiple-choice format more creatively. The idea was not to sort children, for which a speeded format is very helpful. If you want to sort people, make them run a race; if you want to see if kids can "do it," them give them adequate time to "do it."

At the same time, CAP had introduced something called "Direct Assessment of Writing." Instead of asking where the comma goes in a sentence, which can be done in a multiple-choice format, or about subject-verb agreements, the CAP test made students write — which was amazing and revolutionary at the time. In response, teachers started to ask students to write, using the seven CAP genres (persuasive, expository, etc.). And, student writing improved.

One could argue that the professional development programs like EQUALS or the California Writing Project (which was behind the direct assessment of writing) contributed to improved student performance. Or, one could argue that CAP scores improved due to a variety of other factors such as familiarity with the test, better instructional materials, improved nutrition, etc. Whether the focus for teachers on the CAP released items caused the improvement cannot be determined, but the fact is that many students spent considerable time improving their responses to tasks of this sort. That was not a bad way for them to spend their time.

The mathematics community was envious of what was taking place in writing and tried to take advantage of their success. We embraced the extended multiple-choice items and used them in our workshops with teachers, but behind the scenes we said to the CAP folks, "We'd really like to see how we can push the limits."

One of my favorite items from this era was this:

James knows that half of the students from his high school were accepted to the public university nearby, half were accepted to the local private college. He thinks this adds up to 100% so he'll surely be accepted at one or the other. Explain why James might be wrong.

One can determine from the context that this was a high-school item, though the mathematics is clearly elementary. Therefore, it was amazing how many students proved that James was correct. The vast majority of the students wrote "one half plus one half equals one, which is the same as 100%." When this result was reported, even though it was an experimental item and didn't count for anything official, the teachers whose students had been happily adding probabilities without regard for context were horrified. And, assuming that problems like this would, once scoring issues were sorted out, count in the official assessment, they started to give students problems like this to solve. Thus, good assessments can have a beneficial curricular impact.

The extended multiple-choice items, like the triangle problem, and constructed-response items, like James being accepted at college, were released in a California Department of Education publication called *A Question of Thinking* [California 1989]. This was presented as simply a collection of problems that teachers might want to try — like CAP, low stakes, high yield.

Around the same time, the California Mathematics Council (CMC, the state affiliate of the National Council of Teachers of Mathematics) published *Alternative Assessment in Mathematics* [Stenmark 1989]. It showcased items from CAP, the Shell Centre, and other sources. Some 30,000 copies were distributed to CMC members, at conferences, and through other professional networks of teachers. This teacher-to-teacher communication, basically saying, "Look at all these cool ways that we can find out what is going on in our children's mathematical thinking," was — again — low (in fact, no) stakes, high yield.

California seems to prefer the rollercoaster model of educational policy. In the 1980s, under State Superintendent Bill Honig, the curriculum frameworks called for more authentic, extended work in every discipline. In addition to reading and writing in English Language Arts and problem solving in mathematics, the frameworks called for hands-on experimentation in science and working with primary source documents in history.

According to Honig, CAP had intentionally been designed as a low-level assessment because the state's urban superintendents wanted a measure that wouldn't show much differentiation. Honig used the bully pulpit of the state superintendency to get those urban superintendents to see that, in fact, using a low-level test perpetuated inequity; he argued that the test had to be ratcheted up to test what was really valued in order to expose the magnitude of current inequities. The revised state testing program, the California Learning Assessment System (CLAS), was designed to use the new item types, extended multiple-choice and constructed response, in addition to multiple-choice items, to assess the more demanding curricular goals that could not be assessed adequately with multiple-choice items.

Unfortunately, CLAS backfired. Tests that, like CLAS, use constructed-response formats depend on the training of the scorers. In the United States, teachers score only one large-scale test, the New York State Regents' exams. The Advanced Placement exams involve only a small group of elite teachers. Thus, the vast majority of California teachers had no experience evaluating constructed-response items outside of their own classrooms.

Teaching anybody to score with a rubric rather than their own personal standards or judgment is hard work. There is a whole technology to make sure that the scores are accurate, including scorer training and qualification procedures, table leaders and room leaders, and read-behinds. Installing that technology and training the leadership in that technology takes time; the constant monitoring of scorers' accuracy is a major culture change for teachers who are accustomed to grading papers in isolation.

Where would a California education story be if it didn't have some politics? The CLAS program was initially intended to be just like CAP, designed to give curricular information to schools, districts, and the state. Midstream, the governor ordered the program to produce individual student scores. The teachers union had been arguing for individual student scores on the premise that, in exchange for the students' time, you owed them scores on the assessment. Because the time for constructed-response items exceeded the time needed for the multiple-choice tests, the union's increased demands for individual student scores dovetailed nicely with the governor's increased interest in accountability. Unfortunately, neither the governor nor the union understood a psychometric rule of thumb: You need 30 or more data points to get a reliable score. If your goal is to get the score at the school level, you can get lots of data points about lots of subscores. If your goal is an individual student score and you don't have unlimited testing time, then you have to narrow the domain of what you test in order to get a reliable and valid score. A test that was designed for one purpose — school-level scores — was asked suddenly to fulfill a different purpose: individual student scores. It didn't have the right design to accomplish the new purpose and the technology for the scoring wasn't yet robust enough to assign accurate scores to individual students.

To add insult to injury, the CLAS writing assessment asked students to write a brief essay based on their reading of a passage written by Alice Walker in which the protagonist wondered whether or not to get married. Despite the lack of personal freedom that she would experience in a conservative religious tradition, she decided to go ahead with the marriage for the sake of her children. The excerpt was published in the *Orange County Register* and the Eagle Forum became unglued: "Questioning marriage is not an appropriate topic for eighth graders!" (This was despite the fact that the protagonist decided to get married.)

The governor had a huge political problem. He used a psychometric argument concerning the accuracy of the prematurely released individual scores as an excuse for not defending CLAS. Bill Honig, the reform superintendent who had set the more demanding frameworks and assessments in motion, had stepped on the toes of the State Board of Education by getting ahead of them, and he had been indicted for the appearance of conflict of interest, so he was in no position to defend CLAS. Without further ado and with no rational discussion whatsoever, CLAS went down in flames.

When CLAS was in place, just as with the Direct Assessment of Writing, teachers were giving students opportunities to construct responses to challenging questions in mathematics, science, and social studies. For many students, particularly for some of our most neglected students, it was the first time anybody ever asked them, "What do you think?" (I think that question is the most profound assessment of a classroom; if you visit a classroom and don't hear someone say to someone else, "What do you think?" in a whole class period, then those students are being shortchanged.)

Because of the political debacle, Governor Wilson, through his State Board of Education, asked the superintendent, "What shall we use as the state test?" Ninety percent of the local districts were using McGraw-Hill's CTB at that time. The state superintendent (a Democrat) recommended CTB, so the governor (a Republican) picked the Harcourt Assessment's Stanford Achievement Test, Ninth edition (SAT 9), making lots of people in San Antonio, Texas very happy.

Hardly anybody knows that Harcourt, Riverside Publishing, and McGraw-Hill produce practice exams that show the format of the items but not the level of difficulty or the range of mathematics that is assessed. They don't show the differences between what they call "skill problems," "concept problems," and "problem-solving problems." Because teachers don't know what's on the test and it doesn't get released, they drill on arithmetic. That is all they can be certain will be on the test. That is what they practice and the scores go up because practice pays off.

Using the SAT 9 was only an interim solution, because it was not aligned with California standards, so the plan was to add California standards items gradually to the test. There is a blueprint of the test available that shows how many items will be on which standards. But, in 2004, at the time of the MSRI assessment conference, there were no released items available, so teachers had to imagine what they might look like.

An interesting note is that after three years of administering California Standards Tests in mathematics, released items were posted recently [California 2006]. There are 65 items on the fifth grade test: 17 about operations on fractions and decimals, 17 about algebra and functions. An interesting example of the latter is:

What value for z makes this equation true?

$$8 \times 37 = (8 \times 30) + (8 \times z)$$

A. 7   B. 8   C. 30   D. 37

A student might recognize the equation as an instance of the distributive property. Or, the student might try the answer choices to see which number makes the equation true.

Until recently, the only clue that teachers had at their disposal to predict what would be on the test was the "key standards." Responding to initial reactions to the mathematics standards, that there were too many topics on the list to teach in depth, about half of the standards were identified as "key" standards. The test blueprint was designed so that 80% of the test addressed these key standards. When school districts developed pacing plans that covered only the key standards, the percentage was dropped to 70% to encourage teachers to teach all of the standards. Key standards-only pacing plans persist in at least one large California district.

Released items from California state assessments, to the extent that they have been available, have been more influential than the standards or frameworks that they are supposed to exemplify. It's not what the framework or standards say or intend. It is the teachers' perceptions of what counts that afford the students the opportunity to learn.

References

[California 1982] California State Board of Education, *Mathematics framework and the 1980 addendum for California public schools, kindergarten through grade twelve*, Sacramento: Author, 1982.

[California 1989] California Department of Education, *A question of thinking: A first look at students' performance on open-ended questions in mathematics*, Sacramento: Author, 1989.

[California 2006] California Department of Education, 2003, 2004, and 2005 CST released test questions, January 2006. Available at http://www.cde.ca.gov/ta/tg/sr/css05rtq.asp. Retrieved 21 Apr 2006.

[Stenmark 1989] J. K. Stenmark, *Assessment alternatives in mathematics*, Berkeley, CA: University of California, 1989.

Assessing Mathematical Proficiency
MSRI Publications
Volume **53**, 2007

Epilogue
What Do We Need to Know?
Items for a Research Agenda

Eight working groups at the 2004 MSRI conference "Assessing Students' Mathematics Learning: Issues, Costs and Benefits" were charged with formulating items for a research agenda on the topic of the conference. The moderators of the working groups were:

Linda Gojak, John Carroll University

Hyman Bass, University of Michigan at Ann Arbor

Bernard Madison, University of Arkansas

Sue Eddins, Illinois Mathematics and Science Academy

Florence Fasanelli, American Association for the Advancement of Science

Emiliano Gomez, University of California at Berkeley

Shelley Ferguson, San Diego City Schools

Hugh Burkhardt, University of Nottingham

The following items were identified by these working groups. Any item mentioned by more than one group appears only once.

1. On the topic of productive disposition, as used by Robert Moses and others (e.g., *Adding It Up*), the following questions could be studied. How does one actually define this? How does one measure it? How does knowledge of a student's position on a "productive disposition" scale affect assessment? What are the symptoms of an unproductive disposition? (For example, perhaps a student response to a request to work on a task by saying, "I am not working on the task. I am waiting to be told how to do it.") Is an unproductive disposition a learned behavior? If so, what can be done to change that? What items (e.g., classroom setting, teacher, materials) are important in developing productive dispositions?

2. Confirm or deny the following. Hypothesis 1: Productive disposition, as described in Item 1, declines from the early grades to the middle grades. Hypothesis 2: The decline has become worse over time.

3. How do students interpret questions on assessments? How can assessment tools be designed so that the assessor and student interpretations coincide?

4. Results from a high-stakes assessment and an intensive effective assessment (e.g., a well-designed interview conducted by an expert interviewer) could be compared and contrasted.

5. Practical tests could be designed to give results that are similar to "impractical" methods (e.g., interviews). This is the thrust of some work by Ed Dubinsky and others.

6. Treating mathematics as a language: How students learn and practice this language and become proficient in it, and how one can assess this linguistic mastery, should be researchable issues.

7. Is it appropriate to teach for mastery before proceeding? If so, how does one assess actual mastery?

8. What could be learned from extensive interviews with students and their teachers, with the interviews of the groups either held separately or interspersed?

9. Although we know what many people say students are being taught, do we really know what skills and concepts are actually being taught, and are our assessment tools up to the challenge of identifying this? Opinions seem to differ greatly.

10. How can teachers identify and use information about individual students' backgrounds before attempting to teach them? Some method is needed for identifying background, and attaching that information to assessment and achievement, if potential is to be identified.

11. How do students view the need for precision, and how can they be expected to achieve it? How can we assess whether they have done so?

12. What is the connection between the precision used by the teacher and that used by the student? How precise does the teacher have to be for students at each grade level? Are students being taught in precise enough ways to be able to be precise themselves? How do we then actually assess precision?

13. The ability to form generalizations is an important aspect of a student's mathematical development. How does one assess this? How much does a child have to "know" to generalize mathematics concepts, and how "solid" a background is needed to assure that they are not forgotten? How can this background knowledge and solidity be assessed?

14. What is the actual role of assessment of student knowledge in the improvement of teaching and learning, in both theory and practice? How can a large-scale assessment be productively used to improve teacher development?

15. How much of a large-scale assessment should be published ahead of time?

16. How can standards and assessment be aligned, and how much emphasis should actually be placed on such alignment?

17. How can assessment be individualized so that at various points in a student's career an educational plan for the student's further development can be created?

18. It would be interesting to collect a set of assessment tasks on a particular topic and map these to a conceptual space of understanding that might serve as a diagnostic tool. A series of studies could be done on this, perhaps across groups (universities, leadership groups in education, teachers, and so forth).

19. At the high school level, what components are needed to motivate under-performing students to be successful in college preparatory mathematics in order to be accepted and successful in college, and what is the role of assessment in this? Some possible components that were suggested are resources, teacher quality, class time, class size and load, curriculum and pedagogy, assessment, administrative support, student engagement, parental involvement, and transferability of innovation into the classroom.

20. A longitudinal study is needed to determine the impact on student learning of various types of summative assessment. What impact do teacher knowledge, and various curricula have on student performance on high-stakes testing? What impact does high-stakes testing have on student learning? What about the same sorts of questions for formative assessments (interviews, quizzes, embedded assessments, Freudenthal Institute methods, and so forth)?

21. There is a need for comparative studies of a variety of assessment and accountability systems which are promising for the development of mathematical proficiency for *all* student populations.

22. There is a need for comparative studies of conceptions of mathematical proficiency. How are these conceptions reflected locally, nationally, and internationally in assessments, curriculum, professional standards, and practices?

23. What assessment practices (and more generally what pedagogical practices) contribute to a dislike of mathematics? While on the subject of dislike of mathematics: Is stating such a dislike in class by the students acceptable, or does it affect class attitude?

24. How can one make effective comparative assessments across grade levels of difficult (to many students) mathematical concepts such as equivalence relations?

About the Authors

Michèle Artigue is a professor in the Mathematics Department and a member of the Doctorate School in Epistemology, Philosophy and Didactics of Sciences at the University Paris 7. After earning a Ph.D. in mathematical logic, she became involved in the activities of the Institute of Research for Mathematics Education at the University Paris 7; her research interests moved progressively towards the new field of didactic research in mathematics. Like most researchers in this area, she worked first at the elementary level on pupils' conceptions of mathematics. However, she soon got involved, jointly with physicists, in experimental programs for undergraduate students at university and thus in didactic research at more advanced levels. Artigue's main contributions deal with the didactics of calculus and analysis, the integration of computer technologies into mathematics education, and the relationships between epistemology and didactics. This research has led to her editing or co-editing of nine books and more than fifty articles or chapters in international journals and books. She is a member of the scientific committee or editorial board of several international journals and a member of the European network of excellence, Kaleidoscope. Since 1998, she has been vice-president of ICMI, the International Commission on Mathematical Instruction.

Current address: Equipe Didirem, Case 7018, Université Paris 7, 75251 Paris Cedex 05, France (artigue@math.jussieu.fr)

Richard Askey retired after teaching mathematics at the University of Wisconsin for forty years. His work in mathematics was on special functions, and this has had an impact on his work in education which can be seen in two papers on Fibonacci numbers in *Mathematics Teacher*. He is interested in curriculum and the knowledge teachers need to be able to teach well. Both of these can be studied by looking at what is done in other countries, and what was done in the United States in the past. His interest in teaching mathematics started early as illustrated by his starting a tutoring program in high school which was run by the Honor Society.

Current address: University of Wisconsin – Madison, Mathematics Department, 480 Lincoln Drive, Madison WI 53706-1388, USA (askey@math.wisc.edu)

Deborah Lowenberg Ball is Dean of the School of Education and William H. Payne Collegiate Professor and Arthur F. Thurnau Professor at the University of Michigan. Ball's work draws on her many years of experience as an elementary classroom teacher and teacher educator. Her research focuses on mathematics instruction, and on interventions designed to improve its quality and effectiveness. Her research groups study the nature of the mathematical knowledge needed for teaching and develop survey measures that have made possible analyses of the relations among teachers' mathematical knowledge, the quality of their teaching, and their students' performance. Of particular interest in this research is instructional practice that can intervene on significant patterns of educational inequality in mathematics education. In addition, she and her group develop and study opportunities for teachers' learning. Ball's extensive publications, presentations, and Web site are widely used.

Ball is a co-principal investigator of the Study of Instructional Improvement, a large longitudinal study of efforts to improve instruction in reading and mathematics in high-poverty, urban elementary schools. Ball was a member of the Glenn Commission on Mathematics and Science Teaching for the Twenty-First Century. She is a co-director of the NSF-funded Center for Proficiency in Teaching Mathematics, a research and development center aimed at strengthening professional education of mathematics teachers. She chaired the Rand Mathematics Panel on programmatic research in mathematics education, co-chaired the international Study on Teacher Education sponsored by the International Commission on Mathematical Instruction, and she is a trustee of the Mathematical Sciences Research Institute in Berkeley.

Current address: 4119 School of Education, 610 E. University, University of Michigan, Ann Arbor, MI 48109-1259, USA (dball@umich.edu)

Hugh Burkhardt is a theoretical elementary particle physicist and educational engineer. He has been at the Shell Centre for Mathematical Education at the University of Nottingham since 1976, as director until 1992. Since then he has led a series of international projects including, in the U.S., Balanced Assessment, Mathematics Assessment Resource Service (MARS), and its development of a Toolkit for Change Agents. He is the Project Director of MARS with particular responsibility for project processes and progress, and a visiting professor at Michigan State University.

He takes an "engineering" view of educational research and development — that it is about systematic design and development to make a complex system work better, with theory as a guide and empirical evidence the ultimate arbiter. His other interests include making mathematics more functional for everyone through teaching real problem solving and mathematical modeling, computer-aided mathematics education, software interface design, and human-computer interaction.

He remains occasionally active in elementary particle physics.

Current address: Shell Centre, School of Education, Univ. of Nottingham, Jubilee Campus, Nottingham NG8 1BB, UK (Hugh.Burkhardt@nottingham.ac.uk)

Claus H. Carstensen is a research psychologist working principally on the topic of item response theory (IRT) applied to educational research. His publications focus on multidimensional item response models for psychological and educational measurement and applied assessments. His research is carried out mainly in the context of the OECD PISA project. In the past, Professor Carstensen has worked for the international PISA consortium at the Australian Council for Educational Research (Melbourne, Australia), and now is working with the German PISA consortium (Kiel, Germany). In this context, he is mainly involved in the development of assessment instruments, the scaling and linking of data from large-scale-assessment studies, and the analyses of data from complex samples.

Current address: IPN - Leibniz Institute for Science Education, University of Kiel, Olshausenstrasse 62, 24098 Kiel, Germany (carstensen@ipn.uni-kiel.de)

Lily Wong Fillmore recently retired from the faculty of the Graduate School of Education at the University of California, Berkeley. She is a linguist and an educator: much of her research, teaching and writing have focused on issues related to the education of language minority students. Her research and professional specializations are in the areas of second language learning and teaching, the education of language minority students, and the socialization of children for learning across cultures. Over the past thirty-five years, she has conducted studies of second language learners in school settings on Latino, Asian, American Indian and Eskimo children. Her research and publications have focused on social and cognitive processes in language learning, cultural differences in language learning behavior, sources of variation in learning behavior, primary language retention and loss, and academic English learning.

Current address: School of Education, LLSC, Tolman Hall #1670, University of California, Berkeley, CA 94720-1670, USA wongfill@berkeley.edu

Linda Fisher is the Director of the Mathematics Assessment Collaborative (MAC). She oversees mathematics performance test development for grades 3–10, helps develop rubrics, and supervises scoring training. Linda also analyzes student work and misconceptions as demonstrated on the performance assessment; her analysis is included in the teacher resource, *Tools for Teachers*. She was formerly a staff developer for the Middle Grades Mathematics Renaissance and the Mathematics Renaissance K-12 Project. She was also formerly a middle school mathematics teacher in Berryessa School District.

Current address: Noyce Foundation, 2500 El Camino Real, Suite 110, Palo Alto, CA 94306-1723, USA (lfisher@noycefdn.org)

David Foster is the Mathematics Director of the Robert N. Noyce Foundation. He oversees and directs the Silicon Valley Mathematics Initiative, which comprises thirty-five member districts in the San Francisco Bay Area. Foster is the primary author of *Interactive Mathematics: Activities and Investigations*, an innovative mathematics program for middle school students, grades 6 through 8. His other works included *Exploring Circles*, published by Glencoe in 1996 and *Computer Science One*, published by Coherent Curriculum in 1988. He was a Regional Director for the Middle Grade Mathematics Renaissance, the mathematics component of the California Alliance for Math and Science.

Foster taught mathematics and computer science at middle school, high school, and community college for eighteen years. In 1989 he was selected Outstanding Mathematics Teacher by an affiliate of the California Mathematics Council. He is Co-Director of the Santa Clara Valley Math Project and Co-Chair of the advisory committee of the Mathematics Assessment Resource Service. David was founding member of the California Math Project Advisory Committee and the California Coalition for Mathematics. He was a California Curriculum Consultant and has served on the California Teaching Credential Advisory Committee for Mathematics.

Current address: Noyce Foundation, 2500 El Camino Real, Suite 110, Palo Alto, CA 94306-1723, USA (dfoster@noycefdn.org)

Jan de Lange is past chairman/director of the Freudenthal Institute and full professor at University of Utrecht in the Netherlands. The Freudenthal Institute is part of the Faculty of Mathematics and Computer Science and has as its task innovation in mathematics education by research, implementation, dissemination and professionalization.

Following graduate work in mathematics at both the University of Leiden in the Netherlands and Wayne State University in Detroit, he received his Ph.D. at the University of Utrecht. His dissertation, "Mathematics, Insight and Meaning," is on the development of a new applications-oriented curriculum for upper secondary mathematics and more specifically the assessment problems that come with such a curriculum. His research still focuses on modeling and applications and assessment issues and has broadened to multimedia and on issues related to implementation.

De Lange is member of the Mathematical Sciences Education Board. He was a member of the National Advisory Board of the Third International Mathematics and Science Study (TIMSS) and currently chairs the Mathematical Functional Expert Group of the OECD Programme for International Student Assessment (PISA).

Current address: Director, Freudenthal Institute, Faculty of Mathematics and Computer Sciences, Utrecht University, P.O. Box 9432, 3506 GK Utrecht, The Netherlands (J.deLange@fi.uu.nl)

Bernard Madison is a mathematician and mathematics educator with forty years' experience in research, teaching, university administration, and science policy. Following over two decades as department chair or college dean, his recent work has focused on assessment in college mathematics, education for quantitative literacy, and the mathematical education of teachers. He directs two major national NSF-funded college faculty professional development projects for the Mathematical Association of America.

In addition to publishing over twenty mathematics research papers and approximately fifty articles on education and policy issues, he has recently edited two books, one on quantitative literacy and one on assessment. He is now serving as the first President of the National Numeracy Network.

He developed and teaches a course entitled Mathematical Reasoning in a Quantitative World using recent newspaper and magazine articles as source material. He is a native of Kentucky and a first generation college student who began school in one of the last one-room elementary schools in the Commonwealth. He holds a B.S. from Western Kentucky University, and an M.S. and Ph.D. from the University of Kentucky.

Current address: Department of Mathematical Sciences, 301 SCEN, University of Arkansas, Fayetteville, AR 72701, USA (bmadison@uark.edu)

William McCallum is a University Distinguished Professor at the University of Arizona, whose principle interests are arithmetical algebraic geometry and mathematics education. In the latter area he has worked on undergraduate curriculum development, particularly calculus reform, and more recently has become interested in K-12 education. He has worked on a number of projects promoting collaboration between mathematicians and educators, most recently as Director of the Institute for Mathematics and Education at Arizona.

Current address: Department of Mathematics, The University of Arizona, 617 N. Santa Rita Avenue, P.O. Box 210089, Tucson, AZ 85721-0089, USA (wmc@math.arizona.edu)

R. James Milgram is Professor of Mathematics in the Department of Mathematics at Stanford University, a position he has held since 1969. His research in mathematics centers around algebraic and geometric topology, and in recent years his earlier results in these areas have been applied in the areas of robotics and protein folding. His work in mathematics education started with helping to write the 1998 *California Mathematics Standards* with three colleagues at Stanford and the 1999 *California Mathematics Framework* with H.-H. Wu. Since that time he has been involved in helping to write the Achieve, Massachusetts, Michigan, and Georgia Mathematics Standards as well. He is a senior author on a middle school mathematics series, and was the principal investigator on an FIE grant to study the mathematics that pre-service and in-service teachers need to know. More recently he has been developing a course at Stanford with Liping Ma on the mathematics that pre-service elementary school teachers need to know. He is a member of the National Board for Education Science that oversees the Institute of Educational Sciences at the U.S. Department of Education and is on the board that advises Congress on education research. He is a member of the NASA Advisory Council (NAC), and a member of the NAC Committee on Human Capital. He is one of the directors of the Comprehensive National Center for Instruction, and is one of the principal investigators on a recent FIE grant to construct a national arithmetic exam.

Current address: Department of Mathematics, Bldg. 380, Stanford University, 450 Serra Mall, Stanford, CA 94305-2125, USA (milgram@math.stanford.edu)

Judit Moschkovich is Associate Professor of Mathematics Education in the Education Department at the University of California Santa Cruz (UCSC). Before joining the faculty at UCSC in 1999, Dr. Moschkovich worked at TERC in Cambridge, Massachusetts, at the Institute for Research on Learning in Menlo Park, California, and was a lecturer in the mathematics and education departments at San Francisco State University. Her research publications examine students' conceptions of linear functions, conceptual change in mathematics, mathematical discourse practices in and out of school, and bilingual mathematics learners. She has conducted research for over ten years in secondary mathematics classrooms with large numbers of Latino students. Spanish is her first language and she immigrated to the U.S. from Argentina as an adolescent. She is a past member of the Journal for Research in Mathematics Education and Journal of the Learning Sciences editorial panels and was Co-Chair of the

Research in Mathematics Education Special Interest Group for the American Educational Research Association from 2004 to 2006. She was the Principal Investigator of the NSF project Mathematical Discourse in Bilingual Settings: Learning Mathematics in Two Languages and is one of the principal investigators for the Center for the Mathematics Education of Latinos funded by NSF (2004–2009).

Current address: Education Department, 1156 High Street, University of California, Santa Cruz, CA 95064, USA (jmoschko@ucsc.edu)

Pendred Noyce, M.D. was Co-Principal Investigator of the NSF-funded Massachusetts State Systemic Initiative Program and of PALMS, a $16 million dollar NSF-funded State Systemic Initiative to improve mathematics, science and technology education in Massachusetts. She was also Co-Principal Investigator of the Massachusetts Parent Involvement Project. Currently, Penny serves on the Advisory Board at the Center for Study of Mathematics Curriculum at Michigan State University and the Board of Directors of COMAP. She is actively on the Board of Directors of TERC, Concord Consortium, Massachusetts Business Alliance for Education, and a Trustee for the Boston Plan for Excellence and the Libra Foundation. She is also on the Dean's Council at Radcliffe Institute for Advanced Study. She has been a Trustee of the Noyce Foundation since its inception in 1990.

Current address: Noyce Foundation, 2500 El Camino Real, Suite 110, Palo Alto, CA 94306-1723, USA (pnoyce@noycefdn.org)

Judith A. Ramaley is President of Winona State University (WSU) in Minnesota. Prior to joining WSU, she held a presidential professorship in biomedical sciences at the University of Maine and was a Fellow of the Margaret Chase Smith Center for Public Policy. She also served as a Visiting Senior Scientist at the National Academy of Sciences in 2004. From 2001 to 2004, she was Assistant Director, Education and Human Resources Directorate at The National Science Foundation. Dr. Ramaley was President of the University of Vermont and Professor of Biology from 1997 to 2001. She was President and Professor of Biology at Portland State University in Portland, Oregon for seven years between 1990 and 1997.

Dr. Ramaley has a special interest in higher-education reform and institutional change and has played a significant role in designing regional alliances to promote educational cooperation. She also has contributed to a national exploration of the changing nature of work and has written extensively on civic responsibility and partnerships between higher education and community organizations as well as articles on science, technology, engineering, and mathematics.

Current address: President's Office, Somsen 201, Winona State University, P.O. Box 5838, Winona, MN 55987, USA (JRamaley@winona.edu)

Alan Schoenfeld is the Elizabeth and Edward Conner Professor of Education and an affiliated professor of Mathematics at the University of California, Berkeley. Schoenfeld has served as President of the American Educational Research Association and as Vice President of the National Academy of Education. He is a Fellow of the American Association for the Advancement of Science and a senior advisor to the Educational Human Resources Directorate of the National Science Foundation.

Schoenfeld's research focuses on mathematical thinking, teaching, and learning. His book *Mathematical Problem Solving* characterizes what it means to "think mathematically" and describes a research-based undergraduate course in mathematical problem solving. Schoenfeld led the Balanced Assessment project, which developed alternative assessments for K-12 mathematics curricula. He has worked on modeling the process of teaching, and on issues of equity and diversity in mathematics education.

Current address: School of Education, EMST, Tolman Hall #1670, University of California, Berkeley, CA 94720-1670, USA (alans@berkeley.edu)

Susan Sclafani is a managing partner of the Chartwell Education Group, LLC. She recently retired as U.S. Assistant Secretary of Education for Vocational and Adult Education. She also served as Counselor to the Secretary of Education, where she was the U.S. representative to the Organization for Economic Cooperation and Development, and the Asia-Pacific Economic Cooperation. Among the highlights of Dr. Sclafani's term at the Department was the leadership role she played in the creation of the Mathematics and Science Initiative (MSI) to focus attention on the importance of mathematics and science in the education of all students. MSI emphasized the need for teachers knowledgeable in math and science at every level of schooling and the importance of further research in both areas. Her international work led to her leadership of the joint E-Language Learning Project with the Chinese Ministry of Education. She also led the Department's High School Initiative to better prepare students for twenty-first century education, training and the workplace. Prior to serving at the Department, Dr. Sclafani was Chief Academic Officer of one of the nation's largest urban school districts, and in that capacity perfected her diverse skills focusing on technology, curriculum development and construction management. She also has extensive state education and business experience.

Current address: Chartwell Education Group LLC, 1800 K Street, NW, Washington, DC 20006, USA (sclafani@chartwelleducation.com)

Ann Shannon is a mathematics educator with two decades of experience specializing in assessment and urban education. She works as an independent consultant helping states, districts, and schools to better serve English learners and others underrepresented in mathematics.

Dr. Shannon was employed as a Research Fellow at the Shell Centre for Mathematical Education, University of Nottingham, England before moving her work to the University of California. At the University of California, she developed performance assessment materials for the NSF-funded Balanced Assessment project and the New Standards Project. Her 1999 monograph, *Keeping Score*, was published by the National Research Council and drew from her work on Balanced Assessment and with New Standards; it addresses a variety of issues that are relevant to developing, administering, and scoring mathematics assessments. Dr. Shannon subsequently directed the mathematics component of the U. S. Department of Education's Office of Educational Research and Improvement-funded Learning to Think, Thinking to Learn curriculum development project at the University of California (Office of the President), provided mathematical guidance in the development of the California State Mathematics Standards for Preschoolers developed at the University of California at Berkeley's Evaluation and Assessment Research Center, and currently provides support as a specialist to the NSF-funded Diversity in Mathematics Education project.

Current address: 4344 Fruitvale Avenue, Oakland, CA 94602, USA (Annshannon@earthlink.net)

Sara Spiegel is the Director of Administration at the Palo Alto, California office of the Noyce Foundation. She manages human resources and the office in general. In addition, she supports the programs of the Foundation. She also directs the Silicon Valley Scholars program. Sara was previously a Program Associate at the Bay Area School Reform Collaborative. She joined the Noyce Foundation staff in 2001.

Current address: Noyce Foundation, 2500 El Camino Real, Suite 110, Palo Alto, CA 94306-1723, USA (sspiegel@noycefdn.org)

Elizabeth K. Stage is the Director of the Lawrence Hall of Science at the University of California at Berkeley. With a bachelor's degree in chemistry from Smith College, she taught middle school science and mathematics before earning a doctorate in science education from Harvard University. She then spent ten years at the University of California at Berkeley, primarily at the Lawrence Hall of Science, where she conducted research and program evaluation; led professional development programs; and directed public programs in mathematics and computer education. Work on California state frameworks, adoptions, and assessments in mathematics and science led to service at the National Research Council, working on the development of the National Science Education Standards. At the University of California Office of the President, she worked on science assessment based on national standards ("New Standards") and coordinated professional development in mathematics, science, and physical education. She has served on the U.S. Steering Committee for the Third International Mathematics and Science Study and the Science Expert Group for the OECD Program in International Student Assessment; and she currently serves as an advisor to the Center for Assessment and Evaluation of Student Learning. She continues her research interest in understanding the factors that increase the success of students who are underrepresented in the sciences.

Current address: Lawrence Hall of Science, 1 Centennial Drive, University of California, Berkeley, CA 94720-5200 (stage@berkeley.edu)

Elizabeth Taleporos is the Director of Assessment for America's Choice. Prior to joining America's Choice, she managed large-scale test development projects in English Language Arts and Mathematics for several major national test publishers. Before that, she directed the assessment efforts in New York City, managing the efforts in test development, psychometrics, research, analysis, administration, scoring, reporting, and dissemination of information.

Current address: America's Choice, 39 Broadway, Suite 1850, New York, NY 10006, (etaleporos@americaschoice.org)

Mark Wilson is a professor in the Graduate School of Education at the University of California at Berkeley. His interests focus on measurement and applied statistics. His work spans a range of issues in measurement and assessment from the development of new statistical models for analyzing measurement data, to the development of new assessments in subject matter areas such as science education, patient-reported outcomes, and child development, to policy issues in the use of assessment data in accountability systems. He has recently published three books: *Constructing Measures: An Item Response Modeling Approach* (Erlbaum) is an introduction to modern measurement; the second book (with Paul De Boeck of the University of Leuven in Belgium), entitled *Explanatory Item Response Models: A Generalized Linear and Nonlinear Approach* (Springer-Verlag), introduces an overarching framework for the statistical modeling of measurements that makes available new tools for understanding the meaning and nature of measurement; and the third, *Towards Coherence Between Classroom Assessment and Accountability* (University of Chicago Press and National Society for the Study of Education), is an edited volume that explores the issues relating to the relationships between large-scale assessment and classroom-level assessment. He has recently chaired a National Research Council committee on assessment of science achievement: *Systems for State Science Assessment* (with Meryl Bertenthal of the NRC). He is founding editor of a new journal: *Measurement: Interdisciplinary Research and Perspectives.*

Current address: School of Education, POME, Tolman Hall #1670, University of California, Berkeley, CA 94720-1670, USA (markw@berkeley.edu)

Assessing Mathematical Proficiency
MSRI Publications
Volume **53**, 2007

Subject Index

Author Index

Assessing Mathematical Proficiency
MSRI Publications
Volume **53**, 2007

Task Index